Rural Property Valuation

Readers of this text may be interested in the following publications from the Appraisal Institute:

- *The Appraisal of Real Estate*
- *Appraising Conservation and Historic Preservation Easements*
- *Appraising Industrial Properties*
- *The Dictionary of Real Estate Appraisal*

Appraisal Institute®

Professionals Providing Real Estate Solutions

Rural Property Valuation

Appraisal Institute • 200 W. Madison • Suite 1500 • Chicago, IL 60606 • www.appraisalinstitute.org

The Appraisal Institute advances global standards, methodologies, and practices through the professional development of property economics worldwide.

Chief Executive Officer: Frederick H. Grubbe, MBA, CAE
Director of Professional Services and Resources: Evan R. Williams, CAE, IOM
Senior Manager, Publications: Stephanie Shea-Joyce
Senior Book Editor/Technical Writer: Michael McKinley
Senior Technical Book Editor: Emily Ruzich
Manager, Book Design/Production: Michael Landis

For Educational Purposes Only

The materials presented in this text represent the opinions and views of the developers and reviewers. Although these materials may have been reviewed by members of the Appraisal Institute, the views and opinions expressed herein are not endorsed or approved by the Appraisal Institute as policy unless adopted by the Board of Directors pursuant to the Bylaws of the Appraisal Institute. While substantial care has been taken to provide accurate and current data and information, the Appraisal Institute does not warrant the accuracy or timeliness of the data and information contained herein. Further, any principles and conclusions presented in this publication are subject to court decisions and to local, state, and federal laws and regulations and any revisions of such laws and regulations.

This book is sold for educational and informational purposes only with the understanding that the Appraisal Institute is not engaged in rendering legal, accounting, or other professional advice or services. Nothing in these materials is to be construed as the offering of such advice or services. If expert advice or services are required, readers are responsible for obtaining such advice or services from appropriate professionals.

Nondiscrimination Policy

The Appraisal Institute advocates equal opportunity and nondiscrimination in the appraisal profession and conducts its activities in accordance with applicable federal, state, and local laws.

Library of Congress Cataloging-in-Publication Data

Names: Appraisal Institute (U.S.), issuing body.

Title: Rural property valuation / Appraisal Institute.

Description: Chicago, IL : Appraisal Institute, [2017] | Includes index.

Identifiers: LCCN 2016044536 | ISBN 9781935328681

Subjects: LCSH: Farms--Valuation. | Real property--Valuation. | Land use, Rural.

Classification: LCC HD1393 .R864 2017 | DDC 333.33/52--dc23 LC record available at https://lccn.loc.gov/2016044536

Table of Contents

Foreword

The American landscape is changing and rural communities are changing as well. While the need for agricultural products to feed and clothe the population may be relatively stable, the demand for land for recreational and residential use continues to expand. Understanding the uses, characteristics, and value of rural land can be challenging for those who work the land and for the real estate professionals who value it.

The Appraisal Institute is proud to present *Rural Property Valuation*, a long-awaited reference textbook to help both experienced valuers and novice appraisers successfully complete challenging rural valuation assignments. This new text provides up-to-date information on the many forces that affect agricultural land and the commodities it yields. The book lays the groundwork for a broad understanding of rural properties and offers specific data and analytical tools that appraisers can use in their daily work. Appraisal basics such as property rights, financing, market analysis and highest and best use, site and building improvements, and the approaches to value are all presented from a distinctly rural perspective. Additional topics unique to rural property such as federal and state assistance programs, soil analysis, ranch management, and commodity pricing are also covered in depth. The final chapters of the text are devoted to specialized rural properties (livestock ranches, dairy farms, permanent plantings, and timberland) and include valuable insights and information that cannot be found elsewhere.

In short, *Rural Property Valuation* is a landmark text that addresses the needs of practicing appraisers and fills a void in the current literature of real property appraisal.

The Appraisal Institute owes a debt of gratitude to the many volunteers who contributed their expertise to the development of this text, including Stephanie Coleman, MAI, SRA, AI-GRS, AI-RRS; John Evans, MAI; Brian Flynn, MAI, AI-GRS; Steven J. Herzog, MAI, AI-GRS; S. Warren Klutz, MAI, SRA, AI-GRS; Richard LaMont, MAI; Michael S. MaRous, MAI, SRA; Michael Morris, MAI; Doug Nelson, MAI, SRPA, SRA; Herb Pritchett, MAI, SRPA, SRA; John A. Schwartz, MAI; Lee B. Smith, MAI; and Brad Swinney, MAI, AI-GRS. Special thanks are due to the final reviewers, who joined me in reading the complete manuscript to ensure that the text presents a balanced, accurate picture of rural appraisal today: Jim Amorin, MAI, SRA, AI-GRS; Darrell Bullock, MAI, SRA; Stephen Crosson, MAI, SRPA, SRA; Douglas Defoor, MAI, AI-GRS; W. West Foster, MAI, AI-GRS; Brian Linnemeyer, MAI; and Michael Tankersley, MAI, SRA, AI-RRS.

The information in this new text will help real estate practitioners understand rural property markets and expand their expertise, competence, and overall professionalism. It should prove useful to all appraisers, whether they value rural properties once a year, once a month, or every day.

Scott Robinson, MAI, SRA, AI-GRS
2016 President
Appraisal Institute

Preface

The publication of *Rural Property Valuation* rectifies a long-standing gap in the professional literature and provides an up-to-date examination of valuation issues unique to rural property. *The Appraisal of Real Estate* and other foundational valuation texts tend to have an urban focus. This new text addresses the need for guidance on the specific challenges of appraising real property in rural areas.

The first step, of course, is to scrutinize the most important word in the book's title–*rural*. Where is the line between a property deemed "rural" and a property considered "urban"? Is a rural area, by definition, a "sparsely developed area where the land is primarily used for farming, forestry, resource extraction, very low-density residential uses (one unit per 10 acres or less), or open space uses,"[1] or is it a "sparsely developed area, with a population density of fewer than 100 persons per square mile, where the land is undeveloped or primarily used for agricultural purposes"[2]?

Perhaps it is clearer to contrast a rural area with what is classified as urban. The US Census Bureau defines an urban area as "a densely settled core of census tracts and/or census blocks that meet minimum population density requirements, along with adjacent territory containing non-residential urban land uses as well as territory with low population density included to link outlying densely settled

1. Michael Davidson and Fay Dolnick, *A Planner's Dictionary* (Chicago: American Planning Association, 2004), s.v., "rural."
2. Harvey S. Moskowitz and Carl G. Lindbloom, *The Latest Illustrated Book of Development Definitions* (New Brunswick, N.J.: Center for Urban Policy Research at Rutgers University, 2004), s.v., "rural area."

territory with the densely settled core. To qualify as an urban area, the territory identified according to criteria must encompass at least 2,500 people, at least 1,500 of which reside outside institutional group quarters."[3] The same source uses "rural" to describe "all population, housing, and territory not included within an urban area."

Given these definitions, a shocking 80% of the US population lived in urban areas in 2016. However, the land area classified as rural at that time encompassed almost 97% of the country's total land area. Even at the broader county level, a comparison of "metropolitan" and "nonmetropolitan" counties classifies almost two-thirds of the country as nonmetropolitan areas, with 72% of the land area and 15% of the country's population.

So, in terms of land area, urban sprawl is still a long way away from absorbing the rural core of the United States. Extending the de-

Metropolitan and Nonmetropolitan Counties

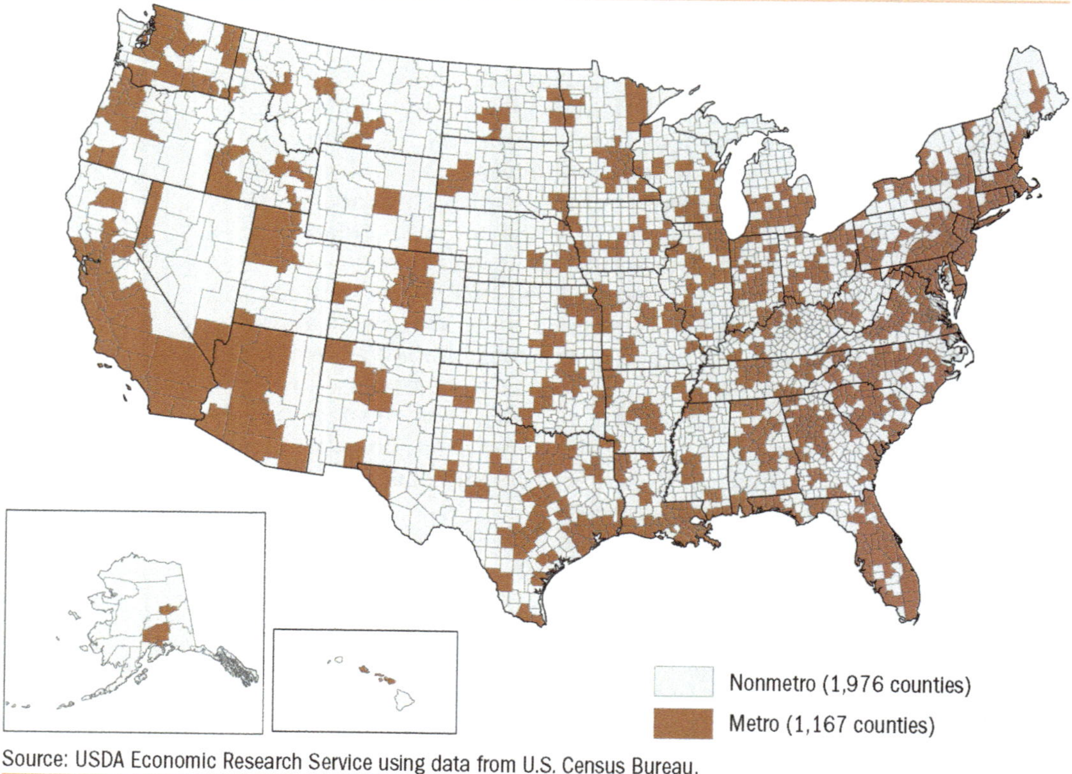

Source: USDA Economic Research Service using data from U.S. Census Bureau.

Image source: https://www.ers.usda.gov/data-products/chart-gallery/gallery/chart-detail/?chartId=58274/ (2013)

3. US Census Bureau, 2010 Census Urban and Rural Classification and Urban Area Criteria, available online at https://www.census.gov/geo/reference/ua/urban-rural-2010.html/.

mographic measures to include employment figures paints a different picture of nonmetropolitan areas. The agricultural industry in the United States, which less than a century ago employed almost 25% of US workers, employs just over 1% of workers in the twenty-first century. Still, the agricultural sector of the US economy (i.e., agriculture and agriculture-related industries) remains sizeable–5.7% of the gross domestic product in 2014.

The value of rural land is tied directly to the various uses of that land. For example, cropland tends to sell for higher prices than pasture land, which is not surprising given that the costs involved in preparing and maintaining land to grow a particular crop are higher than the costs to simply produce natural forage for livestock. Likewise, the income potential of rural land affects the value of land used for various forms of agriculture, with land values tending to track shifts in the prices of agricultural commodities (e.g., corn, soybeans, timber, dairy products). Demographic trends in nearby urban areas can also have a significant influence on the value of rural land in areas in transition from agricultural use to a new, more intensive use of the land. In fact, a common valuation assignment in rural areas is an appraisal of a proposed residential or commercial development on a site formerly used as farmland. Increases

Value Added to GDP by Agriculture and Related Industries

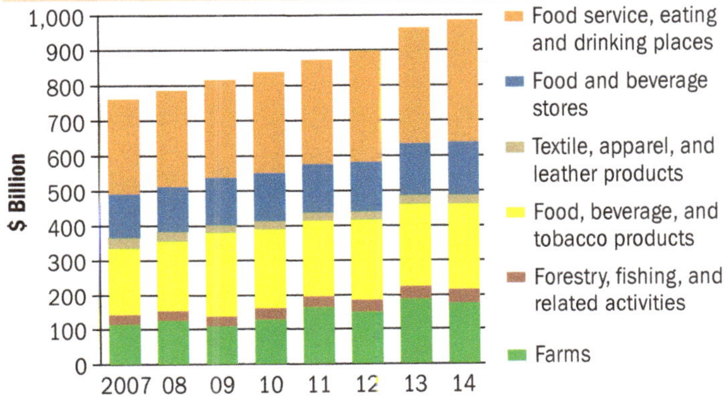

Note: GDP refers to gross domestic product.

Source: USDA, Economic Research Service using data from U.S. Department of Commerce, Bureau of Economic Analysis, Value Added by Industries series.

Image source: https://www.ers.usda.gov/data-products/ag-and-food-statistics-charting-the-essentials/ag-and-food-sectors-and-the-economy.aspx/ (October 14, 2016).

Average US Rural Land Values

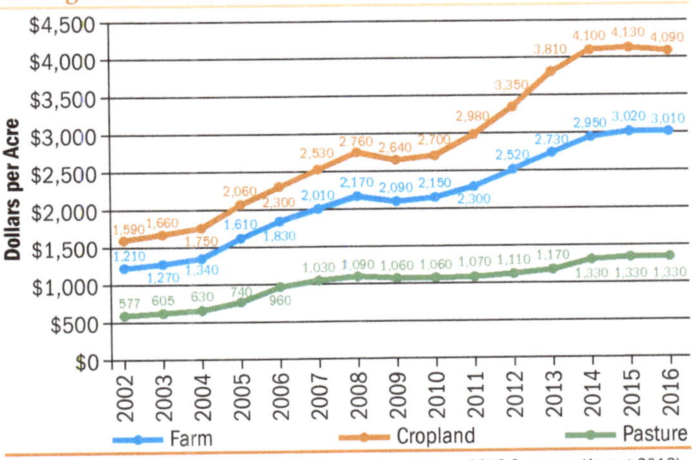

Source: USDA, National Agricultural Statistics Service, *Land Values: 2016 Summary* (August 2016).

Gross Farm Income, Production Expenses, and Net Farm Income, Inflation Adjusted, 2000-2016F

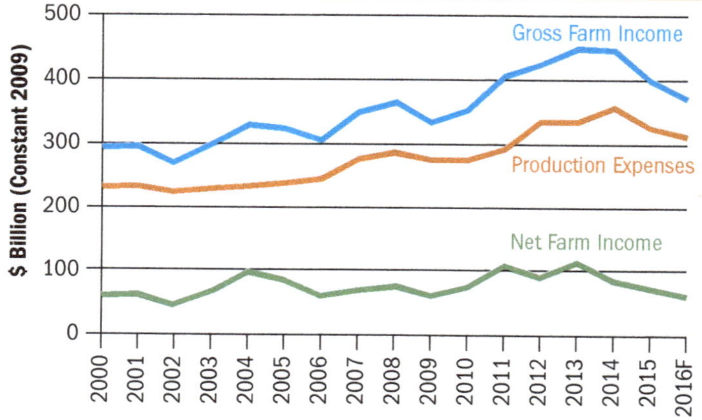

Note: F = forecast. Values are adjusted for inflation using the chain-type GDP deflator, 2009 = 100.

Source: USDA, Economic Research Service, Farm Income and Wealth Statistics. Data as of November 30, 2016.

Image source: https://www.ers.usda.gov/data-products/chart-gallery/gallery/chart-detail/?chartId=76952

US Agricultural Output, Inputs, and Total Factor Productivity

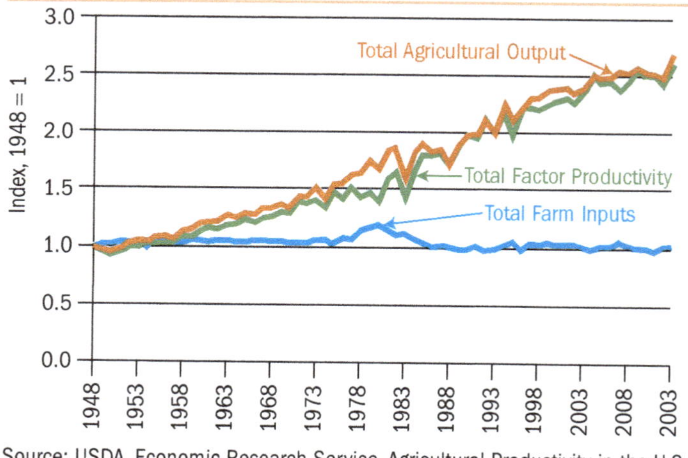

Source: USDA, Economic Research Service, Agricultural Productivity in the U.S. data product. Data as of December 2015.

Image source: https://www.ers.usda.gov/data-products/chart-gallery/gallery/chart-detail/?chartId=58284

in demand for urban (or suburban) development spilling out from the boundaries of urban areas often drive change in land prices for nonagricultural land uses in neighboring rural areas.

The modern six-step process of market analysis for real property valuation, which includes a formal quantification of supply and demand, has been conspicuously absent from the literature on the valuation of rural properties. Formal market analysis based on demographics and household income can be difficult to apply when rural land is, by definition, sparsely populated. Without a doubt, measuring demand for agricultural land uses can be a challenge for appraisers who specialize in the valuation of rural property, particularly in an era of declining competition among a smaller number of market players. Industry consolidation has affected many agricultural market areas dominated by specialized property types, e.g., livestock ranches in the West, dairy farms in the Midwest, concentrated animal feeding operations such as extremely large hog farms in Iowa.

Despite the decline in rural population and the number of workers, farm productivity has risen dramatically since the mid-twentieth century as agricultural practices have improved and technology has eased the burden on farm labor. But in the twenty-first century, gross farm income, production expenses, and net farm income have continued to follow cyclical trajectories. In spite of the increasing efficiency of larger and larger facilities served by sophisticated equipment and systems, agricultural markets still remain tied to market cycles.

Many modern farmers are vigilant market observers, always seeking opportunities to expand their operations by leasing or buying additional land when it is available and affordable and take advantage of economies of scale. Appraisers who can reconcile the increasing value of land with the tightening profit margins of many agricultural enterprises have information that is of great value to lenders, investors, and other market participants who need to know about the health and continued prosperity of agricultural enterprises ranging from the embattled American family farms to large-scale livestock operations.

Chapter 1

Fundamentals of the Appraisal of Rural Property

Much can be said about the basic importance of land resources and real estate in the modern world. They provide people with living space, with raw materials necessary for filling material needs, and with opportunities for satisfactions dear to the heart of man. People look to land for their physical environment, for the food they eat, for fibers and the other materials needed to clothe their bodies, and to provide housing and manufactured goods, for building sites, for recreation opportunities, and for scenery and open space.

Raleigh Barlow, *Land Resource Economics: The Economics of Real Estate*, 4th ed.

Throughout recorded history, people have lived in close association with the soil, fields, forests, and fishing areas that provide food, fiber, shelter, and products for trade. Over the course of the twentieth century, population and capital shifted away from rural areas and concentrated in urban areas. According to United Nations estimates at the time, the global population reached the tipping point in 2007, and now more people live in urban areas than in rural areas. In many parts of the world, rural and agricultural economies cannot match the economic opportunities offered in cities and towns. Agricultural practices and technology have changed too, requiring far fewer people to farm and ranch, and globalization has revolutionized the distribution of agricultural products from primarily local markets to international markets.

The productivity of agricultural and forest land is measured primarily in terms of its capacity to produce commodities, although rural land also furnishes its owners with amenities in the form of housing and recreation. Real estate accounted for about 81% of the assets in the farm sector in 2011.[1]

The Nature of Property

"Whose is the land, his it is, to the sky and the depths." This ancient maxim is the basis for the following legal definition of *land*:

> [Land] includes not only the ground, or soil, but everything that is attached to the earth, whether by course of nature, as are trees and herbage, or by the hand of man, as are houses and other buildings. It includes not only the surface of the earth but everything under it and over it. Thus in legal theory, the surface of the earth is just part of an inverted pyramid having its tip, or apex, at the center of the earth, extending outward through the surface of the earth at the boundary lines of the tract, and continuing on upward to the heavens.[2]

Land provides the foundation for the social and economic activities of people. It is both a tangible physical commodity and a source of wealth. Because land is essential to life and society, it is the subject of many disciplines, including law, economics, finance, sociology, and geography. Within the vast domain of the law, issues relating to the ownership and use of land are considered. In economics, land is regarded as one of the four agents of production, along with labor, capital, and entrepreneurial coordination. Land also provides the natural elements that contribute to a nation's wealth. In finance, the principles of economics are applied within a market economy that furnishes capital for the exchange of property. Knowledge of these principles helps market participants act prudently. From a sociological perspective, land is a resource to be shared by all people and a commodity that can be owned, traded, and used by individuals, corporations, partnerships, and other entities. Geography focuses on describing the physical elements of land and the activities of the people who use it. Permanent, natural features of land include soil, rocks, lakes, streams, trees, plants, and subsurface deposits of gravel, oil, gas, minerals, and water.

Real Estate, Real Property, and Personal Property

In appraisal, the terms *real estate*, *real property*, and *personal property* are widely used but are not always clearly defined. As a term of art, real estate refers to land and the appurtenances affixed to the land, i.e., permanent improvements and attachments such as buildings and fences. Improvements and attachments to land are man-made and, unlike land, they are depreciable. Improvements that are generally consid-

1. US Department of Agriculture, *Agricultural Income and Finance Outlook* AIS-91, December 2011, available online at www.ers.usda.gov/publications/ais-agricultural-income-and-finance-outlook/ais91.aspx.

2. Raymond J. Werner and Robert Kratovil, *Real Estate Law*, 10th ed. (Englewood Cliffs, NJ: Prentice-Hall, 1993), 4.

ered a part of the land, called *site improvements*, include fences, roads, drainage systems, permanent irrigation systems, and irrigation wells.

In contrast, *real property* refers to the interests, benefits, and rights inherent in the ownership of physical real estate. An *estate* in land is the degree, nature, or extent of the interest held.

Personal property refers to movable items of property that are not permanently affixed to, or part of, the real estate. Therefore, personal property is not endowed with the rights of real property ownership. Livestock, machinery, and household furniture are examples of personal property. Personal property is also sometimes known as *tangible personal property*.

Some agricultural properties include intangible business assets that may be a factor in a valuation and may need to be considered for the appraisal to be compliant with appropriate appraisal standards and federal regulations. For instance, an established grain elevator or agricultural fertilizer supply property may be valued using the unit of comparison of throughput (i.e., the price divided by the yearly flow of product through the property) in the application of the sales comparison approach. If the indicated value of the property is substantially higher than the value indicated by the cost approach, the difference in values may be attributable to entrepreneurial profit or to factors related to the operation and the goodwill of the existing business. If the value difference is attributable to goodwill or the operation of the business, then an appraiser may need to allocate and assign a value to this goodwill separate from the value of the real property alone. Appraisals of real property that also include the valuation of non-realty assets require appraisers to have additional training and expertise for professional competency. Identifying the different asset types to be appraised and allocating value to each is a complex valuation assignment.

Other examples of properties with intangible assets would be poultry or swine facilities that have contracts in place with the end users. In the valuation of a concentrated animal feed operation (CAFO) property, if the indicated value of the property derived in the sales comparison approach is higher than the depreciated value of the improvements plus land derived in the application of the cost approach, an appraiser must further analyze the reasons behind this difference. The difference may be attributable to the contract between the farm operator and the integrator who is receiving the finished animal at the contracted price. In some instances, a change in the highest and best use of the property would result if the contracts were terminated, largely because the market would place little or no value on the improvements without the contract in place. As a related example, consider operators of an orchard who choose to sell their produce on their own in their own markets. In this case, appraisers should examine what part of the resulting income stream is attributable to the real estate, what part is attributable to the operation of the retail selling business, and how this income should be handled in an appraisal.

Property Rights and Ownership

At the start of any valuation assignment, an appraiser faces two tasks:

1. identifying and describing the real property or personal property, and
2. determining what property rights are included in the appraisal.

An appraiser must be aware of the differences between the real property and the personal property that may be involved in the proposed appraisal and must understand what items are to be included in the real property to be appraised. Appraisers must also determine the ownership, possession, and control of property to be able to define the appraisal problem. It is important to clearly identify the type of estate involved. (Rural property rights are discussed in depth in Chapter 2.)

The four powers of government–police power, taxation, eminent domain, and escheat–serve as tools for government at all levels to influence the use of land. For example, the government can encourage agriculture uses through lower taxation policies to retain the land in agriculture. Conversely, high taxes can be used to accelerate a change in land use from agriculture to urban uses.

Eminent domain can be exercised by a government agency or by an entity acting under governmental authority, such as an irrigation district, public utility, housing authority, school district, utility company, redevelopment district, or right-of-way agency. The taking of privately held land for public use is the most direct–and often the most controversial–governmental intervention in land use.

The police power of the government has become more prevalent in recent years. As population density has increased and society has become more conscious of environmental and conservation issues, land and resource regulations have proliferated at all levels of government. Under police power provisions, local, state, and federal governments are using regulations to shape and limit the uses permitted on rural properties. Many landowners have brought litigation arguing whether these regulations are a legitimate exercise of the state's police power or are "takings," but the fact remains that government can, through the use of regulations, limit the exercise of property rights on rural properties. As the US population continues to grow, there will be increased competition for land and growing conflicts of interest concerning the most appropriate uses of land.

Private Ownership Rights

In addition to public limitations on land use, a property owner's rights may be restricted by private limitations, which can limit the use or development of a property and may constrain the manner in which ownership can be conveyed. These limitations may be created by easements, deed reservations, or restrictive covenants. Conservation easements and utility easements (e.g., for natural gas pipelines) are common encumbrances on rural land. For example, a utility company may hold an easement that allows it to run its lines across a property and gives its workers access to service those lines. A deed may include restrictions regarding mineral rights or rights of way, or it may specify that the property remain in agricultural use for a specified period of time. For example, these restrictions may limit the division of the property into tracts no smaller than 500 acres or may restrict timber harvests. These restrictions may or may not impact the highest and best use. Other restrictions, such as an unrecorded agreement relative to water rights, may be nearly impossible for an appraiser to discover.

Concepts of Value

The economic concept of value is not inherent in the commodity, good, or service to which value is ascribed. Value is created in the minds of people who make up the market for that commodity, good, or service. A number of important principles support society's understanding of the concept of value. These principles provide the rationale for market behavior and, therefore, help explain changes in the value of real property in rural markets. As with other types of real property, four economic factors create value: (1) utility, (2) scarcity, (3) desire, and (4) effective purchasing power. The complex interaction of these four factors is reflected in the basic economic principle of supply and demand.

Rural property can provide economic benefits to a property owner in a variety of forms:

- as the site of shelter and amenities (i.e., for a residential use)
- as the site of an agribusiness operation (i.e., for use as a farm, orchard, dairy facility, livestock ranch)
- as the source of valuable natural resources that the property owner can use or sell (e.g., oil or gas, minerals)

The property's utility to the property owner serves as a component part in the creation of value.

Scarcity can affect the value of rural property both positively and negatively. For example, insufficient water or drainage affects the utility of land and has a major influence on the value of rural property. That is, the absence of the proper conditions and resources for growing crops or providing pasture for livestock can require costly business expenses (e.g., digging irrigation or drainage ditches) that would affect the cash flow generated by agribusiness operations on that land. On the other hand, land that supports the production of a particularly rare and desirable commodity–e.g., a vineyard with unique environmental conditions–can gain value in the marketplace through the scarcity of alternative sources for the commodity that is produced on that land. The relationship between utility and scarcity has been explored for decades.

Scarcity can be tied to desirability when land–or a product of that land, such as a specific varietal grape–has few substitutes in the marketplace. For example, land with particularly fertile soil would have a productive advantage over an otherwise comparable plot of land and thus would be more desirable to farmers (i.e., the likely buyers of the land) as a site for row crops.

The combination of utility, scarcity, and desire works more effectively to create value when market participants have adequate effective purchasing power to put the most productive land use into action. The availability (or lack) of financing to the farm sector can have a significant influence on the feasibility of putting agricultural land to use (by, say, adding irrigation improvements such as new wells or improving drainage) and thus on the value of that land.

Influences on Value

The value of real property is affected by the interplay of social, economic, governmental, and physical forces, which are continually changing, often in a cyclical pattern. (See examples of the four forces in Table 1.1.) Although the value of real property may seem relatively stable in comparison to the value of stocks or commodities, it is still

Table 1.1 Four Forces that Influence the Value of Real Property

Force	Examples
Social trends	• population characteristics
Economic circumstances	• demand-side indicators • supply-side economic indicators
Government controls and regulations	• public services • zoning and building codes • tax policies • real estate laws
Environmental conditions	• climate • topography • soil • natural barriers to development • the availability of transportation systems • the suitability of the location in terms of linkages • the accessibility of the location

subject to the many pressures and influences created in this interplay. The interaction of the four forces influences the value of every parcel of real estate in the market.

Demographic trends have a strong influence on the value of all types of real property. A change in use from agriculture to part-time farm, rural residential, or recreational use changes the social structure of communities.

In the valuation of rural properties, appraisers need to understand the economic forces that affect the specific community in which the subject property is located. Factors such as typical crops, lease rates, cash rents or cropshare leasing, population, demographics, and historical trends in commodity prices are all important economic considerations.

Appraisers must also diligently identify and examine the potential value influences of

- public services such as fire and police protection, rural water suppliers, natural gas, cable, sewer service, and electric cooperatives
- local zoning, land use regulations, groundwater regulations, chemical application restrictions, odor abatement rules, waste disposal regulations, and endangered species acts
- government agricultural programs such as Farm Service Agency commodity programs, lending policies, and other cost-sharing programs, including the Conservation Reserve Program
- special legislation that influences imports or exports, property taxation, environmental and water quality control, limitations on foreign ownership, and emergency disaster relief programs pertaining to agricultural properties

Environmental forces that may be analyzed for real estate appraisal purposes include

- climatic conditions such as snowfall, rainfall, temperature, and humidity
- topography and soils
- water availability
- primary transportation systems including federal and state highways, railroads, airports, and navigable waterways
- the nature and desirability of the immediate area surrounding the property
- the presence of toxic contaminants such as asbestos, radon, and PCBs (polychlorinated biphenyls)
- natural barriers to development such as rivers, mountains, lakes, and oceans

Principles of Economic Theory

The characteristics of land and real property, which are the focus of appraisal activity, reflect society's attitudes and beliefs about the relative value of specific economic activities. To analyze the many dynamic and interactive factors that influence people's attitudes and beliefs about value, a number of fundamental economic concepts must be addressed.

Anticipation

The principle of *anticipation* is the perception that value is created by the expectation of benefits to be derived in the future. In the market for agricultural properties, both owner occupancy and income production contribute to the purchase decision. Although a farm or ranch property may be purchased as an investment with the expectation of future income benefits, it is also likely to provide a home for its owner. For farm, forest, and ranch properties, anticipation is tied to productivity, which will determine the income stream that is derived from the operation. This also applies to the anticipation of an investor who holds land in expectation of an increase in value.

Real property appraisers must be aware of real estate trends that affect the perceptions of buyers and sellers and their anticipation of the future. Historical data relating to the subject property or a particular market is relevant only insofar as it helps interpret current market anticipations. An agricultural property's historical crop and forage production may reveal factors that will affect the property's ability to produce a certain net income in the future.

Change

The dynamic nature of the social, economic, governmental, and environmental forces that influence real property value accounts for *change*. The physical deterioration and obsolescence of property are other manifestations of the inevitability of change.

During the early 1980s, a combination of factors affected the market for farmland in the United States, and the result was a significant decline in land values. As economic conditions improved and governmental influences changed in the late 1980s, land values began to increase. After 1985, improving plant genetics, rising commodity prices, and research on best farming practices led to more revenue per acre flowing to the farmer, which supported rising prices for farmland. In 1985, the average corn yield nationwide was 118 bushels per acre with an average price received of $1.57 per bushel, for an average revenue of $185 per corn acre planted. By 2014, this average corn acre revenue had increased more than threefold. In 2014, the average corn yield nationwide was 171 bushels per acre with the average price of $3.65 per bushel (down from a high of $6.43 in 2012), for

Figure 1.1 Average Cropland Value in the United States

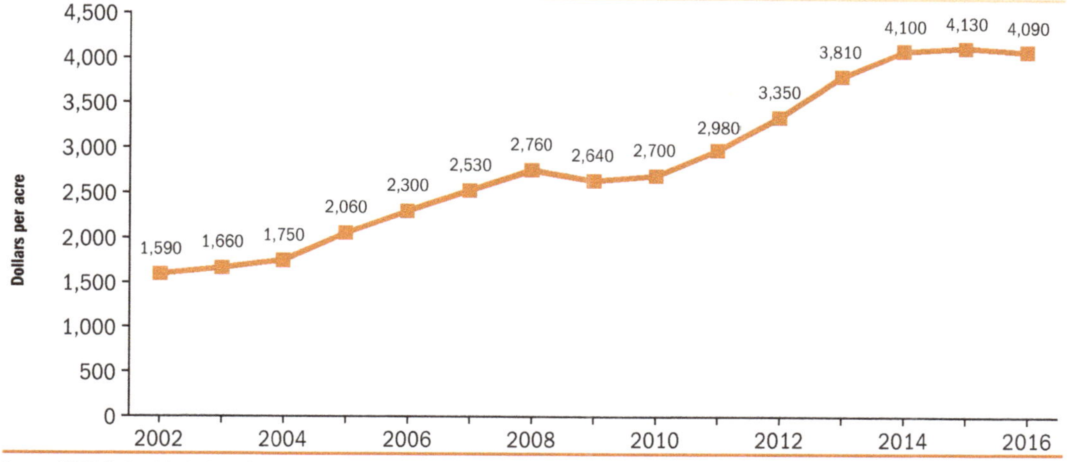

Source: USDA National Agricultural Statistics Service (August 6, 2016), www.nass.usda.gov/Charts_and_Maps/Land_Values

an average revenue of $624 per corn acre planted. It is therefore not surprising that the average value of cropland in the nation increased from $713 per acre in 1985 to almost $4,100 per acre in 2016.

Property values can also change if commodity prices are affected by environmental regulations. For instance, in the early 2000s the mandate for gasoline to contain a certain percentage of ethanol coupled with federal incentives for the building of ethanol plants pushed the demand for corn to levels where more corn was used for ethanol purposes than for animal feed. This, in turn, pushed up the price of corn, which led to an increase in the price of farmland used to grow corn.

Value is also affected by the improvements on a property. Because obsolescence can begin in the design phase and deterioration may start while a building or improvement is still being constructed, the different types of deterioration and obsolescence found in a property have unique implications in appraisal.

An acceleration in price levels for rural land may come from two sources. In the agricultural community, agriculturalists may have a strong demand for more land. At the fringes of urban communities, farmland may be desired for subdivision development. In either case, the supply of land will likely increase because the capital necessary to buy the land, either for agricultural development or urban expansion, is available in sufficient quantities to bring the land to market.

Supply and Demand

The principle of *supply and demand* states that the price of a commodity varies directly with demand and inversely, but not necessarily proportionately, with supply. Theoretically, the supply of and demand for commodities tend toward equilibrium, where market value, price, and cost are equal.

Land is both a commodity and the source of commodities, so the supply of rural land refers to the amount of land suitable for a specific use rather than simply the total number of available acres. The supply and demand of a farm product directly affect the price that a farmer may receive for that product in the market. Indeed, many economics textbooks note that the markets for agricultural products come closest to fitting the characteristics of an efficient market. If supply and demand tends to increase or decrease the prices received by farmers for their products (and thereby their income), then over time the value of farmland would also be expected to increase or decrease, but not by the same percentage or even at the same time as prices and income increase or decrease. Value is defined as the present worth of the future benefits to be derived. Therefore, the value of income-producing farmland is based much more on expectations of future income and prices than current income and prices.

Competition

Competition between buyers or tenants represents the interactive efforts of two or more potential buyers or tenants to make a purchase (or to secure a lease). Competition between sellers represents the interactive efforts of two or more potential sellers to effect a sale. Competition is fundamental to the dynamics of supply and demand in a free-enterprise, profit-maximizing economic system.

Competition for real property used for agricultural production is intricately related to national and international markets for agricultural commodities. When demand and prices for commodities increase, an investment in agricultural production is perceived as profitable. Thus, competition for land that can be used for commodity production tends to increase, with a resultant increase in agricultural land values.

The phenomenon of ruinous competition is easily observed in agricultural support businesses such as suppliers of production needs and commodity-processing and marketing facilities. When the price of a commodity begins to rise, more farmers plant their lands in that commodity, which increases the supply and eventually drives the price for the commodity down. This cycle plays out year after year. The prices for soybeans and corn have been at historical highs for a number of years, but in 2015 crops were sold at prices that were below the breakeven point for many farmers. This is a result of waning demand for grains and falling prices. In 2013, the prices of peanuts increased significantly. Farmers in the Mississippi Delta, which is not known as a large peanut-producing region, began buying equipment and planting acreage in peanuts. This phenomenon only lasted one growing season because the producers delivered large quantities to the market and the prices of peanuts decreased to the point that most producers returned to their traditional crops the following season.

Substitution

The principle of *substitution* states that when several similar or commensurate commodities, goods, or services are available, the one giving the greatest value for the best price attracts the greatest demand and widest distribution. For example, if two farms offer approximately the same advantages, the prospective buyer will select the one with the lower price. Soil, topography, climate, size, and location are some of the factors that are considered in determining if a farm property is equally desirable.

Opportunity Cost

Opportunity cost is the cost of options foregone or opportunities not chosen. An investor who selects one type of investment forgoes the opportunity to invest in other types of available investments. Harkening back to the value factor of scarcity, a person's capital to invest in real estate is relatively scarce when compared to the universe of available purchasing alternatives. Even Warren Buffet and Berkshire-Hathaway do not have enough capital to pursue all the purchase opportunities available to them. They pursue a few select investment opportunities. When relatively scarce capital is employed to purchase real property, it is not available for another purchase, and the "real" cost to the buyer of that purchase is the loss of the capital to pursue other purchases. For example, the opportunity cost of purchasing an acre of rural land at $10,000 per acre means that the capital is no longer available to purchase another equally desirable farm at $9,995 per acre if it were to come on the market. A well-informed investor will select the investment that seems likely to provide the highest rate of return commensurate with the risk associated with that investment. The lack of liquidity that the investor experiences over the term of the investment poses a potential opportunity cost if alternative investments at comparable levels of risk outperform the investment chosen. Opportunity costs are especially significant in estimating the rates of return required to attract capital.

Balance

The principle of *balance* holds that real property value is created and sustained when contrasting, opposing, or interacting elements are in a state of equilibrium. This principle applies to relationships among various property components as well as the relationship between the costs of production and the property's productivity. The principle of balance governs the related principles of diminishing returns, contribution, surplus productivity, and conformity.

Figure 1.2 Point of Balance

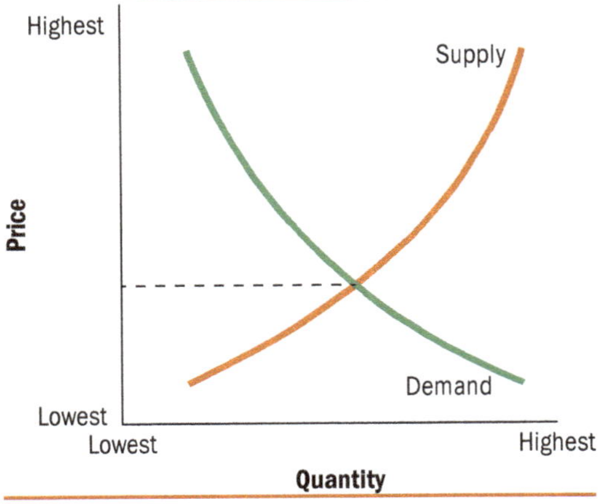

The fertilization of farmland provides a simple example of the law of diminishing returns. Applying fertilizer to land increases crop yields only up to a point, beyond which the additional fertilizer will produce no further increase in the marginal output of the acreage. The optimum amount of fertilization is achieved when the value of the increment in yield resulting from the last unit of fertilizer equals the additional expenditure on fertilizer. This is the point of balance (see Figure 1.2).

Contribution

The cost of an item does not necessarily equal its value. For example, the construction of an expensive residence on a farm or ranch does not necessarily cause the property value to increase in direct proportion to the improvement's cost. The installation of an irrigation system or the development of a specialty crop or orchard may never create value in the market equal to the cost of development.

Surplus Productivity

The value of any part of an enterprise must be economically justified by its effect on productivity and utility. For example, the capital outlay for an irrigation system is justified only if the resulting increased productivity will increase property value by an amount that exceeds the expenditure.

Conformity

The styles and uses of the properties in a given area may conform for several reasons, including economic pressures and the shared preferences of owners for certain types of structures, crops, amenities, and services.

Externalities

Externalities may refer to the use or physical attributes of properties located near the subject property or to the economic conditions that affect the market in which the subject property competes. For example, an increase in the purchasing power of farm owners in the area where a grain elevator or milk plant is located will likely affect the sales potential of the property. As another example, the low interest rates in the marketplace and the resulting low opportunity cost of capital investment after the 2007 financial meltdown fueled the rise of farmland prices.

Highest and Best Use

The concept of *highest and best use* is predicated on the assumption that buyers and sellers set prices for properties based on their conclusions about the most profitable use of the site or property. Highest and best use is the reasonably probable use of property that results in the highest value. The use must be legally permissible, physically possible, financially feasible, and maximally productive to result in the highest estimate of land value. Highest and best use is determined separately for the land or site as though vacant and available for an anticipated use and for the property as improved. The related concept of *consistent use* holds that land cannot be valued based on one use while improvements are valued based on another.

In market value appraisals of improved property, appraisers consider a number of alternative uses of the existing improvements:

- Demolish the existing improvements and redevelop the site.
- Convert, renovate, or alter the existing improvements to enhance the current use or change the use of the property to a more productive use.
- Retain the existing improvements and continue the current use. The existing improvements may represent an interim use that helps defray the cost of carrying the property and demolition costs until all approvals have been obtained and actual construction may begin.

For example, a large equipment storage structure on an orchard property could continue to be used to store equipment or it could be converted for fruit storage or housing laborers. Even when an existing improvement does not represent the highest and best use of a site as though vacant, it should not necessarily be demolished. Demolition should be considered when the land value as though vacant exceeds the value of the property as improved, plus the cost of demolishing the improvements, hauling costs, and dumping fees.

When appraising agricultural properties, appraisers must consider whether the agricultural use is a long-term or interim use. If it is long-term, the type of operation that will provide the greatest return to the land must be identified. For example, conversion of the land use from dry to irrigated cropland or from cropland to vineyard might result in an increase in the net return to the land.

The existing agricultural use may not be the highest and best use for the property in its entirety, especially in an area undergoing change to a more intensive use such as residential development. Nevertheless, the existing suboptimal use will continue until conversion becomes financially feasible.

Certain land uses may have negative value if substantial costs would be incurred to remove them. For example, consider a single operating farm in an area where subdivision development is intense. Use of the property as a farm would seem to be the exact opposite of highest and best use. Nevertheless, when a financial feasibility analysis is completed and all factors are considered, such a use may be the most profitable interim use if the supply of land already devoted to subdivision development exceeds current demand.

Agricultural use may be an interim use for transitional properties until the ultimate highest and best use is achieved. During this period the land may be valued for its prospective highest and best use. If this is done, however, the value indication must be discounted at an appropriate rate for the remainder of the projected interim holding period.

The Valuation Process

The valuation process is a systematic set of procedures an appraiser follows to provide answers to a client's questions about real property value. It is a model that can be adapted to a wide variety of questions that relate to value and that is applicable to valuation assignments involving all types of real property, such as farms, ranches, open land, and other rural property types.

The valuation process is accomplished through specific steps. The number of steps followed depends on the intended use of the assignment results, the nature of the property, the scope of work deemed appropriate for the assignment, and the availability of data. The components of the valuation process are shown in Figure 1.3.

Identification of the Appraisal Problem

The first step in the valuation process is similar for rural properties and other types of real property. A typical problem in the appraisal of agricultural and rural properties, however, is the difficulty of describing and identifying the real property. Several parties may have ownership interests in a given property, and the extent of these interests and their duration must be specified. Whether an appraiser is asked to develop an opinion of the value of the fee simple estate or the value of a partial interest, the specific rights to be valued must be identified at the outset of the appraisal process.

Also, each tract of land differs from every other tract by virtue of its location and physical character. This is true for any parcel of real estate, but the productive capacity of agricultural land is of particular importance in the valuation of property used for agricultural purposes. Thus, the physical character of rural land must be carefully identified, analyzed, and discussed in an appraisal assignment.

Scope of Work

Typically, an appraisal of a vacant dry land farm would require less work than an appraisal of an irrigated, multi-enterprise improved

Figure 1.3 The Valuation Process

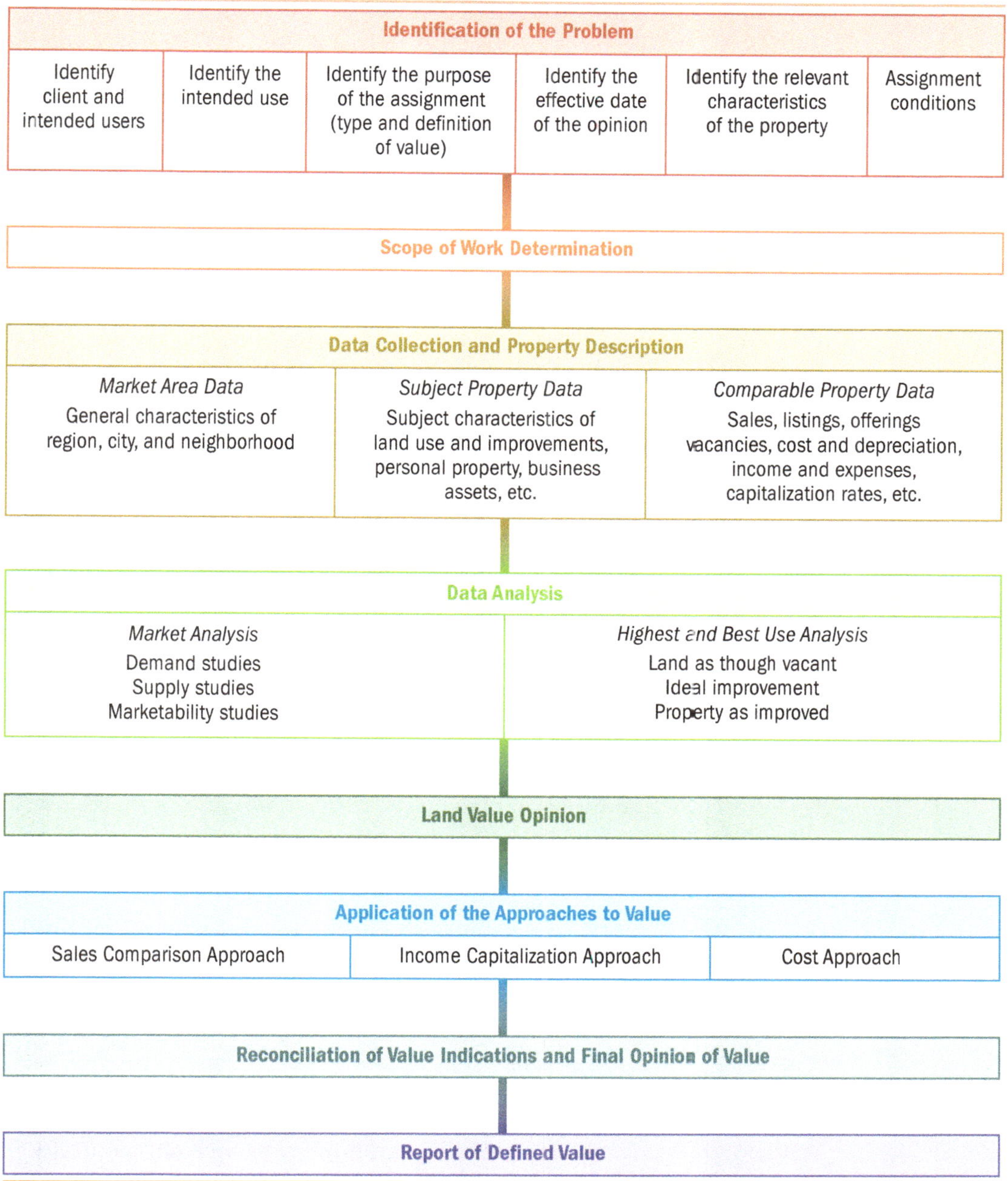

farm. Some appraisals of rural property may be finished within a few days, while others may take considerably longer. For some assignments, it can take weeks or months to gather the required data to complete the analysis and value opinion.

To determine the scope of work of a valuation assignment involving rural property, an appraiser needs to answer the following questions:

1. Does this assignment require the assistance of experts from other fields, such as soil scientists or engineers?
2. Can this assignment be completed by the appraiser alone, or will assistance from other appraisers or staff be needed to complete the project?

What the actions of the appraiser's peers would likely be for the same assignment, given the same level of expertise, and what the market of regularly intended users would reasonably expect under similar circumstances should also be considered.

Data Collection

In the valuation process for rural property, the appraiser collects general data about the marketplace and specific data about the subject property and comparable properties as he or she would in appraisals of other property types and augments this data by collecting certain market and property data that is particularly important to rural property. Relevant general data includes information on

1. social forces such as trends in the size, makeup, customs, and perceptions of the market,
2. economic forces that contribute to the profitability of agriculture,
3. governmental forces such as regulations relating to land use, wetlands, and endangered species, pending and potential litigation, changing water regulations, and tax and lending regulations that affect value, and
4. environmental forces such as weather, growing seasons, groundwater levels, groundwater recharge, and frost potential.

Relevant specific data includes the property's legal description, physical location, address, access, historical and current use, soils, water rights, well and pump capacity, irrigation system, environmental restrictions, land inventory, site improvements, building improvements, zoning, assessments and taxes, special amenities, and leases and rents. An appraiser should also research the presence of potentially hazardous material on the subject property.

Data on comparable properties is important because it helps an appraiser assess how the subject property would compete in the market against other properties. To complete a working farm or ranch appraisal, an appraiser should research the neighborhood to determine its financial health, the availability of funds, and the overall economic outlook. This primary research should help the appraiser determine whether there is demand for the subject property now or in the future. In the appraisal of a recreational property, however, out-of-town buyers may be influenced more by changes in their areas than changes in the subject property's market area. For example, wealthy, out-of-state buyers may decide not to invest in a Wyoming ranch because Congress passed a new tax law that reduces the amount of depreciation that could be claimed. At the same time, local ranchers may see the ranch as a good investment because they know that calf prices are at an all-time high.

Market Analysis and Highest and Best Use Analysis

Even the simplest valuation assignments must be based on a solid understanding of prevalent market conditions and the highest and best use of the real estate. The two forms of analysis are related. In fact, an appraiser's investigation into trends affecting the economic base of the market area leads directly to the determination of highest and best use.

Land Value Opinion

For most appraisals of agricultural property, the land component is the primary contributor to the total value of the property. The traditional techniques of land valuation–sales comparison, allocation, ground rent capitalization–are generally applicable in appraisals of rural property when adequate market evidence is available for analysis.

Application of the Approaches to Value

In a rural property appraisal assignment, all recognized approaches to value that apply should be used. Sometimes all three approaches are pertinent and provide reliable indications of value. In other cases, only one or two approaches may be applicable.

In the mountainous recreational areas of the Wyoming ranch market, the income capitalization approach would not provide a reliable value indication because the market is being driven by recreation potential, aesthetics, and emotion rather than concern over potential earnings. Similarly, the sales comparison approach may not be applicable for appraising a cattle feedlot in a market where there have been no feedlot sales.

Additional considerations may be involved in applying specific approaches to rural property. In the application of the sales comparison approach, the price per *gross* acre and the price per *cultivated* acre can vary. A farm on the urban fringe may be marketed based on the price per gross acre when urban influences affect value. The same farm may revert to the price per cultivated acre when the urban market recedes.

The application of the cost approach may also require additional considerations. For example, in an appraisal of a farm that has many older improvements, the results of the cost approach might not be given as much weight as the results of the sales comparison approach. Also, older improvements typically can be used for storage until the structure ultimately falls down. Because it is difficult to determine accurate replacement costs for older improvements and the amount of depreciation to deduct, the reliability of the cost approach estimate may be weakened.

Reconciliation of Value Indications

The reconciliation step in the valuation process involves the analysis of different value indications to arrive at a value opinion. The essential considerations of reconciliation–e.g., the appropriateness, accuracy, and quality and quantity of market evidence–are as applicable to an appraisal assignment involving rural property as they are to appraisals of other property types.

The Appraisal Report

Collecting the proper data and deriving a final value opinion are critical to the appraisal process. Presenting an appraiser's final conclusions to the users of the appraisal report in a clear, persuasive manner is also essential. Effective communication through an oral or written appraisal report is well covered in the professional literature.

In reporting the results of an appraisal of rural property, it is often efficient and convenient to use information on file for much of the background data. However, each item of factual data should be carefully examined to ensure that it is relevant to the appraisal problem. The area and neighborhood description section of an appraisal of a vegetable farm with a roadside market requires different information and emphasis than the appraisal of a dairy farm in the same neighborhood. The area and neighborhood descriptions presented in the report must be relevant to the appraisal.

Comparable sale and rent information can be presented in different formats, but the formats should include a recorded document reference, sale date, sale price, and financing terms. Many sales of agricultural properties are package sales that include chattels. Itemization of the chattels and their contributory value to the sale should be described and analyzed. A physical description of an agricultural property needs to include all physical attributes that may affect value, including a land inventory. An accurate land inventory categorizes the different classes or types of land by acreage. The physical description should also include building descriptions, information on water rights and irrigation systems, and a soil description. Leases or rents at the time of sale should be noted, and livestock ranches should include the buyers' estimate of forage production (carrying capacity). Reference should also be made to sale confirmations.

Figure 1.4 shows topics that are often relevant in reporting the appraisal of a rural property.

An appraisal report should include plat maps, aerial maps, soil maps, topographical maps, site characteristic maps, building plats, and the like as appropriate for the appraisal problem. Agricultural properties are diverse, and some appraisals require significantly more data than others. For example, if the majority of the overall property value consists of land, fewer improvement photos will be needed. By contrast, if a significant portion of the total property value

Figure 1.4 Identifying Characteristics of Rural Property

- Size
- Access
- History and current use and operation
- Soils
- Water rights, irrigation systems, stock water, well information, etc.
- Topography including surface conditions, slopes, drainage, elevations, etc.
- Land inventory—an accurate accounting of all of the acreage by class.

  Class may be determined using area standards or by simply identifying the best to the worst land within the subject property based on soils, productivity, water availability, topography, etc. The land description would include soils, irrigation, potential cropping, and other features.
- Crop and forage history, including yields, carrying capacity, rotation, fertilizer program management methods, etc.
- Farm and ranch lease information (e.g., state leases, Bureau of Land Management grazing allotments)
- Government crop program information, grazing permits
- Description of land improvements, including fences, corrals, levees, roads, ditches, etc.
- Description of building improvements, including the condition, quality, size, amenities, and utility of each
- Utilities
- Taxes and assessments
- Zoning
- Easements, encroachments, and deed restrictions
- Wetlands, flood zones, endangered species, hazardous materials
- Detriments
- Special or recreational amenities
- Land use (e.g., acreage in cultivation, acreage irrigated, woodland acreage, minerals)
- Advantages and disadvantages of the property in relation to its market area
- Minerals
- Contiguous uses
- Amenities

is attributable to the improvements, fewer land photos may be included. The appraiser can assist readers in understanding maps, photographs, and other exhibits by adequately describing what is shown.

Extensive analysis will probably not be needed in presenting the conclusions of the highest and best use analysis of a simple agricultural property with a conforming land use in an agricultural area. On the other hand, an agricultural property on the urban fringe may require an in-depth analysis

Types of Value

Because the concept of value is central to appraisal, clear distinctions must be drawn among the related terms *price*, *cost*, and *value*. *Price* refers to the amount a particular purchaser agrees to pay and a particular seller agrees to accept under the circumstances surrounding their transaction. A *market* is a set of arrangements in which buyers and sellers are brought together through the price mechanism. Real estate markets involve the exchange of property rights for other assets, such as money. The market for family-operated dairy farms and the market for orchards are examples of specific real estate markets. *Cost* is used by appraisers in relation to production rather than exchange and can be divided into direct (hard) costs, indirect (soft) costs, and construction and development costs. *Value* is commonly perceived as the anticipation of benefits to be obtained in the future. Cost and price are facts, whereas value is always an opinion rather than a fact.

Because value is an economic concept, the monetary worth of property, goods, or services to buyers and sellers is an expression of value. To avoid confusion, appraisers do not use the word *value* alone. Instead they refer to *market value*, *fair value*, *use value*, *investment value*, *assessed value*, and other specific kinds of value. Market value is the focus of most real property appraisal assignments.

Market Value

A number of different definitions of market value can be found in a variety of sources, including appraisal texts, real estate dictionaries, professional standards, federal and state laws and regulations, licensing laws, and court decisions. Despite differing opinions on individual aspects of the market value definition, it is generally agreed that market value results from the collective value judgments of market participants.

The market value of limited-market properties may be based on their current use or the most likely alternative use. Limited-market properties are properties that are not commonly exchanged or rented and have relatively few potential buyers, such as manufacturing plants, research facilities, food processing facilities, and cold storage facilities. There may be little evidence to support market value estimates of these properties due to limited market interest and the extended market exposure needed to sell them. When a use is so specialized that there is no demonstrable market for it but the use is viable and likely to continue, an appraiser may be asked to develop an opinion of use value, which is discussed below.

Other Types of Value

Market value is not the only type of real property value that can be the focus of an appraisal assignment. Other types of value include

- use value
- investment value
- the value of a going concern
- insurable value
- assessed value and special use value

Use value focuses on the value a specific property has for a specific use without regard to the property's highest and best use. *Investment value* is the value of an investment to a particular investor or class of investors, based on individual investment requirements. The *value of a going concern* is defined as all tangible and intangible assets of an established and operating business with an indefinite life. *Insurable value* is the value of an asset or asset group that is covered by an insurance policy. *Assessed value* refers to the property's value according to the tax rolls in ad valorem taxation, and *special use value* is relevant in the valuation of family-owned farms and ranches for estate tax purposes.

The use value of urban real estate is tied to the improvements and the user's management of the property, while the use value of agricultural properties relates to what can be done with the land. The use value of an agricultural property reflects the income that the property generates under a specific use. The land, not the farm buildings or current farm operation, is the primary consideration in use value appraisals of agricultural properties.

Court decisions and specific statutes may also create the need for special-use value appraisals. For instance, many states require agricultural use appraisals of farmland for property tax purposes (i.e., value based on soil productivity) rather than opinions of value based on highest and best use. The current IRS regulation on estate taxes allows land under an interim agricultural use to be valued according to this alternative use even though the land has development potential.[3] Application of the regulation can be complex, and there are

3. The section on special-use valuation in United States Estate (and Generation-Skipping Transfer) Tax Return (IRS Instructions for Form 706) states: "Under section 2032A, you may elect to value certain farm and closely held business real property at its farm or business use value rather than its fair market value (FMV). Both special-use valuation and alternate valuation may be elected."

penalties for violating the terms of use of the special use classification–e.g., selling the property to an unqualified heir within 10 years, failure to use the property for the qualified purpose.

An appraisal of some type of value other than market value has a significant effect on the scope of work of an appraisal assignment, i.e., in the identification of the type of value and the definition of value relevant in the assignment. Similarly, an appraisal of the leased fee estate in the real property or of the real property along with other types of assets such as personal property and intangible business assets would have a significantly different scope of work than an appraisal of the complete bundle of rights in the real property. The next chapter examines the identification of the rights and interests in real property that are commonly appraised in a valuation assignment involving rural property.

Chapter 2

Identifying Rights and Interests to Be Appraised in Rural Property

As discussed in the previous chapter, the term *real property* refers to a complicated legal concept that is often misinterpreted as the physical parcel of land. Proper usage of the term relates to the benefits and rights inherent in the ownership of the physical real estate. In other words, the property consists not of real estate but of rights to real estate.

The property rights inherent in real estate include the right to occupy, sell, lease, or bequeath it and any other rights or benefits that can be derived from the ownership of the land. These various benefits of ownership are known as the *bundle of rights.* When a property owner has all of these rights, he or she has a complete ownership interest in the property, which is known as the *fee simple estate.* To best illustrate the bundle of rights, imagine a group of sticks. Each one of these sticks represents a right inherent in the ownership of property. The total number of sticks represents fee simple ownership. A lease, an easement, or a deed restriction takes one of the sticks away from the bundle of rights, leaving the property with an encumbrance.

Fee simple title, the complete bundle of rights, can be divided in many ways:

- by a lease (leased fee or leasehold estate)
- through an easement (for a highway, power lines, pipelines, etc.)
- by deed restrictions
- through a life estate

In the appraisal of rural and agricultural properties, one or more property rights are usually encumbered or have been severed from the fee simple estate. An estate may be subject to access easements, utility easements, mineral reservations, or other encumbrances. For example, consider a farm that has had the mineral estate severed. In this case, the property rights appraised are "fee interests in the surface estate." An analysis of sales in the market area reveals that surface interests in farm properties are selling for $6,000 per acre, and fee mineral interests contribute $1,000 per acre. If the property were appraised in fee simple at $7,000 per acre, it would be overvalued by $1,000 per acre.

Possession of a title in fee establishes the fee simple estate–i.e., absolute ownership unencumbered by any other interest or estate, subject only to the limitations imposed by the governmental powers of taxation, eminent domain, police power, and escheat. The fee simple estate includes air rights, subsurface rights, and all other property rights associated with real estate ownership. A fee simple estate that is mortgaged, however, does not include the full bundle of rights. Because an appraiser usually is not valuing the fee simple estate in agricultural properties, a conditional statement clearly describing the estate is often required. In most rural areas, minerals have been partially or totally severed from property rights and many properties are subject to access easements, deed reservations, and other restrictions. The comparable sales used to value a property should reflect the same or similar estates. In some instances, certain rights may not be included in the estate appraised, but these missing rights may have little or no value, meaning that they do not contribute to value in the context of the highest and best use of the property. An appraiser's job is to determine which rights contribute to value and how much value they add.

An interest in real estate is defined by degrees. A fee simple estate denotes complete, but not absolute, ownership. A leased fee estate typically describes a property subject to a lease that has a definite expiration date. A leasehold estate is the interest of the tenant or lessee.

A partial interest is any interest or group of interests that makes up less than the entire bundle of rights. Partial interests can be created in several ways:

- economically
- legally
- physically
- financially

Economic Interests

The most common type of economic interest is created when the fee simple estate is divided by a written lease, such as a farmer who leases lands for crops or for grazing livestock. In that circumstance, the lessor and the lessee each obtain partial interests, which are stipulated in contract form and are subject to contract law. The divided interests resulting from a lease represent two distinct but related interests–the leased fee interest and the leasehold interest.

A leased fee interest is the lessor's, or landlord's, interest. A landlord holds specified rights that include the right of use and occupancy conveyed by the lease to others. The rights of the lessor (the leased fee owner) and the lessee (leaseholder) are specified by contract terms contained in the lease.

Although the specific details of leases vary, holding a leased fee interest generally provides the lessor with the following:

- rent to be paid by the lessee under stipulated terms
- the right of repossession at the termination of the lease
- default provisions

When a lease is legally delivered, the lessor is obligated to surrender possession of the property to the tenant for the lease period and abide by the lease provisions. The lessor's interest in a property is considered a leased fee interest throughout the duration of the lease.

The leasehold interest is the tenant's position. A leasehold is the right to use property for a stated term under the conditions of the lease. When a lease is signed, the tenant acquires the right to possess the property for the leased period and to sublease it subject to the conditions of the lease. In return, the tenant is obligated to abide by the lease terms, pay rent, and surrender possession of the property at the termination of the lease.

In some instances a property is subject to verbal sharecrop arrangements. State laws address the rights of the lessor and lessee with regard to verbal agreements. In some states a verbal sharecrop arrangement must be terminated in writing by the lessor prior to a specified date set by the statute. Cash leases that are unwritten may carry the same termination rights, depending on state law. In some areas, local custom calls for written agreements or a written lease that is extended from year to year. Appraisers should be aware of local customs and state laws relating to agricultural leases and their implied effect on value.

Lease Documents

Leases can be verbal or written contracts. Verbal contracts are normally for less than one year and may or may not be enforceable under state law. Both written and verbal leases obligate the tenant to pay rent to the landlord and may include other agreements between the parties. A lease document should include a legal description of the land and any improvements included with the property.

Leases typically include clauses that specify the following:

- the length of the agreement
- the amount of annual or monthly rent
- the amount of any deposit
- the required insurance on the buildings and liability coverage on the land
- the landlord's right of entry
- provisions for subleasing, maintenance, and repair on all structures
- which party is liable for taxes

Leases may also include an eminent domain clause, direct payment for a possible taking or damages, a default clause, and renewal options.

It is important that appraisers understand the effect leases have on the property interest being appraised. Local custom and state laws are of prime importance. A property's value can be influenced positively or negatively by the terms and conditions of the lease agreement, just as the value of a property involved in a sales transaction may be influenced by the terms of the financing agreement.

Similar to leases are Conservation Reserve Program (CRP) contracts, which are agreements between landowners and the federal

Table 2.1 Payments from Conservation Reserve Programs

Payment Type	Description	Limit	Sign-up Type
Rental payment	Annual payment to participants. Based on soil productivity for each county and the average dryland case rental rate.	$50,000 annually for any person or legal entity	General and continuous sign-up
Cost-share payment	Payment for a percentage of installing or establishing an eligible practice	No more than 50% of the actual or average cost of establishing the practice	General and continuous sign-up
Maintenance incentive payment	Reimburses participants for the average annual cost of certain practice maintenance	$5 per acre per year	Certain continuous sign-up practices
One-time sign-up incentive payment (SIP)	One-time incentive payment made to participants that enroll certain practices	$10 per acre per year enrolled (not to exceed 10 years)	Certain continuous sign-up practices
One-time practice incentive payment (PIP)	One-time incentive payment for eligible installation costs for certain practices	40% of the eligible cost of practice installation	Certain continuous sign-up practices
Other financial incentive	Additional incentives, as part of annual rental payments, for windbreaks, grass waterways, filter strips, and riparian buffers	Up to 20% of the annual rental payment	Certain continuous sign-up practices

Sources: 16 U.S.C. 3834, 7 C.F.R. 1410.40-1410.42, and USDA, FSA, Conservation Reserve Program Continuous Sign-Up, Fact Sheet, July 2010, www.fsa.usda.gov/Internet/FSA_File/crp_contsignup_072610.pdf

government to leave cropland idle for a specified period of time–usually 10 years. For example, the Grassland Reserve Program (GRP) contracts limited future development and cropping uses of the land and retained the right to conduct common grazing practices and operations related to the production of forage and seeding.[1] In most cases, CRP contracts are not recorded, but the landowner is subject to very restrictive language during the holding period and a substantial penalty can be imposed if the lease terms are compromised. Appraisers should research and understand the implication of the language in CRP contracts and assess the quality of the income stream during the holding period.

Rental Payments

The three most common types of agricultural leases are gross leases, net leases, and cropshare leases. Farmland is typically leased on a cash rent or cropshare basis, both of which may or may not involve expense sharing. These leases delineate different responsibilities for the tenant and landowner. In a gross lease, the landlord receives the stipulated rent and is obligated to pay some or all property expenses, including real estate taxes. In a net lease, however, the tenant pays all or most of the property expenses in addition to the stipulated base rent. Property expenses usually include real estate taxes and the costs of insurance, management, maintenance, and structural repairs. In a cropshare lease, the landlord receives a share of the annual gross income and may participate in some of the annual expenses. Again, local customs often outline the level of responsibility ascribed to each position.

Rental payments are typically flat or level payments made throughout the contract term. A step-up or step-down lease provides for a certain rent for an initial period, followed by an increase (or decrease) in rent over stated periods. An *index lease* provides for periodic rent adjustments based on the change in an economic index. Index leases are usually long-term. A *revaluation lease* provides for periodic adjustments to contract rent based on the prevailing market conditions.

Rental payments can be monthly, quarterly, semiannual, or annual, depending on whether the property is subject to a cash lease or a sharecrop arrangement. Cash agricultural leases usually call for annual or semiannual payment in advance or partially in advance. Sharecrop leases are sometimes verbal and typically call for a share of the crop with payment at the time of harvest. Payments can be set on a percentage basis–i.e., one-quarter to one-third of the crop payable to the landowner or a 50–50 split for crop production. The tenant and landlord may also share in any government payments. Under

1. The GRP was repealed in the 2014 Farm Bill, but grassland preservation contracts–similar to what was available under the GRP–are currently eligible under the CRP.

sharecrop leases, the landlord is usually obligated to pay taxes and property maintenance costs, while the tenant provides the labor for maintenance and operation of the property. In analyzing sharecrop agreements, appraisers should look for automatic extensions and consider the perceived risk to the landlord in comparison with a cash lease arrangement.

The most important obligation associated with the rights to use and occupy a leasehold interest is the payment of rent. Contract rent is the actual rental income specified in a lease. The rent is paid by the tenant to the lessor according to the terms in the lease contract. A leasehold has positive value when contract rent is less than market rent, i.e., the rental income that a property would command in the open market. In turn, the leased fee value may be negatively affected by a positive leasehold. A leased fee encumbered with a long-term, below-market, fixed rental rate may be worth less than an unencumbered fee simple estate.

Legal Interests

The most common legal partitions of the complete bundle of rights in rural property are life estates, easements, and transferable development rights.

A life estate is defined as the total rights of use, occupancy, and control of a specified property limited to the lifetime of a designated party. The designated party is generally known as the *life tenant* and is obligated to maintain the property in good condition and pay all applicable taxes during the term of the life estate. Life estates can be created in several ways:

- by operations of law
- by wills
- by deeds of conveyance

For example, a fee owner may leave a will that gives farmland to his widow for her remaining lifetime and, at her death, passes the land on to their children. Thus, the widow acquires a life estate and functions as a life tenant with the children becoming the remaindermen. The income stream flowing to the life tenant is normally the basis for valuing a life estate. An appraiser can estimate the duration of the income stream by consulting actuarial tables, which indicate the expected remaining life of the tenant. Once the duration of the income stream is set and the net operating income from the estate is established, an appropriate market discount rate can be selected and applied to estimate the value of this property interest.

Obviously, the income from a real property interest subject to a life estate would be discounted at a market rate derived from other, similar real estate interests. For example, if actuarial tables indicate a 15-year life expectancy for the life tenant, a 15-year discount period would be applied to the reversionary interest of the estate. Life estates are common in rural areas where parents sell their property to their children and retain a life estate in the buildings and a small acreage around the structures. The value of the whole property depends on when the buildings and acreage are rejoined with the remainder of the property. The contribution of the portion containing the structures and acreage would be discounted at an appropriate market rate and added to the remaining value of the unencumbered property.

An easement is an interest in real property that transfers use, but not ownership, of a portion of an owner's property. Easements usually permit a specific portion of a property to be used for identified purposes, such as access to an adjoining property or as the location of a certain utility. Surface easements are the most common easements on agricultural property. Subterranean and overhead easements are used for public utilities, fiber-optic cables, subways, and bridges. Other easements such as conservation easements and scenic easements may prohibit the owner of the underlying fee simple estate from certain uses of the property without giving the holder of the easement any possessory interest in the real estate.

Clearly a property or entity that enjoys the benefit of an easement gains additional rights, while a property that is subject to an easement is burdened. The easement attached to the property benefitted is referred to as an *easement appurtenant.* The property whose owner acquires an easement is known as the *dominant tenement.* The property that is subject to the easement is called the *servient tenement.*

A conservation easement is a typical example of a contract between private parties, in which a land owner enters into an agreement with, say, a conservation group that limits the future use of a portion of the owner's property, often to ensure that some natural environment will not be developed. The property owner hands over certain specified rights of use and may receive a tax deduction. In exchange, the conservation group may or may not pay compensation (and offer ongoing property tax savings). An example of an easement created by prescription might be a right of access granted to the public who for many years have used a trail as a shortcut through a parcel of privately owned land. Prescriptive easement cases often require proof similar to that found necessary for adverse possession. Open space easements can inhibit the overall use of a property. The basic procedure for valuing these easements is similar to the procedure applied in eminent domain appraisals–i.e., the value of the property before the easement is compared to the value after imposition of the easement.

Transferable development rights (TDRs)–sometimes referred to as *severable use rights* (SURs)–became popular in the real estate industry during the 1970s. A TDR is a development right that is separated from a landowner's bundle of rights and transferred, generally by sale, to another landowner in another location. Some TDRs preserve property uses for agricultural production, open space, or historic buildings. In this arrangement, a preservation (or sending) district and a development (or receiving) district are identified. Landowners in the preservation district are assigned development rights that they cannot use to develop their own land but can sell to other landowners in the development district. The landowners in the development district can use these transferred rights to build at higher densities than zoning laws in the development district would normally permit. Development rights may also be transferred to land trusts or governmental agencies to curtail development and preserve environmentally sensitive properties.

Physical Interests

The partition of real property among physical interests can include vertical interests, such as air rights above the surface of the land, and subsurface rights, such as an ownership interest in an oil or gas deposit below the surface. For rural property that will be used for agricultural purposes, the water rights related to a particular parcel can have a significant effect on the productivity of that land and thereby the value of the real property.

The most common vertical interests in real property are subsurface rights and air rights. A subsurface right is a right to the use of and profits from the underground portion of a designated property. Air rights are the property rights associated with the use, control, and regulation of air space over a parcel of real estate.

The vertical division of real property is significant because engineering advances have dramatically affected land use and, therefore, highest and best use considerations. Congress has declared that the federal government has complete and exclusive sovereignty over the nation's airspace and that every citizen has "a public right to freedom of transit in air commerce through the navigable air space of the United States."[2] Many states restrict the ownership and use of subsurface areas, such as underground aquifers and oil or gas reserves.

Mineral Rights

Minerals include, but are not limited to, several broad categories including sand and gravel, precious metals, stone, clay deposits, gemstones, oil and gas, and quarry types such as building stone. Essentially, *minerals* is a term used to refer to any subterranean commodity that has or may have economic value.

2. The Air Commerce Act of 1926 (formerly 49 USC 171 et seq.), the Civil Aeronautics Act of 1938 (formerly 49 USC 401 et seq.), and the Federal Aviation Act of 1958 (see 49 USC 401).

Mineral rights consist of the right to extract some or all minerals contained in or below the surface of a property. Mineral rights may be granted with surface rights or without surface entry because the "mineral estate" is the dominant tenant in most states. When mineral rights are severed from the surface, the owner of the mineral interest has a right to occupy the amount of surface area that is reasonably necessary for extraction purposes. In these cases, however, the mineral rights holder must pay the market value of the surface acres occupied and any damages. In some areas, the surface damage income is greater than the owner's share of the mineral income from royalties.

A mining right for a specific tract of land is a right to enter and occupy the land for the purpose of extraction, either by underground or surface methods. In some instances, this right may also include the use of vegetative resources, such as timber, on the surface for the extraction process.

Types of Ownership

Substantial mining and quarry properties are owner-operated. The owner may have purchased the property from another private party or by patenting a mining claim. The development or transfer of the mineral rights in privately owned (fee title) land is accomplished within the framework of state laws relating to property titles, sales and conveyances, leases, licenses, and contracts. Private rights are typically subject to federal or state statutes relating to conservation, wildlife and wetlands mitigation, taxation, safety, and other matters.

When fee title is not owned, leases provide the vehicle for mining lands owned by private parties or federal and state agencies for specified types of minerals. Leases between mining operators and private owners have the following characteristics:

- They may include a purchase option.
- Lease payment may take the form of a royalty based on production.
- The rights may be subject to abandonment or forfeiture.
- Written leases may cover subletting.

Leases between mining operators and public owners have the characteristics illustrated in Figure 2.1.

A *royalty* originally was the rent or tax paid for the privilege of mining. In the United States, a royalty is a reservation to the lessor or seller of a portion of the minerals, or proceeds from their sale, at no cost. There have been many sales of mineral properties in which the seller retained a royalty interest even though title changed hands. The royalty is actually part of the production costs (gross royalties are typically deducted from net sales). The royalty should be included in valuing the leasehold estate when it can be accurately determined

Figure 2.1 Characteristics of Leases between Mining Operators and Public Owners

- The state "land commission" or a similar agency has jurisdiction over all mineral operations on state-owned lands or lands that contain minerals reserved by the state.
- Mining operations on public lands are subject to assessment as possessory interests just as possessory rights for other purposes are assessed for tax purposes.
- A permit or lease to prospect for minerals other than oil and gas may be required by the state.
- Leases include a production agreement, which will likely call for reclamation bonds and possible escrow for road work, utilities, and other requirements.
- The right to extract minerals on federal lands may be obtainable by lease or permit from the Secretary of the Interior, as opposed to filing a claim or patenting.
- Federal mineral deposits subject to "leasing" include coal, petroleum, natural gas, phosphates, oil shale, sodium salts, and potash.
- Federal "permits" are available for extracting pumice, clay, sand and gravel, and other "common varieties."

and is not contingent on unknown future commodity prices, profit levels, or other intangible measures.

Valuations of minerals may entail analysis of the following:

- a patented claim
- an unpatented claim
- a mill site
- in-place reserves (i.e., proven and probable reserves)
- structures only (leasehold or improvements held in fee)
- an operating mine

Patented mining claims are legally equivalent to a fee simple title acquired from the United States under the 1872 mining law. A *patent* is the conveyance by the United States of title to a citizen, in a situation in which the property was in the public domain and is being conveyed into private ownership for the first time. A patent is the equivalent of a quitclaim deed from the sovereign. All valuable mineral deposits (i.e., gold, silver, iron, lead, copper, and other metals plus many nonmetallic items like gypsum, limestone, and rare earths) on or in public domain lands, not withdrawn from mineral entry, are open to exploration, occupation, and patent. Patented claims are subject to taxation since they are, in effect, fee property. They can be mortgaged or sold like any other fee interest. Lands not open to mining claims include federal lands not in the public domain–e.g., acquired lands, public domain lands withdrawn from mineral entry, privately held lands, patented town sites, and land deeded to the state for unpaid taxes.

Unpatented claims are, in effect, the right to explore and produce locatable minerals under the 1872 mining law. No application for a patent is necessary to operate a mine on public domain land. In most cases, a legal opinion is required as backup to determine the disposition of unpatented claims. Unpatented claims include mill site claims, lode claims, and placer claims. Lode claims must be based on the discovery of a valuable mineral deposit. Lode claims may include extra lateral rights, if they include the top or apex of the vein or lode. Each lode claim is limited to about 20 acres (600 feet × 1,500 feet, or slightly more than 20 acres) regardless of the number of locators. A placer claim may be located on all other types of valuable and identifiable mineral deposits. The placer location must also be based on a discovery, but a discovery that is not in a vein or lode formation. The maximum size for a placer claim is 20 acres per person. However, an association placer may contain 20 acres per associate, up to a maximum of eight associates, and 160 acres in a single claim.

Mill site claims may be located on a maximum of five acres of nonmineral land in conjunction with either lode or placer claims for the purpose of milling or processing the mineral. If the mining claim is forfeited or abandoned, the right to the mill site may also be lost, unless it is used for custom milling.

Ownership of Oil and Gas Deposits

Oil and gas properties tend to be more complicated than other mineral entities because of the wide range of ownership interests that may be encountered. Ownership can be broken down into several categories. It is necessary to know what type of ownership is being evaluated in the appraisal process. Various ownership interests are described in Table 2.2.

Table 2.2 Ownership Interests

Mineral owner	The person or entity possessing title to the mineral property. The mineral owner may or may not hold title to the associated surface of the land. In this case, the mineral estate is considered severed from the surface ownership.
Leasehold owner	The person or entity holding a lease on the mineral estate with the right to extract the oil and gas reserves present.
Royalty interest owner	That portion of the produced oil and gas retained by the mineral owner as partial compensation for leasing to the leasehold owner. This is normally a percentage of the gross production and is subject only to severance and income taxes.
Overriding royalty interest owner	That portion of the produced oil and gas generally retained by a third party or parties as compensation for services or subleasing of a particular mineral property. This is also normally a percentage of the gross production and is subject only to severance and income taxes. This interest is created from or carved out of the leasehold interest.
Working interest owner	The party who pays 100% of the costs of production. Working interest owners are usually, but not always, the leasehold owners.
Reversionary interest	Sometimes called a *back-in*, a reversionary interest is an interest that will revert normally to the leasehold owner at a predesignated time, such as pay out of the drilling and completion cost of an oil or gas well. This interest is usually an option to continue with a given overriding royalty, or to convert that interest to a larger working interest.
Net revenue interest	The net revenue to the working interest owners after all production costs, royalty, and overriding royalty interests have been paid.

To further complicate many oil and gas properties, vertically segregated mineral ownership may exist. One company may have a leasehold interest in a given mineral property down to a certain depth below ground level, such as 8,000 feet below the surface, while a second company may have the leasehold for all zones below 8,000 feet.

Water Rights

The legal right to water is as important to the value of a property as the physical source of the water. A *water right* is a right to a definite or conditional flow (quantity) of water, generally for use at stated times and in stated quantities, for irrigation, livestock consumption, real estate developments (subdivisions), hydroelectric power development, and other uses. Water rights vary greatly throughout the United States and are affected by numerous compromises, exceptions, and changes. Most water rights, however, are controlled by state law. Many disputes occur because of the fluid nature of water and the fact that laws were passed without scientific knowledge of the interdependencies of the water cycle.

Surface Water Doctrine

Laws regarding surface water rights in the United States are complex primarily because of differences in the availability of water. The *riparian doctrine* is applied in states where water has historically been plentiful. In the most arid states, water rights are based on the doctrine of prior appropriation. Some states employ both doctrines, with the appropriative doctrine being dominant. The major differences between the riparian and appropriative doctrines of surface water are summarized in Table 2.3.

The riparian doctrine of surface water is based on English common law and gives the owner of the land bordering a lake or stream

Table 2.3 Summary of Riparian and Appropriative Rights to Surface Water

Riparian Right	Appropriative Right
Prevails in eastern United States	Prevails in western United States
Is based on English common law	Assumes public ownership of water
Is linked to land ownership and physical contiguity	Is acquired by diverting water for beneficial use and meeting specific local requirements
Entitles user to reasonable use, without materially diminishing quality of watercourse; "right to simple usufructuary while it passes along"	Entitles user to physical consumption of a certain amount of water for specified land
Entitles user to equal sharing	Entitles user to consumption on a seniority basis; "first in time is first in right"
Depends on ownership of land in physical proximity to water and transfers automatically with the land	Is a real property right, separate from right to land and separately transferable

the right to use the water on the contiguous land. The riparian right exists solely because the land is located next to a natural water supply, and the right to the surface water resides in the ownership of the land.

The riparian owners up and down the stream have equal usufructuary rights. The term *usufructuary* refers to the "use of the fruits." In other words, the riparian owners up and down the stream are entitled to share the water for reasonable use without seriously diminishing the quantity or quality of the water used by other owners. The owners' rights are equal, regardless of location along the stream or the time when each property was purchased. The riparian doctrine does not define "reasonable use" or guarantee any particular quantity of water. The resulting uncertainty can lead to legal conflict among the owners along a watercourse. Reasonable use is determined in a particular case only when one party brings suit and the court makes a decision. As long as the practical application of riparian rights is confined to areas where the supply of water generally exceeds the demand, all users are able to share the available water and there is no need to specify quantity or reasonable use.

A riparian right can be lost by selling the land next to the water, leaving the remaining land cut off from the supply of water. Similarly, the right can be lost if the stream shifts or moves away from the riparian land. A riparian right can also be lost by condemnation or by failing to prevent someone else from diverting water for a certain number of years, as set by the statute of limitations (prescription).

Under the doctrine of prior appropriation, water is owned by the state and may be granted for beneficial use through appropriation. Appropriative surface water rights are acquired by use and the performance of certain legal requirements, as directed by the state, the county, or an irrigation district. These rights apply to a specific quantity of water and often to particular times of use, places of use, and methods of diversion. Unlike riparian surface rights, which depend on the ownership of land in physical proximity to water, appropriative surface rights can entitle the owner of the water right to surface water that is not contiguous to the land on which it is used. In addition, the owners of appropriate surface rights are entitled to consume or use the appropriated quantity of water, provided that the use for the water is beneficial. The oldest water right has precedence over more recently acquired water rights. Simply stated, first in time means first in right. This means that in times of scarcity, the holder of the oldest right is entitled to a full quantity of water before any is given to later holders. This principle of seniority clearly differs from the equal sharing of water in the riparian doctrine. Many appropriative water rights can also be transferred independently of the land.

Although the details vary with state regulations, the usual pattern for establishing appropriate water rights is:

1. The landowner files an application to use a certain quantity of water on specifically designated land, taking the water from a designated diversion point.
2. The landowner proves beneficial use for this water.
3. The appropriate authority in that state grants a decree or adjudication.

The water rights may be evidenced by a contract with the Water and Power Resources Service (formerly the US Bureau of Reclamation), with the Secretary of the Interior, or with a public utility water distributor. Water rights may also be given by an individual state certificate or decree, by shares of stock in an irrigation company, or by location in an organized irrigation district. These agencies differ widely in administration and the terms on which water is granted. Appraisers must therefore be aware of the different regulations, local considerations, and any changes in the regulations. If water is obtained from public agencies, the laws governing use can be changed, and that change can put limitations on the number of acres for which federal water can be used. For example, the Reclamation Act of 1902 specified that federal water can be provided for only 160 acres per owner (320 acres for a farm couple). The Reclamation Reform Act of 1982 increased the acreage limitations to 960 acres for a family farm. The act also affects the use of federal water on corporate-owned lands, methods for calculating the cost of water used on excess acres, and the sale of lands in excess of the acreage limitations.

Private agencies can also change the cost of water, the terms of delivery, or the dependability of delivery. For this reason, appraisers should be aware of the management practices of the agency, the condition of indebtedness, the type of payments or assessments, and any shares or voting rights accorded the owners of water rights. The members of a private irrigation company could, for instance, vote to raise the individual assessments on water rights, with a corresponding increase in the expenses of the irrigator dependent on this water. These organizations usually keep records of the amounts, timing, and costs of water used by irrigators, which can be employed by appraisers in estimating the amount, sufficiency, dependability, and expense of the water used for a particular tract of land.

To maintain an appropriative right, the owner must demonstrate "beneficial use." An appropriative right can be lost by condemnation, prescription, abandonment, or nonuse for three to five years, depending on state law. With regard to water, the general rule is, "Use it or lose it."

Groundwater Doctrines

Most state groundwater laws were developed after the surface water laws. Sometimes, groundwater doctrines follow or are combined with the surface water laws of a state.

Resembling the riparian doctrine for surface water, the *absolute ownership doctrine* states that a landowner can take as much water as desired from beneath the owned land. The water is considered part of the landowner's soil. The *reasonable use doctrine* is a modification of the previous principle because use of groundwater can affect the water supplies of other landowners. This modification limits each landowner to reasonable use, recognizing equal rights to groundwater resources. A further modification of the reasonable use doctrine is the *correlative rights doctrine*, which states that the landowner's use must be correlated with the use of others, especially during times of water shortage. When water is in short supply, each landowner is entitled to groundwater in proportion to the percentage of land he or she owns in relation to the other lands overlying that groundwater supply.

Under the doctrine of prior appropriation, groundwater, like surface water, belongs to the people of the state and is subject to appropriation for beneficial use through a priority system similar to that used for surface water. In 1959, Hawaii passed a groundwater use act providing for regulation of groundwater withdrawals from designated areas by a state commission.

Other Water Rights

A neighbor may gain *prescriptive rights* by openly diverting water to which others have prior rights. If the diverted water is used without interruption for the number of consecutive years prescribed by state law and other legal requirements are met, a prescriptive right against the original owner is granted. A prescriptive right may be lost in the same way–i.e., it may be prescribed by someone else openly diverting the water for the specified period of time. A *vested right* is a clause in appropriation law that protects the rights of persons who appropriated water for beneficial use before the passing of appropriation laws. A vested right can also be lost by lack of use.

The water rights in a particular location can affect the value of property in different ways. In an appraisal of irrigated property, an appraiser must know whether the water rights are appurtenant to the land or transferable separately from the land. If they are separate, the legal source, conditions, terms, and value of the water rights should be known and noted by the appraiser. If water rights are not transferred with the land, the property's value may decline significantly and its highest and best use may be changed.

Financial Interests

An exceedingly common division of the complete bundle of rights in real property is the separation created by equity and debt ownership interests–i.e., a property with a mortgage. The availability of capital for investment in rural property has a significant effect on the frequency of sales and consequently on the prices asked and achieved in the market. In individual transactions, the financing terms are of great concern in the valuation process, particularly if the financing terms for transactions of comparable properties differ greatly from the financing terms for the property being appraised, requiring a significant adjustment to the sales prices of the comparable properties in the application of the sales comparison approach.

Chapter 3 examines the complexity of the financing of rural property and the markets for debt and equity interests in detail.

Chapter 3

Financing Rural Property

Rural property valuers need to be aware of how the availability of capital influences the agricultural property market and how the US agricultural industry's capital requirements are met. Capital is the agent of production that brings land, labor, and coordination (the other three agents of production) into use. Farm sector assets in the United States–dominated by real estate–represented around $3.005 trillion in 2015, according to the Congressional Research Service.[1]

Figure 3.1 Real Estate Assets Made Up 81% of Total Farm Assets in 2015

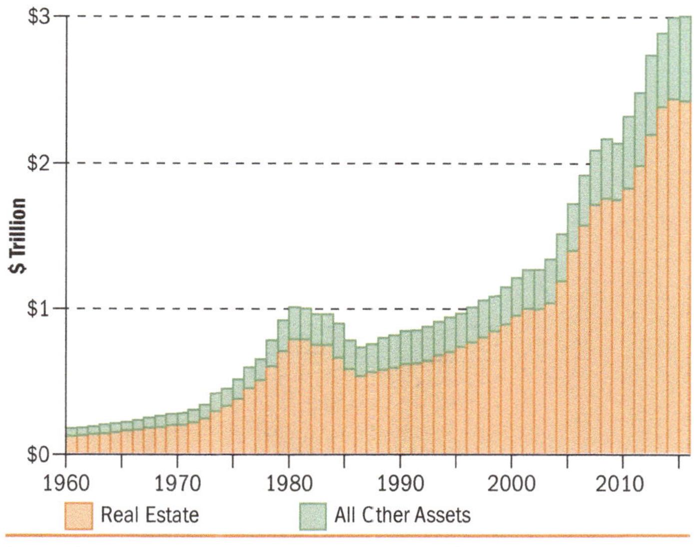

Source: USDA, ERS, "2015 Farm Income Forecast," February 10, 2015. All values are in nominal terms, that is, not adjusted for inflation. 2014 is preliminary; 2015 is forecast.

Notes: Non-real estate assets include financial assets, inventories of agricultural products, and the value of machinery and motor vehicles.

Agricultural property attracts outside capital investment when the return on that investment is competitive with other

1. *US Farm Income Outlook for 2015* (Washington, DC: Congressional Research Service, February 18, 2015). Accessed online on 17 November 2015 at www.fas.org/sgp/crs/misc/R40152.pdf/.

investment opportunities. Local buyers may accept inferior returns on an individual property if it benefits their larger holdings as a whole. In today's market, investment returns are measured in terms of economic, social, governmental, and environmental benefits.

Agricultural property represents an investment of capital that is in some ways similar and in other ways significantly different from investments in the urban sector of the nation's economy. Like commercial property in the urban sector, agriculture requires capital investment to support operations and acquire the equipment, fixtures, and real estate used in production and processing. The capital consumed by agriculture takes the form of equity and debt, and it may represent a short-term (for one crop or season), intermediate-term (for equipment or livestock), or long-term (real estate) investment. In contrast to the urban sector, agricultural asset investments are frequently dependent on the economics of a single commodity, are usually in a property that requires skilled management, and may involve a residential use along with the agricultural use.

Income to provide a return on and recapture of the capital invested in an agricultural property is highly dependent on weather, commodity price and production cost relationships, and, particularly in grain and fiber production, government programs including the Federal Crop Insurance Program. To a far greater degree than in the urban sector, the rural property appraiser must consider the viability, stability, and durability of the agricultural enterprise on the property in the course of completing an appraisal of an operating business. Figure 3.2 outlines the history of agricultural capital markets in the United States. Figure 3.3 illustrates the evolution of agricultural finance in selected international markets.

How the Capital Markets Work

Agriculture requires the investment of capital to support operations and provide the assets used in the production and processing of agricultural goods. The profit and other benefits from this use of capital must be sufficient to reward the investor at a rate at least equal to that offered by alternative investments. Today's agricultural property owners and managers have access to a wide variety of financing sources, but they also face a broader group of competitors for capital. National, regional, and state economies, multinational and national corporations, and the urban, commercial, and residential markets are competitors for capital. Private individuals and companies, community banks and thrifts, retirement funds, hedge funds, publicly traded companies, insurance companies, and money center banks all have the opportunity to channel investment funds to maximize yields through the securities markets, syndications, pooling, and a host of other financing mechanisms. Agricultural lending must also compete with these investment alternatives for capital.

Figure 3.2 History of the Agricultural Capital Markets in the United States

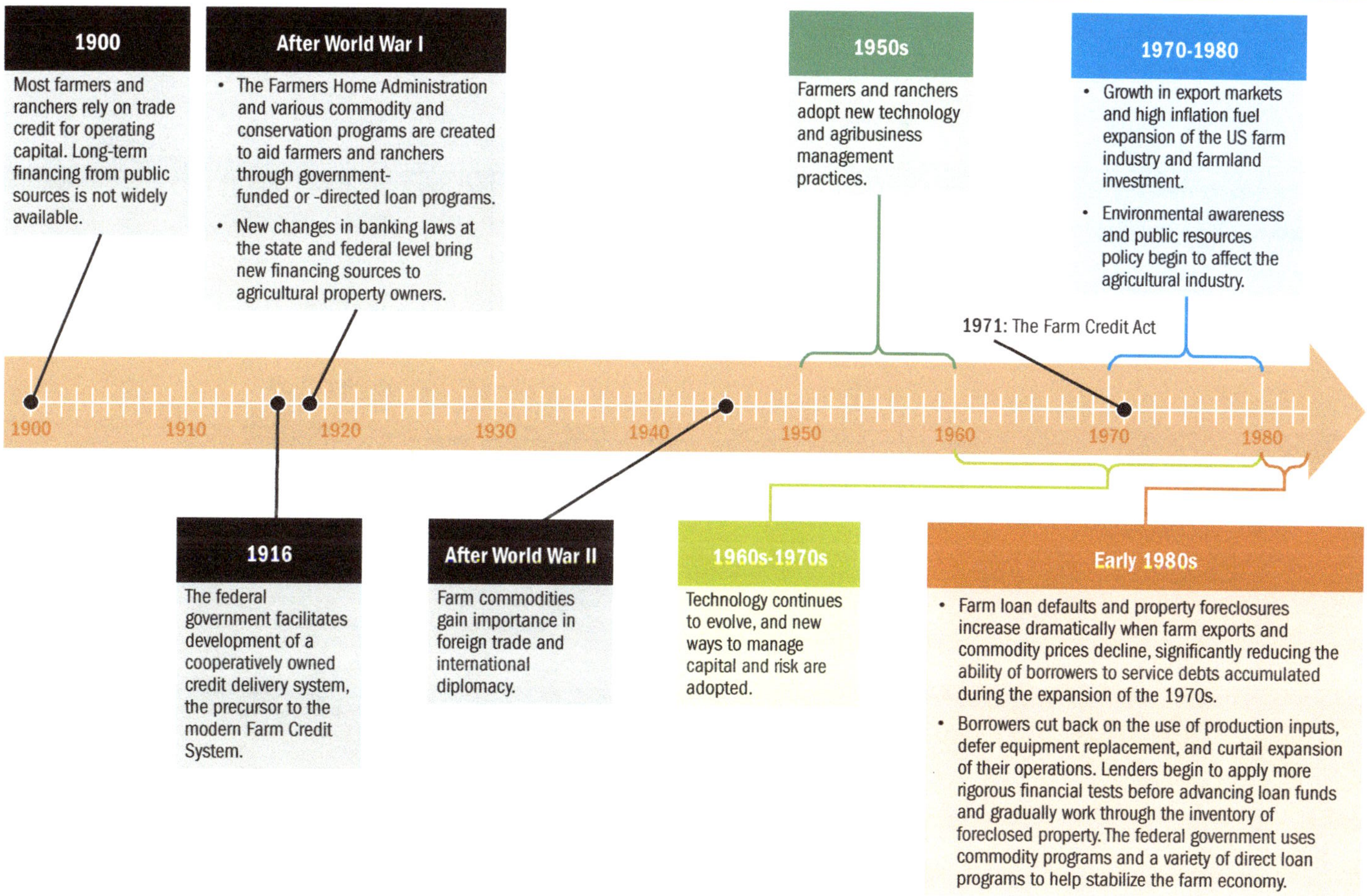

Figure 3.2 History of the Agricultural Capital Markets in the United States *(continued)*

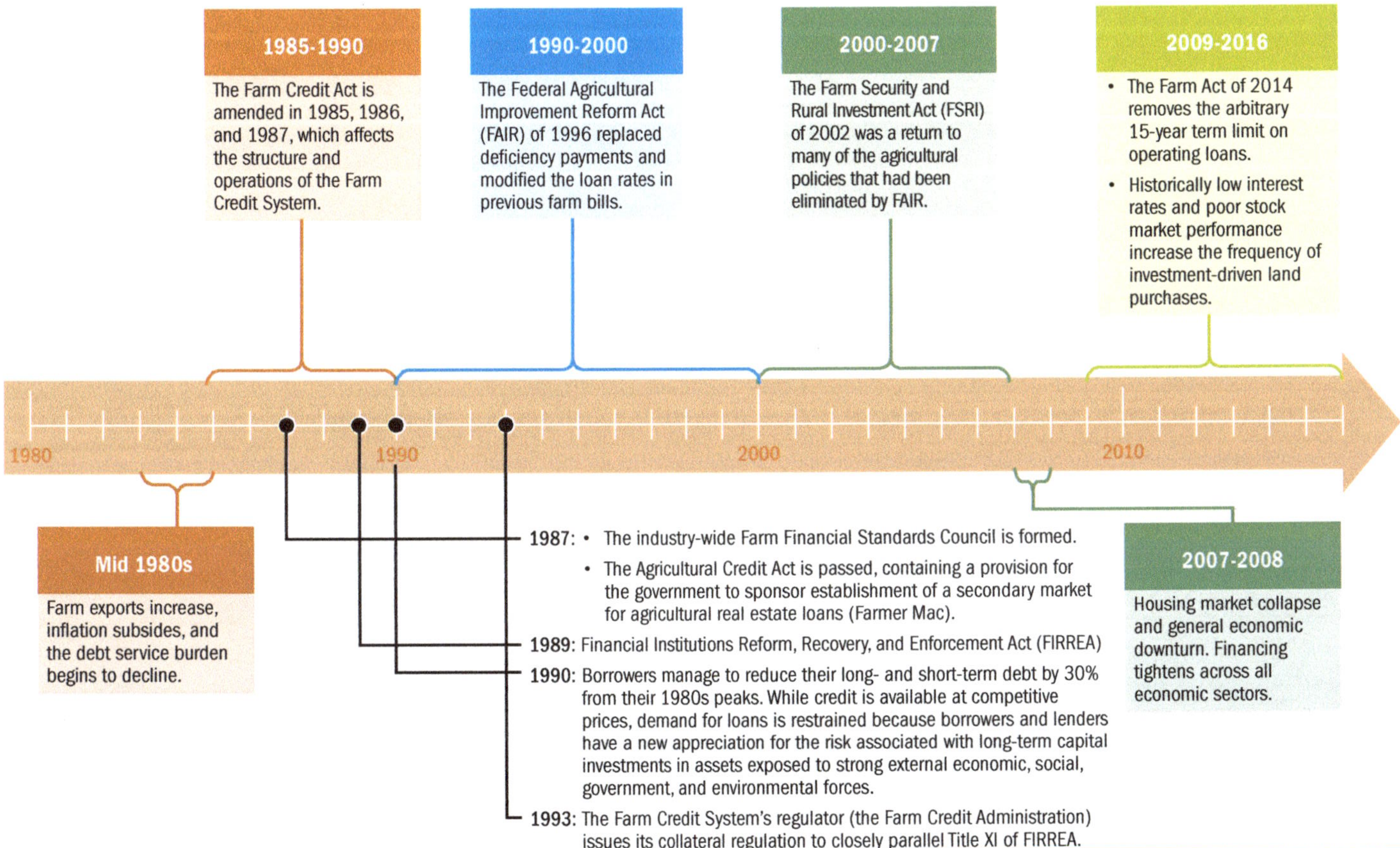

Figure 3.3 Stages of Agricultural Finance Development

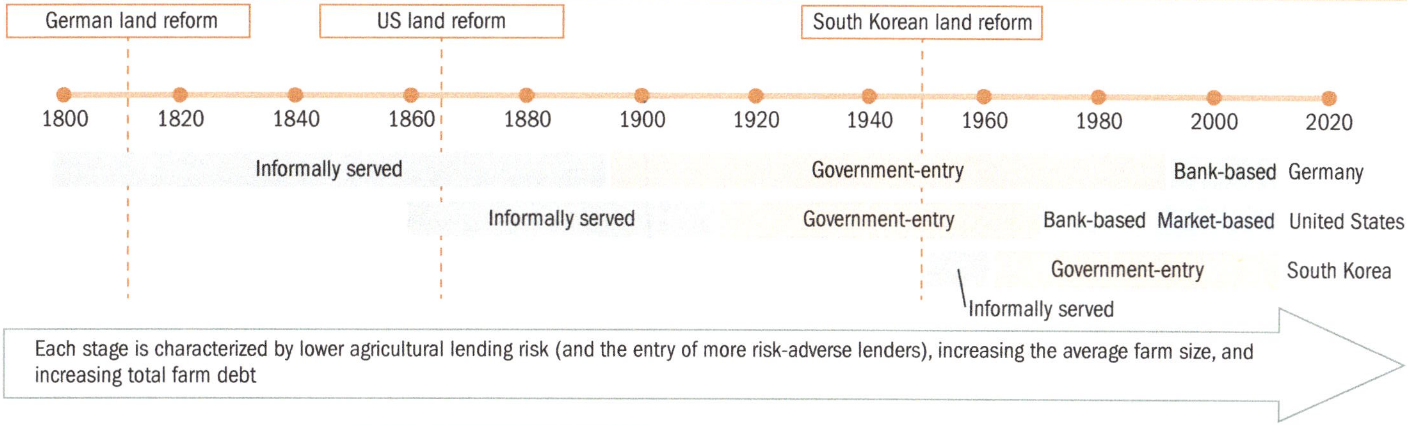

The Informally Served Stage	The Government-Entry Stage	The Bank-Based Stage	The Market-Based Stage
Government role: Little or no involvement	*Government role:* Expanding available credit, working through farm-level organizations, enhancing agricultural productivity more broadly	*Government role:* Oversight and regulation	*Government role:* Deregulation
Sources of farm debt: Non-institutional moneylenders	*Sources of farm debt:* Government-supported community lenders	*Sources of farm debt:* Commercial lenders' share of agricultural debt is greater than or equal to government-sponsored institutions; Informal lenders a minority	*Sources of farm debt:* Banks, securities, and "merchant creditors"
Farm structure: Smallholders and tenant farmers	*Farm structure:* Decrease in smallholders as land consolodation increases	*Farm structure:* Land consolidation intensifies	*Farm structure:* Split between large commercial and smallholder with links to merchants creditors
Farm productivity: Low	*Farm productivity:* Increased input use drives rising productivity	*Farm productivity:* Productivity increases	*Farm productivity:* High

Source: www.raflearning.org/post/role-government-developing-agricultural-finance-look-history-germany-us-and-south-korea

Agricultural property owners and managers evaluate the risk in their operations before committing their future livelihood to debt service. Lenders likewise evaluate the risk inherent in an agricultural property in the course of making a loan commitment.

Lenders typically evaluate the risk in any credit decision by considering five factors:

Character	Character is represented by the applicant's experience, personal history, and management abilities, particularly with the type of agricultural enterprise and business entity associated with the property.
Credit	The character evaluation, together with the applicant's credit history, provides an indication of the applicant's willingness and ability to repay the debt.
Capacity	Capacity refers to the applicant's previous financial performance and well-supported business and enterprise cash flow projections. The evaluation of the cash flow provides an indication of the applicant's ability to repay the debt from business operations or other sources.
Collateral	Collateral is the amount of security the property to be taken as collateral represents. When the character and capacity evaluations result in a need to take collateral as security, the collateral taken must reasonably protect the lender while allowing the applicant to manage the business constructively.
Conditions	The conditions factor refers to loan terms (e.g., advance rate, amortization period, types of security, specific performance requirements) and covenants.

Appraisals completed for lenders do not usually involve character and credit factors.

Although the work of appraisers in connection with lending focuses on the collateral category, consideration of the capacity category is also needed in the course of the appraisal work. The capacity of a property to produce an income flow is a function of physical property characteristics and management, subject to commodity price level fluctuations. The commodity associated with the property establishes a reference point when considering the quantity and stability of the income that a property can be expected to generate.

When developing an opinion of market value, appraisers focus on the level of management that is "typical" for the subject property. An income analysis capitalizing an income stream that results from superior management will tend to overstate the value of the real property. Conversely, capitalizing an income stream that results from inferior management will tend to penalize the property. In other words, when developing an opinion of market value for real property, appraisers should focus on what is typical for the property rather than the results that superior or inferior management achieves. If an appraiser is developing an opinion of investment value or some other value specific to a particular ownership (but not market value), it may or may not be appropriate to consider a specific, atypical management level.

Commodity price changes and climatic cycles since the 1980s have created a renewed appreciation for the need to develop "stabilized" estimates of production and income for use in market analysis and credit decisions.

Rural property includes several primary property types and uses, each with its particular capital needs and sources. The primary categories addressed in this text are

- agribusiness
- production agriculture
- timberland
- rural residential

Rural properties are used for various other purposes, including recreation, open space or ecological preservation, and mineral extraction.

Agribusiness

Agribusiness properties typically are vertically or horizontally integrated units, with at least a portion of the property devoted to packing, processing, or shipping agricultural commodities. Although there are exceptions, most properties of this type are single-purpose by design and have limited market appeal as operating units without the associated business entity and its intangible property (such as franchise rights, trade name, label, or contracts). When these properties are sold as operating units, the sale most often involves a business entity transfer with the price for the tangible real and personal property and intangible property components allocated based on negotiations involving accountants and tax attorneys. When these properties are sold as vacant or nonoperating units, the buyers are typically paying only the price they view as reasonable for the property's productive capacity, net of their cost to reconfigure or rehabilitate the property to suit their needs.

Agribusiness properties are commonly included in corporate finance lending as part of a general security instrument taken by the lender. It is difficult to develop a stand-alone market value for the real or personal property portions of these properties as components in an operating unit. Most often, the value developed represents the contributory value of the property appraised, based on its historical or designed throughput capacity, to the going concern that makes up the related agribusiness entity. When the property appraised is part of a viable business operation, the function of the appraisal is most often to aid the lender in developing a reasonable allocation of the loan principal between short-, intermediate-, and long-term amortization schedules. A real property appraiser may need to enlist the help of personal property and business valuers in determining appropriate allocations.

When the property appraised is not operating due to the financial failure of the ownership entity, the function of the appraisal is most often to aid the owner, the lender, or a court in estimating the amount the property will bring when liquidated.

Production Agriculture

Production agriculture includes a vast array of crop, livestock, and other rural property operations. The capital needs of production agriculture property owners vary with the type of enterprise associated with the property and its ownership. Some properties, such as those held in ownership by one family for generations, have very low or zero long-term debt or have no need for borrowed operating funds, while other operations have complex debt structures, including a full range of loan types and terms.

The enterprise located on the property can produce a single commodity or a broad range of commodities. Integrated units involving several commodities are not unusual. Some properties are the sites of very specialized and intensive operations. Some examples are stand-alone livestock or livestock commodity production units, such as a farrow-to-finish swine facility, a dry-lot dairy, a poultry (broiler or egg) unit, an aquaculture unit, and the like. These single-purpose properties, while perhaps not involving vertical integration, take on the character of an agribusiness property in the farm real estate market and in the type of financing they use.

Appraisals of production agricultural properties for lenders must address the highest and best use and market value of the property under typical ownership and management. In particular, these appraisals must develop a sound basis for estimates of the property's productive capacity and earning ability.

When an appraisal is conducted in connection with a long-term loan, the specific characteristics of the market for the property need to be fully described. The lender and borrower are in the process of making a long-term capital investment decision and need to be aware of the stability and durability of the market for the property well beyond the current production cycle. Although appraisers are not expected to forecast the future, professional rural property appraisers have a unique perspective on the market for these properties, and lenders as well as owners seek the benefit of that perspective when long-term capital investment decisions are made.

Timberland

Financing timberland (exclusive of farm woodlots) is most often handled by insurance companies or Farm Credit System institutions.

Large corporations finance their timberland acquisitions internally or through methods such as private placement bonds. When private individuals require timberland financing, the loan amount is generally dependent on the merchantable timber value, and the lender includes loan covenants requiring partial reduction of the loan as the timber is harvested.

As environmental concerns, particularly over endangered species, have grown since the 1980s, it has become apparent that professional appraisers need to be familiar with the ecology and local forest practice regulations as well as with the economics of the timber industry. Timberland financing can extend well beyond the 20- to 30-year loan term common to other production agriculture property, and future harvest rights can be remarkably different than the present rights. Financing for timberlands is highly specialized, and any appraiser practicing in this market needs to be appropriately qualified and well aware of the market's characteristics.

Rural Residential

The financing of rural residential properties is carried out by a variety of institutions, including those serving production agriculture and urban property markets. Commercial banks and thrifts, Farm Credit System institutions, credit unions, and mortgage brokerage firms are active in financing rural residential properties, and Federal Housing Administration (FHA), Veterans Administration (VA), and Rural Development (RD) loans can be obtained in rural areas. USDA Rural Development also makes loans available for community development in rural areas of a specified size.

The appraisal of rural residential properties other than farm properties generally relies heavily on the use of the sales comparison approach to value. In rural communities, residential and commercial developments are equivalent to the urban use of land in rural settings. In these settings, appraisers must apply the same considerations that are required in a more populated setting, but these appraisers will also need to evaluate how an entirely new type of property in the community will be received by the market.

Although appraisals of rural residential properties have much in common with appraisals of urban residential properties, there are also important differences between the two. Elements such as access, water and sanitation, power and telephone services, and distances and convenience to community services (e.g., schools, medical facilities, churches, markets, fire protection) can cause significant differences in the market for a rural residential property. Other important factors include the property setting, adjacent property uses, linkages to employment centers, community income levels, and local land use or subdivision restrictions. The latter element can be highly significant when the rural residential property involves a parcel that is larger than what is needed for a single dwelling.

Types of Agricultural Property Financing

The unique needs of the agricultural industry have given rise to the expansion of credit sources to meet its increased credit demands. These demands include loans to service short-, intermediate-, and long-term needs.

Short-term credit, generally called *production credit*, provides working capital for annual and seasonal operating costs. The purchase of production inputs such as seed, feed for animals, and fertilizer, as well as other essential operating expenses, are often financed. Loans or advances are repaid from the proceeds from the harvest of annual or seasonal crops, which often serve as the collateral for this type of credit. The customary maximum duration of this type of loan is the crop season or livestock production cycle, i.e., generally one year.

Intermediate-term credit is extended for inputs that are used for more than one season, such as equipment, breeding livestock, and dairy cattle, and that are leased or financed for three to 10 years. Intermediate-term credit may also be used to refinance short-term debts during adverse periods in the farm economy. Personal property is ordinarily used as collateral for this type of loan, but lenders may require mortgages on real estate as additional collateral for short- and intermediate-term loans when borrowers have only marginal security otherwise. However, this use of real estate as additional security for a loan is a result of short- and intermediate-term lending rather than real estate lending.

Long-term credit is used primarily for the purchase of real estate or the consolidation of short-term or intermediate-term debts. Homes in rural areas, as well as farm and ranch land and buildings, are financed through long-term loans. Soil and irrigation improvements may also be financed using long-term credit. In all cases, the loan is secured by a mortgage or deed of trust on the real estate itself. The multiple purposes of long-term credit make a variety of terms necessary, but most real estate loans are made for periods ranging from 10 to 30 years.

Agricultural property owners use a variety of sources for credit. The major sources of short- and intermediate-term credit are

- commercial banks
- Farm Credit System institutions
- the Farm Service Agency (FSA)
- merchants

- dealers
- suppliers
- processors
- installment/land contracts
- owner financing

The major sources of long-term loans for rural properties are

- individuals
- commercial banks
- life insurance companies
- Farm Credit System institutions
- Farmer Mac
- the Farm Service Agency

Processors have increased their long-term financing activity, particularly to help producers establish or expand permanent plantings or make other production unit improvements.

A major share of debt financing is provided by institutional lenders, such as banks, insurance companies, Farm Credit System institutions, Farmer Mac, and certain government agencies–mainly the Farm Service Agency, which was previously the Farmers Home Administration. Private sources, including individuals, trade creditors, and processors, represent a significant share as well. The percentage of debt financing provided by these various sources changes over time.

Commercial Banks

Commercial banks are private corporations chartered by either federal or state governments.[2] For the purpose of this discussion, "banks" includes thrifts (savings and loan companies), although thrifts may have significantly different business structures.

Banks derive their loan funds from money deposited in savings and demand accounts and are required by law to keep a certain percentage of these deposits in reserve. As depository institutions, banks are subject to federal regulation by the Federal Deposit Insurance Corporation (FDIC). Depending on its charter, a bank may also be regulated by state bodies or federal agencies such as the Office of the Comptroller of the Currency (OCC) or Office of Thrift Supervision (OTS). State banks that are members of the Federal Reserve System and bank holding companies or non-bank subsidiaries of bank holding companies are regulated by the Federal Reserve Board (FRB).

2. Additional detail on bank and thrift sources of credit is available in *The Appraisal of Real Estate*, 14th ed. (Chicago: Appraisal Institute, 2013), 129–162.

In the past, commercial banks were not considered a primary source of funds for financing agricultural real estate. Banks historically concentrated on commercial lending to agriculture with short- and intermediate-term loans for production financing and for equipment, fixtures, or property improvements. Real estate mortgages or deeds of trust were taken by banks as additional security on some commercial loans. However, in response to the modern farmer's increased credit needs after World War II, and particularly since the 1970s, banks expanded their long-term and commercial agricultural lending activities. By creating branch offices and participating with urban banks and other production credit lenders through correspondent banking relationships, banks have made more funds available to farmers and ranchers.

In 2013, commercial banks accounted for about 37% of the outstanding farm debt on real estate and almost half of non-real estate debt (see Table 3.1). The interest rate charged by commercial banks is subject to any usury laws in effect, and the loans themselves must be fully amortized. Commercial banks make conventional loans (loans whose security is based solely on the property used as collateral) and loans that are insured or guaranteed by agencies of the federal government, such as the FHA, VA, and FSA.

Commercial banks figure most significantly in the extension of short- and intermediate-term credit to eligible farmers and ranchers, and those institutions are the most active source of loan funds in this area (see Figure 3.6).

Table 3.1 US Farm Sector Debt (in Billions of Dollars)

	2011	2012	2013	2014	2015	2016†
Real Estate	**167.2**	**173.2**	**184.8**	**197.0**	**209.5**	**213.6**
Farm Credit System	75.3	80.3	85.3	NA	NA	NA
Farm Service Agency	3.3	3.7	3.7	NA	NA	NA
Farmer Mac	3.8	3.8	4.1	NA	NA	NA
Commercial banks	52.9	64.5	68.9	NA	NA	NA
Life insurance companies	13.0	11.5	12.0	NA	NA	NA
Individuals and others	18.1	8.7	10.1	NA	NA	NA
Storage facility loans	0.7	0.7	0.7	NA	NA	NA
Non-real Estate	**127.3**	**123.8**	**130.2**	**149.4**	**147.4**	**140.6**
Farm Credit System	41.1	42.7	44.0	NA	NA	NA
Farm Service Agency*	3.5	3.4	2.8	NA	NA	NA
Commercial banks	59.2	59.9	63.7	NA	NA	NA
Individuals and others	23.5	17.9	19.6	NA	NA	NA

* Includes CCC storage and drying facilities loans but excludes debt on operator dwellings and for nonfarm purposes

† Forecast

Source: USDA, www.ers.usda.gov/data-products/farm-income-and-wealth-statistics/data-files-us-and-state-level-farm-income-and-wealth-statistics.aspx

Figure 3.4 US Farm Assets, Debt, and Equity (1960-2016)

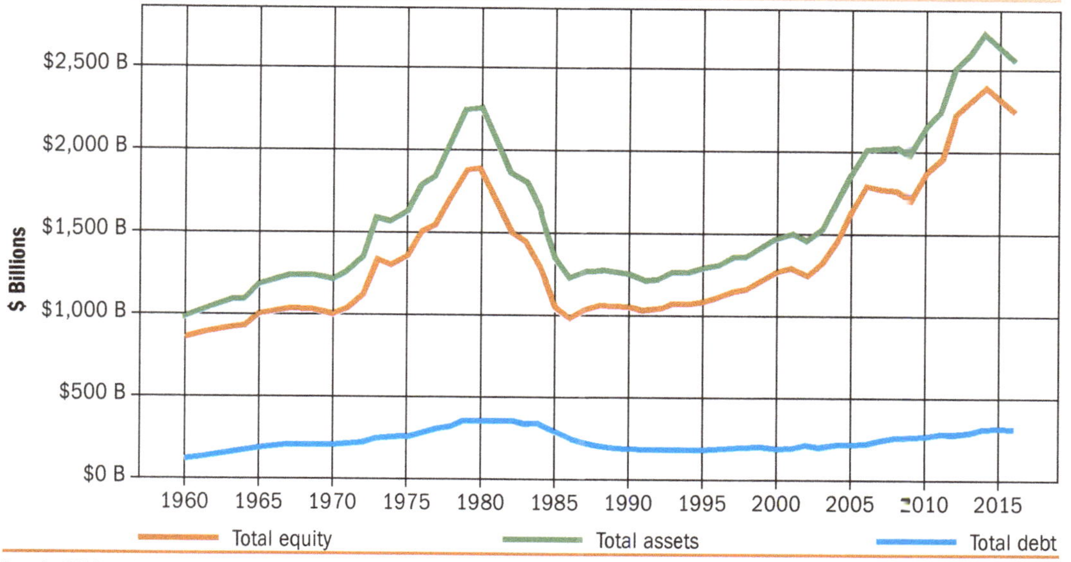

Data for 2016 are forecasts. Data values are rounded to the nearest thousand dollars. Inflation-adjusted values use the chain-type GDP deflator, 2009 = 100 from BEA, DOC.

Source: USDA, www.ers.usda.gov/data-products/farm-income-and-wealth-statistics/charts-and-maps-of-us-farm-balance-sheet-data.aspx

Figure 3.5 Percentage Change in US Farm Assets, Debt, and Equity from Prior Year (1961-2016)

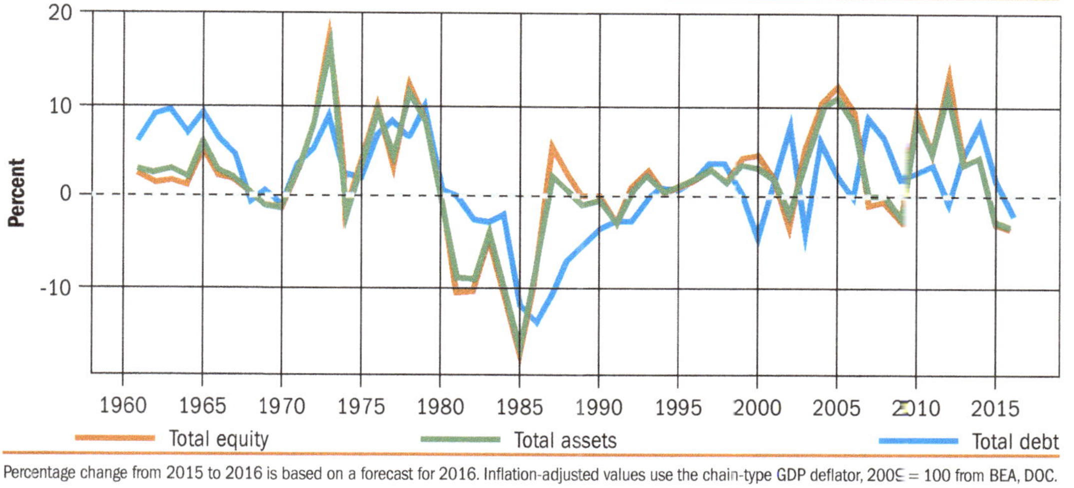

Percentage change from 2015 to 2016 is based on a forecast for 2016. Inflation-adjusted values use the chain-type GDP deflator, 2009 = 100 from BEA, DOC.

Source: USDA, www.ers.usda.gov/data-products/farm-income-and-wealth-statistics.aspx, released August 30, 2016.

Figure 3.6 Farm Loan Growth

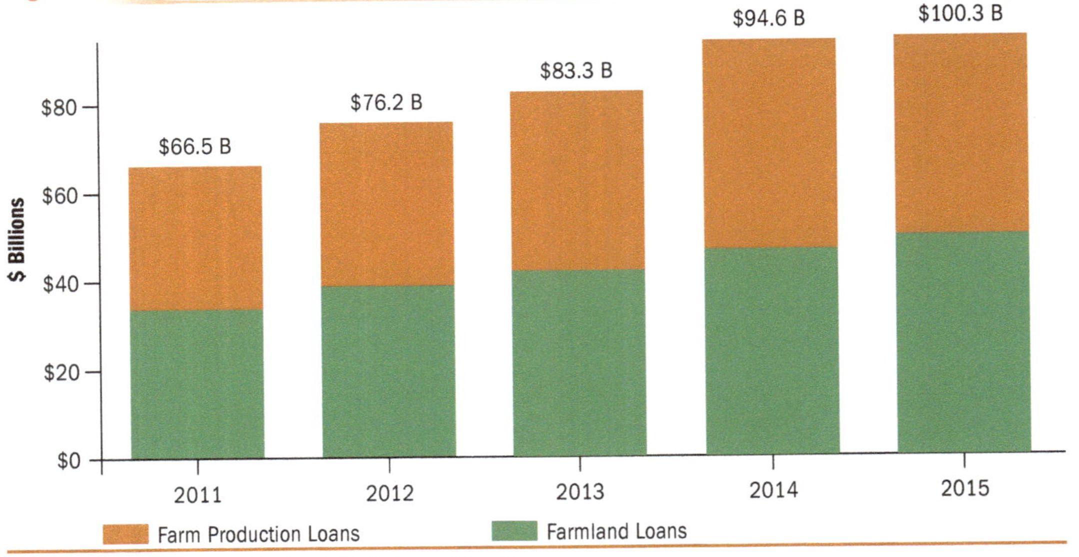

Source: www.agriculture.senate.gov/imo/media/doc/Testimony_Wolfe.pdf

Life Insurance Companies

Life insurance companies are governed by both the state in which they are incorporated and the states in which they operate. For many years, life insurance companies were the largest institutional lenders involved in farm real estate loans. As recently as the mid-1960s, their share equaled that of the Farm Credit System (Federal Land Bank). In 1981, life insurance companies were the second most active institutional lenders, holding 14% of outstanding farm real estate loans. However, in recent years their share of farm real estate loans has decreased. The Farm Credit System and commercial banks have largely absorbed this lost share by dramatically increasing their market share since 1976.

Life insurance companies have traditionally invested heavily in real estate. Their activities include both mortgage lending and property ownership. Life insurance companies may acquire full ownership of real estate for their own investment accounts or for the separate investment accounts that they manage for pension trusts. Although real estate ownership may amount to only 3% to 5% of a company's assets, a large dollar investment is involved because major life companies may have billions of dollars in assets.

Insurance companies are also substantial mortgage lenders for their own accounts, under contracts with pension managers, and as managers of separate accounts. Many portfolio managers, especially those of the larger companies, view agricultural mortgages and equities as safe and rewarding investments for a diversified portfolio.

The mortgage investments of life insurance companies cover the full range of realty types (e.g., residences, apartments, offices, shopping malls, hotels, and industrial properties) as well as all types of agricultural real estate (e.g., row crops, permanent plantings, timber, livestock ranches, and other agribusiness). Although their principal rural investment activity is in loans for farm real estate, they also invest in a limited amount of loans to agricultural support businesses such as feedlots, dairies, poultry facilities, hog facilities, and grain elevators. Because many companies have great financial resources, they have been important in mortgaging large income-producing properties.

Life insurance companies have consistently invested in safe, high-yield enterprises. Although a large percentage of their funds is invested in commercial and industrial real estate loans, life insurance companies also invest in agricultural properties that present little risk and offer reliable profit. Much of their agricultural lending activity originally took place in specific geographic areas such as the north central United States and in the southern Plains states. Large commercial farms located in these areas, where soil and weather conditions are predictably good, attracted life insurance company loans. More recently, because of investor requirements for higher yields, there has been an increase in agricultural investments in the western and southeastern states, where invested funds obtain a higher return.

Seller Financing

Seller financing has consistently been an important source of credit for rural properties. This source of credit remains strong. The importance of seller financing may be seen as an effect of general economic trends in the United States as well as trends that affect agricultural property value and the nature of the industry. As recent experience has clearly illustrated, the value of agricultural real estate can decline as well as increase. Lenders are restricted in the amounts they may lend and the interest rates they may charge, whereas private owner-lenders are not under constraints (other than usury laws). When market conditions cause lenders to reduce the supply of funds available for real estate loans, the importance of seller financing tends to increase.

Another trend in agriculture that influences the use of seller financing is the need for funds to expand productive capacity, particularly when expansion involves unproven technology or adding a new commodity to the borrower's operation. With extensive cash outflow and increased risk, many borrowers seek ways to expand capacity with as little cash investment as is both wise and feasible. Seller financing is often a good alternative when the risk exceeds an institutional lender's limits.

Farm and ranch land transfers often occur between relatives. Parents who wish their children to take over the management and production of their operation often begin the process of transfer long before they expect it to be fully accomplished. In this manner, the younger farmer's or rancher's credit credentials become established, and the final transfer of ownership is facilitated. Appraisers are often involved in estate appraisals even before the estate is created, for the purpose of aiding in the transfer of ownership.

The primary instrument used when a farm or ranch owner finances the sale of property is the land contract for sale (also known as the *contract for deed*, the *land contract*, and the *contract for sale*). In financing by land contract, the buyer makes a relatively low down payment for the land, although the sale price for the property may be higher than it would be if the property was financed through an institutional lender. In exchange for providing the buyer with the benefit of a lower initial capital outlay, the seller receives the prospect of higher future gains. The seller holds the deed until the terms of the contract are fulfilled or until the buyer has enough equity in the property to renegotiate for a conventional mortgage. In this type of transaction, the buyer takes possession of the property and pays all taxes, assessments, and insurance. The advantage to the seller, in addition to the higher sale price, is that payment of taxes on the yield from the sale may be spread out over a number of years.

A second instrument that may be used in seller financing is the purchase-money mortgage, through which a seller conveys the property to the buyer by means of a conventional deed but takes back a mortgage on the property as security for the unpaid balance of the purchase price. This method is used when the purchaser has a down payment and wants to receive the interest, which may be higher than savings rates.

The Farm Credit System

The Farm Credit System (FCS) came into being gradually between 1916 and 1933. Some of its parts were mandated by Congress and others by executive order. In its efforts to help farmers become self-sufficient owner-operators with greater economic strength, the US Government developed a system for evaluating the credit needs of the American farmer, providing a stable source of funds for those needs, and creating the means through which credit needs could be met throughout the normal up-and-down production and economic cycles that affect the agricultural industry.

Farm credit organizations provide more than a third of the credit used by American farmers, ranchers, and cooperatives, in addition to serving the credit needs of rural homeowners, electric and telephone cooperatives, and rural water systems. The Farm Credit System is a nationwide network of borrower-owned lending institutions and specialized service organizations. Components of the system include

- the Farm Credit Administration
- Farm Credit Banks
- the Federal Farm Credit Banks Funding Corporation (the Funding Corporation)
- the Farm Credit System Insurance Corporation
- System associations (ACAs, PCAs, FLCAs, FLBAs)
- the Farm Credit Leasing Services Corporation
- the Farm Credit Council

The Farm Credit System is an agency of the federal government, and it is independently regulated by the Farm Credit Administration (FCA). The FCA was created by executive order in 1933 as the supervising and coordinating agency for the entire Farm Credit System. The role of the FCA was changed to that of an arm's-length regulator by the 1985 and 1987 amendments to the Farm Credit Act of 1971. Located in McLean, Virginia, the FCA is directed by a three-member board nominated by the President and confirmed by the Senate. The FCA is responsible for examining and ensuring the safety and soundness of all FCS institutions.

The Federal Farm Credit Banks Funding Corporation manages the sale of FCS securities in the nation's money and capital markets through a network of investment dealers and dealer banks in New York. Farm Credit Consolidated Systemwide Debt Obligations are the primary source of loan funds for the FCS and are the joint and several liability of system banks. The funding corporation provides the investment community and the system with an efficient method of funding the loans made to the nation's farmers, ranchers, and cooperatives. Loans are priced on the basis of the current market, with the borrower's rate reflecting the nature of the credit, its risk and term, and the cost of providing the credit services.

Typical services include operating and real estate loans with fixed, adjustable, or variable rates; rural home mortgages; credit life insurance and crop insurance; and various financially related services such as farm record keeping and financial planning.

The Farm Credit System is divided into districts covering the United States and Puerto Rico. The FCS districts (as shown in Figure 3.7) are

- CoBank, ACB
- AgFirst Farm Credit Bank
- AgriBank, FCB
- Farm Credit Bank of Texas

Figure 3.7 Bank Chartered Territories of the Farm Credit System

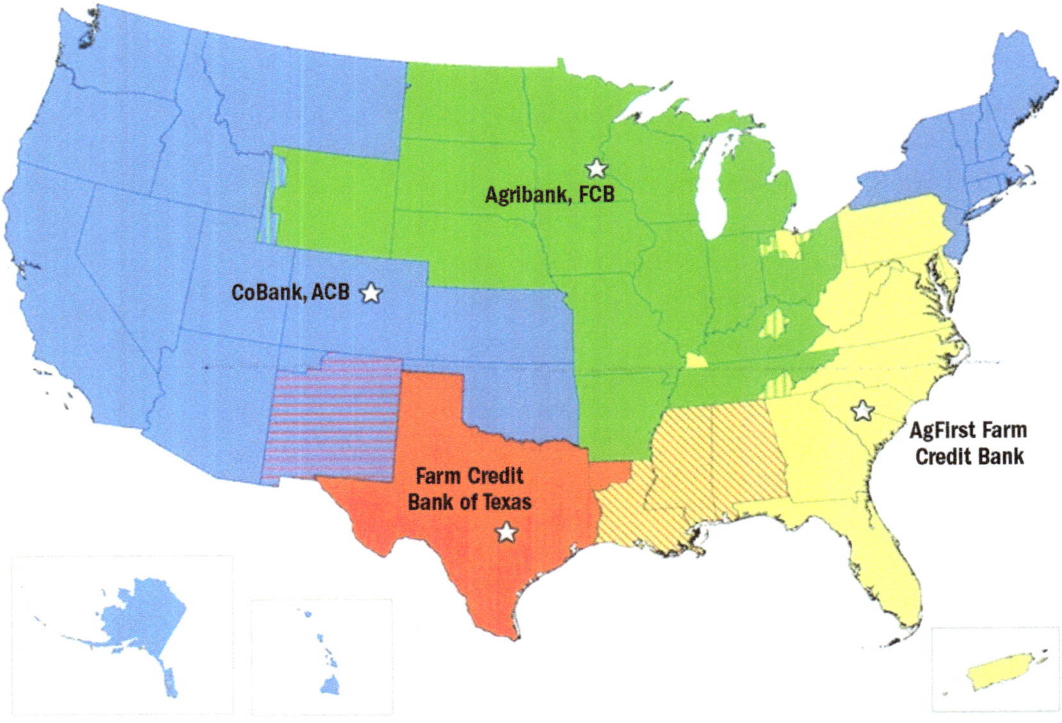

- Funded by AgFirst Farm Credit Bank
- Funded by AgriBank, FCB
- Funded by CoBank,ACB
- Funded by Farm Credit Bank of Texas
- Funded by CoBank, ACB and AgriBank, FCB
- Funded by Farm Credit Bank of Texas and CoBank, ACB
- Funded by AgFirst Farm Credit Bank and AgriBank, FCB
- Funded by AgFirst Farm Credit Bank and Farm Credit Bank of Texas

NOTE: As of January 1, 2015, CoBank, ACB, funds 26 associations in the indicated areas and serves cooperatives nationwide; Farm Credit Bank of Texas funds 14 associations; AgriBank, FCB, funds 17 associations; and AgFirst Credit Bank funds 19 associations. The Farm Credit System contains a total of 80 banks and associations.

Source: www.fca.gov/info/directory.html

As is common in the financial services industry, there is ongoing potential for change (through mergers and affiliation) in the current FCS district and association alignments.

The various FCS institutions make loans to farmers and ranchers through local associations. With the exception of the federal land bank associations (FLBAs), the local associations are direct lenders that make loans using funds generated through retained earnings and a credit line from the FCB. FCBs make long-term mortgage loans through the FLBAs that are agents of the FCB. PCAs make short- and intermediate-term loans. ACAs, created by the merger of one or more FLBAs and PCAs, have received a transfer of authority from their FCB to make and participate in long-term real estate mortgage loans as well as make direct short- and intermediate-term loans. FLCAs are associations (formerly Federal Land Bank Associations) to which direct long-term real estate lending authority has been transferred from their FCB. In some districts, FLBAs, PCAs, ACAs, and FLCAs operate under joint management and are commonly referred to as Farm Credit Services.

The Farm Credit Leasing Services Corporation (FCL), headquartered in Minneapolis, is owned by FCS banks and has sales offices located throughout the nation. FCL provides leasing services to eligible borrowers, including agricultural producers, cooperatives, and rural utilities.

The Farm Credit Council (FCC) is the national trade association representing the interests of the Farm Credit System. The council provides an important mechanism for grassroots involvement in the development of FCS legislative positions and policies while also providing its membership with a wide array of system-related services. Membership in the FCC is made up of Farm Credit District Councils and CoBank. These member councils, in turn, draw their membership from Farm Credit Banks and associations, as well as other cooperative farm lenders and associations.

The USDA and Other Government Sources

The Farmers Home Administration (FmHA) was conceived and organized during the 1930s to provide credit to individuals who were unable to obtain adequate credit from commercial lenders. The FmHA has never been in competition with commercial lenders; rather, it has tried to supplement credit that is available from traditional lending sources. Congress dissolved the agency in 1996. The responsibility for the farm loan programs of FmHA was given to the Farm Service Agency (FSA). To qualify for an FSA loan, the applicant must meet both income and geographic requirements and must be unable to qualify for all the needed credit at reasonable rates and terms from conventional lenders or other loan sources.

Through the FSA's program, borrowers receive loans and supervision designed to aid them in improving their financial situations. The loans are intended to be intermediate in nature, their duration only until the borrower's financial condition warrants transfer to other sources of credit.

Two categories of loans are made either by or through the FSA: (1) direct loans and (2) guaranteed loans. Funds for loans made directly by or guaranteed by the FSA are derived from annual appropriations by Congress. Since 1988, emphasis has shifted from the FSA making direct loans to the Guaranteed Loan Program. Commercial lenders supply funds for FSA-guaranteed loans. Under this program, up to 90% of the loan may be guaranteed by the FSA. Unlike direct loans that are

Table 3.2 Types of FSA Loans

Type of Loan	Maximum Loan Amount	Rates and Terms
Direct farm ownership	$300,000	· Term: Up to 40 years · Interest rate: fixed†
Direct down payment	The lowest of the following: · 45% of the farm or ranch purchase price · 45% of the appraised value · $225,000	· Term: Up to 20 years · Interest rate: fixed†
Direct operating	$300,000	· Term: 1 to 7 years · Interest rate: fixed†
Direct emergency	The lowest of the following: · 100% actual or physical losses · $500,000	· Term: 1 to 7 years (possibly up to 20 years) for non-real estate purposes · Up to 40 years for physical losses on real estate · Interest rate: fixed†
Guaranteed farm ownership	The maximum loan amount is adjusted annually for inflation*	· Term: Up to 40 years · The interest rate is negotiated by the borrower and lender
Guaranteed operating	The maximum loan amount is adjusted annually for inflation*	· Term: 1 to 7 years · The interest rate is negotiated by the borrower and lender
Guaranteed conservation	The maximum loan amount is adjusted annually for inflation*	· Term: Up to 20 years · The interest rate is negotiated by the borrower and lender
Land contract guarantee	A maximum purchase price of $500,000 on a new land contract	· Term: Contract payments must be amortized for a minimum of 20 years with equal payments during the term of the guarantee. The guarantee period is 10 years. · Interest rates must be fixed and can not exceed the direct farm ownership interest rate plus 3 percentage points.

* To determine the maximum loan limit for the current year, check with your local FSA loan officer at http://offices.sc.egov.usda.gov/locator/app or www.fsa.usda.gov/Internet/FSA_File/guaranteed_farm_loans.pdf.

† Direct loan interest rates are adjusted monthly and are posted online at www.fsa.usda.gov/FSA/webapp?area=home&subject=fmlp&topic=fir/ and at the local FSA office. Once a loan is closed, the interest rate is fixed at the rate in effect on the date of loan approval or loan closing, whichever is lower.

Note: FSA offers other types of loans, such as marketing assistance loans and farm storage facility loans. These are not part of FSA's Farm Loan Programs and, therefore, may have different application or eligibility requirements.

Source: Farm Service Agency, *Your Guide to FSA Farm Loans* (June 2012), 12-13. Available online at www.fsa.usda.gov/Assets/USDA-FSA-Public/usda-files/Farm-Loan-Programs/pdfs/fsa_br_01_web_booklet.pdf

originated, processed, and serviced by FSA personnel, the commercial lender performs these functions under the Guaranteed Loan Program.

Farm ownership loans are extended to qualified family farmers who obtain a substantial portion of their income from farming and are considered owner-operators even if part of their income is derived from off-the-farm labor. The loans are extended for purchasing land, refinancing debts, constructing or repairing farm buildings, improving farmland, developing water facilities, and establishing farm-based business enterprises to supplement farming income.

Under the category of farm operating loans, the FSA extends short- and intermediate-term credit to eligible family farm operators. These loans may be made for purchasing livestock or equipment and supplies needed for farm production, refinancing debts, making minor real estate improvements, and establishing non-farm enterprises to supplement farm income.

Farm emergency loans that cover losses that are not otherwise compensated are sometimes made available in cases of natural disasters.

The Rural Housing and Community Program loans of the old Farmers Home Administration (FmHA) were folded into a new USDA agency named Rural Development (RD). The RD agency is involved in an extensive variety of loan types. Rural Development directs its major lending activities toward rural communities, businesses, and low-income rural families and individuals who do not currently own safe and adequate housing. Qualified veterans are given preference in the processing for RD loans, but they must meet the same qualifications as nonveterans. An RD loan can be made on up to 100% of the property's market value, but there are maximum amount limits that vary according to the loan category. Interest rates on some direct loans made by RD may be subsidized based on the borrower's income. Rates on guaranteed loans are set by the lending institution that carries the loan.

Rural housing loans are made for the short- and long-term needs of low- to moderate-income families to build, purchase, or improve homes in rural areas. Towns with populations below specific levels qualify as rural areas for RD loans. Families with above-moderate income living in these areas are eligible for RD rural housing loans under the Guaranteed Loan Program. Short-term loans are made for the repair and rehabilitation of substandard rural properties.

Several rental housing loan programs available through RD provide rental units for low- to moderate-income rural residents and farm laborers. Loans are made to eligible individuals, trusts, associations, partnerships, limited partnerships, state or local public agencies, consumer cooperatives, and profit and nonprofit corporations. The maximum repayment period for most of these multifamily housing loans is 50 years.

The Small Business Administration (SBA) was created by Congress in 1953 as a permanent and independent government agency.

Its primary purpose is to protect the interests and serve the needs of small business concerns in the United States, particularly by providing loans in cases where those funds are necessary and appropriate. In 1976, legislation was passed to include farms among small business concerns that the SBA may serve. However, because the FSA is still the principal government agency serving the credit needs of the farm industry, the SBA's role in this area is secondary to that of the FSA.

The SBA basically guarantees loans funded by lending institutions, with the SBA agreeing to cover up to 90% of the loss in the case of default. The lender's interest rate on guaranteed loans is established by lending institutions but is subject to maximum limitations set by the SBA. Interest rates also vary according to the type and purpose of the loan.

Although the SBA does make agribusiness loans to farmers, the principal responsibility for these loans now rests with RD or the FSA. Most SBA loans are made to farmers under the administration's regular guaranteed business loan program. Under its regular business loan programs, the SBA can guarantee or make loans for both the short- and long-term needs of rural business ventures. The loan maturity varies according to the purpose of the loan, usually five to seven years for working capital and up to 25 years for real estate or equipment. However, the SBA expects the borrower to repay the loan as quickly as possible, and maximum loan terms are not automatically available in every case.

The Federal Agricultural Mortgage Corporation (Farmer Mac)

The Federal Agricultural Mortgage Corporation (Farmer Mac) is a federally chartered agency of the United States established by the Agricultural Credit Act of 1987, which amended the Farm Credit Act of 1971. Farmer Mac was established primarily to attract new capital for the financing of agricultural real estate and rural housing loans and to provide liquidity to agricultural real estate and rural housing lenders. Farmer Mac is intended to aid the development of a secondary market for agricultural real estate and rural housing loans (referred to as *qualified loans*) made by participating *originators*, and pooled by eligible agricultural mortgage marketing facilities (referred to as *poolers*). These loans are secured by first liens on agricultural real estate, including rural housing, by guaranteeing the timely payment of principal and interest on obligations (Farmer Mac I securities) backed by loans and securities that represent interests in loans (the Farmer Mac I program).

Farmer Mac also issues certificates (Farmer Mac II securities) representing interests in pools of portions (i.e., guaranteed portions) of farm ownership, operating loans, and certain other loans guaranteed by the United States acting through the Farm Service Agency (the Farmer Mac II program).

The Agricultural Credit Act authorizes Farmer Mac to borrow up to $1.5 billion from the Secretary of the Treasury, subject to certain conditions, to enable Farmer Mac to fulfill its guarantee obligations. The debt created by this borrowing will bear interest at a rate determined by the Secretary of the Treasury, taking into consideration the average rate on outstanding US marketable obligations as of the last day of the calendar month ending before the date of purchase of the obligations.

FARMER MAC

Chapter 4

Government Intervention in Agricultural Markets

In the modern era, agriculture has been one of the most highly regulated industries in the United States. From regulations governing the use of herbicides and pesticides to limitations on the production and international trade of certain crops, governmental actions can have a great influence on the economic sector and, by extension, on the market for agricultural properties. In addition to the federal Agricultural Foreign Investment Disclosure Act of 1978—which requires all foreign persons to report transfers of interests in rural property, with significant penalties for failure to disclose a transaction—several states have laws in place that restrict the ownership and purchase of farmland by foreign investors. The flip side of regulatory restriction is the support that government often provides, in a variety of forms, for agriculture as an industry. The previous chapter covered monetary policy and government-supported financing for rural property, whereas this chapter focuses on governmental subsidies for agricultural operations and other programs that can be important factors in the profitability of agricultural enterprises.

Appraisers must stay current on government commodity programs and other programs that affect agriculture in the local market. Appraisers may find that government programs affect the highest and best use of the property being appraised. That is, these programs may partially determine which crop will yield the greatest net return. For example, a farmer may be able to obtain a greater net return (through target prices or loans) on a crop for which the government provides support than a crop without government support. Appraisers should investigate (1) whether government support for the crop affects the projected income of the farm and (2) whether acreage allotments or market quotas transfer with the land or with the producer. The effect of those allotments or quotas can then be related to property value.

Government programs may have a positive, negative, or nominal effect on an area's production, income, and property values. It is the responsibility of appraisers to analyze the local situation, in light of national developments and periodic changes, and integrate the findings into the application of the approaches to value when relevant. National regulations are subject to change by Congressional acts and certain decisions of the Secretary of Agriculture. The local county office remains the best source for information on specific crops and annual changes.

By collecting information on the relationship of government price supports to land values over a period of years, appraisers may be able to uncover useful trends. It may be useful to note what percentage of farmers in the area are participating in government commodity programs. For example, what percentage of the total wheat crop in the area is under price supports? Has the government-supported price on wheat influenced the number of acres planted (or not planted) with other crops? Does the market perceive government programs as important?

Since World War I, the federal government has taken a variety of measures to ensure the safety of consumers and the economic health of the farming community. Despite the long history of governmental subsidies to farmers dating back to the 1930s (see Figures 4.1 and 4.2), economists continue to debate the efficiency of governmental payments to agricultural producers. The current Farm Bill provides for the specific implementation of government intervention in the agricultural marketplace.

The amount and type of government support to agricultural producers varies by region and over time, as the effectiveness and economic efficiency of different programs are tested and refined. Political fashion and economic need based on market shifts also influence the types of government expenditures made from year to year. (See Figures 4.3, 4.4, and 4.5.)

The Farm Bill

Most current farm support programs have their origins in the national crisis of the Great Depression. That depression struck farmers especially hard and many farms were lost to foreclosure. In 1932, the net income of farmers dropped to less than a third of what their income was in 1929. The prices for the products farmers were selling declined much more than the prices for the goods and services they were buying. The nationwide financial crisis contributed to the passage of the Agricultural Adjustment Act of 1933, which introduced the concept of *parity*. The goal of parity is to provide "an equality of exchange relationship between agriculture and industry or between

Figure 4.1 Important Dates for US Farm Bill Policy and Selected Related Laws

1862	Agriculture Act of 1862 established the US Department of Agriculture
1862, 1890	Morrill Land Grant Acts (7 USC §§ 301 et seq., 7 USC §§ 321 et seq.)
1887	Hatch Experiment Season Act (7 USC §§ 361a-361i)
1890, 1906	Federal Meat Inspection Act (21 USC §§ 601 et seq.)
1914	Smith-Lever Extension Act (7 USC §§ 341 et seq.)
1916	Federal Farm Loan Act (PL 64-158)
1921	Packers and Stockyards Act (7 USC §§ 181-229)
1933	Agricultural Adjustment Act of 1933 (PL 73-10)
1933	The Farm Credit Act (PL 73-75)
1933, 1939	Creation of the Commodity Credit Corporation (15 USC §714)
1935	Section 32 of the Act of August 24, 1935 (7 USC §612)
1935	Soil Conservation and Domestic Allotment Act (PL 74-46)
1938	Agricultural Adjustment Act of 1938 and Federal Crop Insurance Act (PL 75-430)
1946	Richard B. Russell National School Lunch Act (PL 79-396)
1946	Agricultural Marketing Act of 1946 (7 USC §1621)
1948	Agriculture Act of 1948 (PL 80-897)
1949	Agriculture Act of 1949 (PL 81-439)
1954	Agricultural Trade Development and Assistance Act (PL 480) (7 USC §§ 1691 et seq.)
1954	Agriculture Act of 1954 (PL 83-690)
1956	Agriculture Act of 1956 (PL 84-540)
1956	The Soil Bank Act (1956 Farm Bill)
1957	Poultry Products Inspection Act (21 USC 451 et seq.)
1961	Agriculture Act of 1961 and Consolidated Farm and Rural Development Act (PL 87-128)
1964	Food Stamp Act of 1964 (PL 88-525)
1965	Food and Agriculture Act of 1965 (PL 89-321)
1966	Child Nutrition Act of 1966 (PL 89-642)
1970	Agriculture Act of 1970 (PL 91-524)
1970	Egg Products Inspection Act (21 USC 1031 et seq.)
1972	Consolidated Farm and Rural Development Act of 1972 (PL 92-419)
1973	Agriculture and Consumer Protection Act of 1973 (PL 93-86): considered to be the first "omnibus farm bill"
1976	Farmer-to-Consumer Direct Marketing Act of 1976 (PL 94-463)
1977	National Agricultural Research, Extension, and Teaching Policy Act (7 USC §§ 3101)
1977	Food and Agriculture Act of 1977 (PL 95-113)
1980	Federal Crop Insurance Act (PL 96-365)
1981	Agriculture and Food Act of 1981 (PL 97-98)
1985	Food Security Act of 1985 (PL 99-198)
1985	Animal Welfare Act (1985 Farm Bill, 7 USC §§ 2131-2159)
1988	Disaster Assistance Act of 1988 (PL 100-380)
1990	Food, Agriculture, Conservation and Trade Act of 1990 (PL 101-624)
1990	Organic Food Production Act (1990 Farm Bill, 7 USC §§ 6501-6528)
1996	Federal Agriculture Improvement and Reform Act of 1996 (PL 104-127)
1998	Agricultural Research, Extension, and Education Reform Act (PL 105-185)
2000	Agriculture Risk Protection Act (PL 106-224)
2000	Biomass Research and Development Act of 2000 (PL 306-224)
2002	Farm Security and Rural Investment Act of 2002 (PL 107-171)
2000, 2002	Plant Protection Act (7 USC §7701 et seq.) and Animal Health Protection Act (7 USC §8301 et seq.), consolidating previous laws
2003	Healthy Forests Restoration Act of 2003 (PL 108-148)
2004	Specialty Cross Competitiveness Act of 2004 (PL 108-465)
2005	Energy Policy Act of 2005 (PL 109-58)
2008	Food Conservation and Energy Act of 2008 (PL 110-246)
2014	Agricultural Act of 2014 (PL 113-79)

Source: Congressional Research Service, *What is a Farm Bill?* (2014).

Figure 4.2 Milestones in US Agricultural Policy

1933 Agricultural Adjustment Act: First "farm bill" established the New Deal mix of commodity-specific price and income support programs.

1936 Soil Conservation and Domestic Allotment Act: First direct links created between soil conservation and commodity programs.

1949 Agricultural Act: Established policy of high, fixed-price supports and acreage allotments as permanent farm policy. Programs revert to the 1949 provisions should a new farm bill fail to pass.

1954 Agricultural Act: Introduced flexible price supports to commodity programs.

1956 Agricultural Act: Established the Soil Bank program, which introduced the use of conservation reserve in addition to acreage control for supply management. The program ended after only two years.

1965 Food and Agriculture Act: Introduced new income support payments in combination with reduced price supports and continued supply controls.

1970 Agriculture Act: First inclusion of the title for Rural Development in a farm bill.

1973 Agriculture and Consumer Protection Act: Introduced target prices and deficiency payments to replace price supports, coupled with low commodity loan rates, to increase producer reliance on markets and allow for free involvement of commodities at world prices.

1985 Food Security Act: Introduced marketing loan provisions to commodity loan programs to reduce forfeitures by allowing for the repayment of loans at a lower rate when market prices fell, with the intention of aiding in reducing government-held surplus grain. Reestablished a conservation reserve.

1996 Federal Agriculture Improvement and Reform Act: Replaced price support and supply control program with program of direct payments based on historical production. Introduced nearly complete planting flexibility.

2002 Farm Security and Rural Investment Act: Introduced counter-cyclical payments program triggered when current prices fell below a target level, but paid based on historical production. Introduced working-lands conservation payments through the Conservation Security Program. Continued planting flexibility and program of direct payments based on historical production, allowing updating of historical base acres and adding historical soybean acres.

Source: Compiled by Economic Research Service USDA. The complete texts of US farm bills from 1933 to 2002 are available on the website of the National Agricultural Law Center (www.nationalaglawcenter.org/farmbills).

Figure 4.3 Total Farm-Related Government Payments in 2014

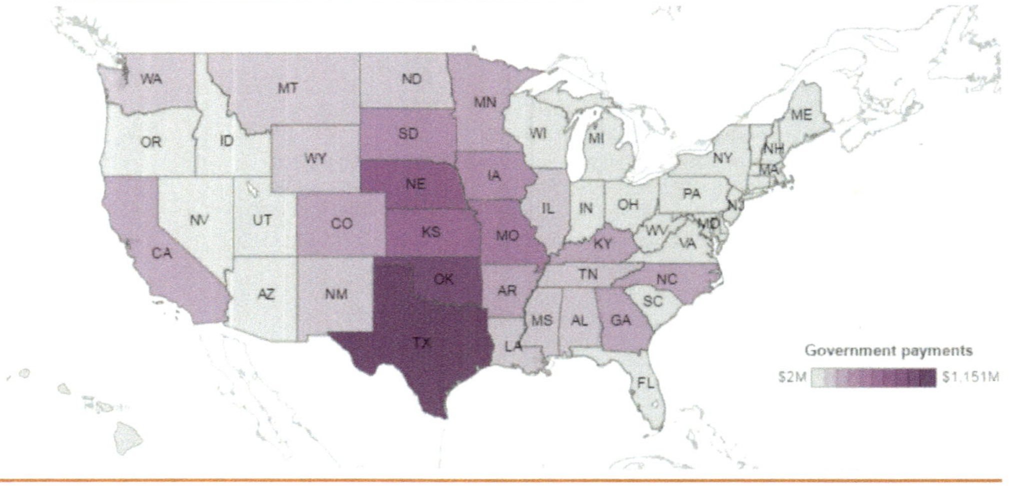

Source: USDA Economic Research Service, www.ers.usda.gov/data-products/farm-income-and-wealth-statistics/charts-and-maps-of-us-farm-income-statement-data.aspx

Figure 4.4 US Government Farm Support, Direct Outlays (1996-2016)

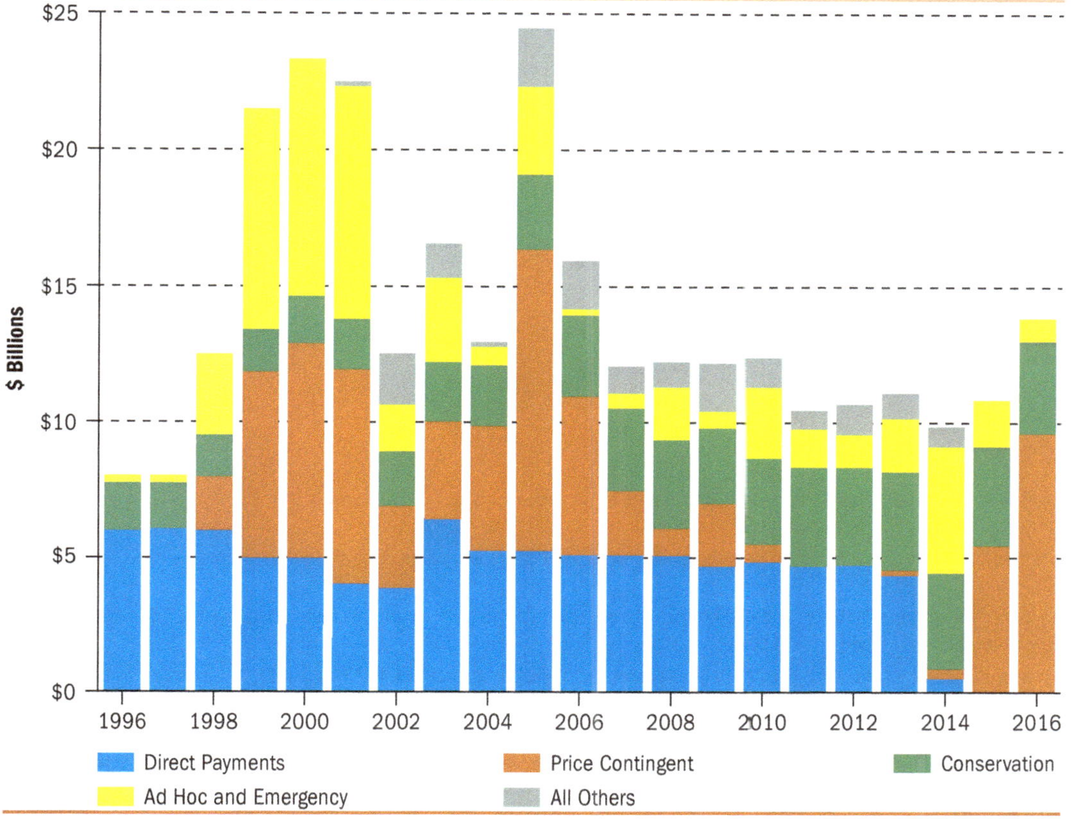

Notes: All values are nominal, that is, not adjusted for inflation. 2016 is a forecast. Data are on a fiscal year basis and may not correspond exactly with the crop or calendar year. "Direct Payments" include production flexibility contract payments enacted under the 1996 farm bill and fixed direct payments of the 2002 and 2008 farm bills; "Price Contingent" outlays include loan deficiency payments, marketing loan gains, counter-cyclical payments, and ACRE payments; "Conservation" outlays include Conservation Reserve Program payments along with other conservation program outlays; "Ad Hoc and Emergency" includes emergency supplemental crop and livestock disaster payments and market loss assistance payments for relief of low commodity prices; and "All Other" outlays include peanut quota buyout payments, milk income loss payments, tobacco transition payments, and other miscellaneous expenditures.

Source: USDA, ERS, "2016 Farm Income Forecast," February 9, 2016. www.fas.org/sgp/crs/misc/R40152.pdf

persons living on farms and persons not on farms."[1] *Parity* means that the price for a unit of an agricultural commodity would have the same purchasing power that it did in a previous period. For example, if the price of a bushel of wheat was equal to the cost of a shirt in the base period, then today a bushel of wheat should buy a shirt at

1. The word *parity* did not appear until the Agricultural Adjustment Act of 1938, which replaced the Agricultural Adjustment Act of 1933 after it was declared unconstitutional in 1936. See Wayne D. Rasmussen and Gladys L. Baker, *Price-Support and Adjustment Programs from 1933 through 1978: A Short History*, USDA Agricultural Information Bulletin No. 424 (Washington, DC: USDA, Economics, Statistics, and Cooperatives Service, 1979), 1-4.

Figure 4.5 FSA Loans Made and Payments to Farmers and Ranchers (Fiscal Year 2015)

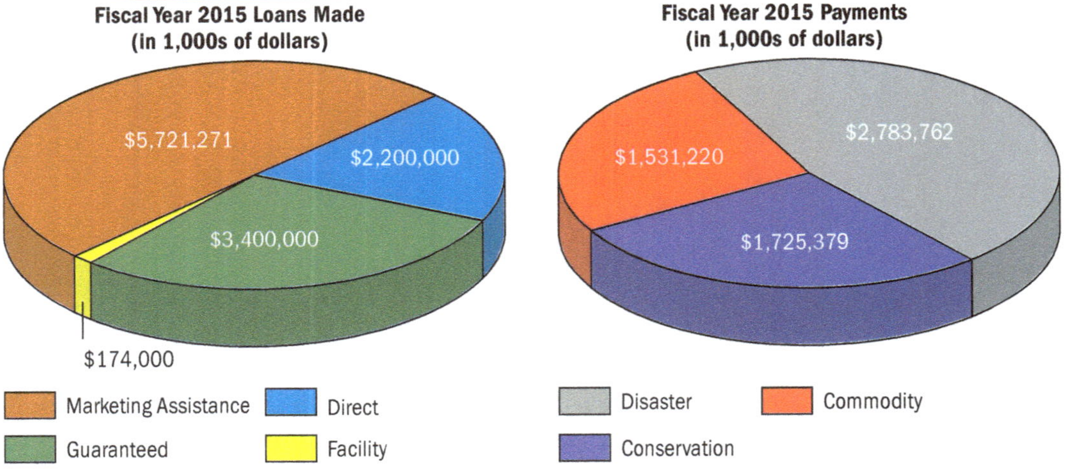

Direct and guaranteed farm loans: Provide farmers and ranchers access to credit when they are unable to obtain financing from a commercial source at reasonable rates and terms.

Farm storage facility loans: Offer low-interest financing to producers seeking to improve their storage and handling facilities.

Marketing assistance loans: Provide producers interim financing to meet cash flow needs and enable more consistent marketing of commodities throughout the year.

Disaster assistance programs: Provide financial support to crop and livestock producers affected by drought, flood, hurricanes, and other natural disasters.

Commodity programs: Assist farmers and ranchers in managing financial risks associated with commodity price and revenue fluctuations. Commodity program payments are low in fiscal year 2015 because Price Loss Coverage and Agricultural Risk Coverage payments totaling around $6 billion for crop year 2014 were issued at the beginning of fiscal year 2016.

Conservation programs: Assist farmers and ranchers in applying conservation measures to enhance and maintain the quality of soil, water, and related natural resources and wildlife.

Source: *USDA, Farm Service Agency Impacts: Selected Accomplishments 2015.*

current market prices. Calculating parity was accomplished by using an index for prices paid for goods and items from 1909 to 1914. (After 1948, the period was based on 1910 to 1914 prices.) This base period was selected because it was considered a time of relatively normal and stable prices.

Two of the methods used for stabilizing prices in the Agricultural Adjustment Act of 1933 were (1) direct payments to farmers who voluntarily reduced acreage planted with certain crops, and (2) marketing regulations through voluntary agreements with associations of producers, processors, and other handlers of agricultural products. The commodities affected were wheat, cotton, corn, hogs, rice, tobacco, milk, and milk products. This list was expanded to include rye, flax, barley, grain sorghums, peanuts, and cattle by the Jones-Connally Act of 1934. The Commodity Credit Corporation (CCC) was established in 1933 to make nonrecourse loans to farmers. Some of the original commodities covered have been dropped from the programs, and many regulations have changed, but payments, loans, and limitations on planted acreage or production are still important methods used in the operation of current programs.

Although the first Agricultural Adjustment Act (1933) was declared unconstitutional, it was soon followed by the Soil Conservation and Domestic Allotment Act of 1936 and the more comprehensive Agricultural Adjustment Act of 1938. Acreage allotments were set up for corn, cotton, rice, tobacco, and wheat. State and local committees were directed to administer the programs. This legislation made specific provisions for the allocation of payments for price support. Nonrecourse loans on corn, wheat, and cotton made the "ever-normal granary" plan (i.e., price stabilization measures) possible. The loans served to create a floor under farm prices when they were declining and also to finance the storage of commodities until they were needed and demand raised the prices. Marketing quotas and crop insurance plans were also established.

During World War II, the large stocks of wheat, corn, and cotton resulting from the CCC loans were an important military reserve. The post-war years were marked by continuing controversy over the levels and types of support to be given to farm production. The Soil Bank, established in 1956 but later discontinued, was an unsuccessful attempt to take enough farmland out of production (usually on contracts of 5 to 10 years) in order to balance the supply of and demand for agricultural products.

Direct payments–i.e., a program in which producers were paid regardless of whether they incurred losses–were eventually eliminated from the Farm Act and replaced in 2014 with several new revenue support programs. In particular, the Price Loss Coverage (PLC) program in the 2014 Farm Act replaced the Counter-Cyclical Price (CCP) program. The PLC program has a set of reference prices based on substantially higher support levels for most crops covered by the program. Also, the Agriculture Risk Coverage (ARC) program was established as an alternative to the PLC program. Reference prices for the ARC program are based on moving averages and thus fluctuate over time, in contrast to the fixed prices of the PLC program. For example, the reference prices in the 2014 Farm Act range from 8% to 88% higher than PLC target prices mandated in the earlier farm bill, raising the threshold at which government price support payments are activated.

Commodity Programs

The agricultural commodities covered in the 2014 Farm Act include wheat, corn, grain sorghum, barley, oats, long-grain and medi-

Table 4.1 2014 Farm Act Reference Prices

Covered Commodities	Reference Prices
Wheat	$5.50 per bushel
Corn	$3.70 per bushel
Grain sorghum	$3.95 per bushel
Barley	$4.95 per bushel
Oats	$2.40 per bushel
Long-grain rice	$14.00 per hundredweight
Medium-grain rice	$14.00 per hundredweight
Soybeans	$8.40 per bushel
Other oilseeds	$20.15 per hundredweight
Dry peas	$11.00 per hundredweight
Lentils	$19.97 per hundredweight
Small chickpeas	$19.04 per hundredweight
Large chickpeas	$21.54 per hundredweight
Peanuts	$535.00 per ton

Source: USDA Economic Research Service, *Agricultural Act of 2014, Title I*, www.ers.usda.gov/agricultural-act-of-2014-highlights-and-implications/crop-commodity-programs.aspx.

Table 4.2 CCP Target Prices vs. 2014 Farm Act (PLC) Reference Prices

Commodity	Unit	CCP Target Prices	Reference Prices	Percentage Change
Wheat	Bushel	$4.17	$5.50	31.89%
Corn	Bushel	$2.63	$3.70	40.68%
Grain sorghum	Bushel	$2.63	$3.95	50.19%
Barley	Bushel	$2.63	$4.95	88.21%
Oats	Bushel	$1.79	$2.40	34.08%
Upland cotton	Pound	$0.71	NA	NA
Long-grain rice	Hundredweight	$10.50	$14.00	33.33%
Medium-grain rice	Hundredweight	$10.50	$14.00	33.33%
Peanuts	Short ton	$495.00	$535.00	8.08%
Soybeans	Bushel	$6.00	$8.40	40.00%
Other oilseeds	Hundredweight	$12.68	$20.15	58.91%
Dry peas	Hundredweight	$8.32	$11.00	32.21%
Lentils	Hundredweight	$12.81	$19.97	55.89%
Small chickpeas	Hundredweight	$10.36	$19.04	83.78%

Note: Upland cotton is not a covered commodity under PLC.

Source: USDA Economic Research Service, *Agricultural Act of 2014, Title I*, and *Food, Conservation, and Energy Act of 2008, Title 1*, www.ers.usda.gov/agricultural-act-of-2014-highlights-and-implications/crop-commodity-programs.aspx.

um-grain rice, soybeans and other oilseeds, dry peas, lentils, small and large chickpeas, and peanuts. The list can change over time as certain commodities are added or dropped from the relevant programs.

For some crops, the FSA limits the amount of production on an individual farm by using either a marketing quota or an acreage allotment. A *marketing quota* limits or allots the amount of production in bushels or pounds that can be marketed and for which the farmer is eligible for price support. The quota or allotment is initially established on the basis of the farm's history of production during a base period. The quota or allotment is then adjusted yearly in light of projected supply and demand. The term *allotment* is often used to mean either the marketing quota or the acreage limitation. The state FSA office can provide information about local regulations and usually a historical record for each farm, typically listed by farm number.

Crop Insurance

Other government programs that may benefit an appraised property and affect land values are various types of crop insurance and disaster payments. The programs are designed to compensate farmers during times when they are prevented from planting or when yields are drastically reduced because of drought, flood, natural disaster, or other conditions beyond their control. Some farmers use additional private insurance for greater protection. The premiums for crop insurance may be subsidized through a government program. If pertinent, the conditions and expenses of the crop insurance for the particular property being appraised should be investigated.

Appraisers should be aware of the relative importance of insurance in the area and the frequency and severity of damage due to local weather. Of course, the farms in a particular area can be viewed as subject to very similar risks. If the risks for the subject property and comparable properties are similar enough, this factor is not a useful variable for comparison. Insurance or disaster payment programs are yearly and, thus, probably would not affect long-term values. Federal crop insurance does not transfer with a sale, although it can transfer with an inheritance or the dissolution of a partnership. A producer who is selling a farm might be due indemnity from crop insurance, but the indemnity would go to the producer rather than the new buyer.

Foreign Trade and Tariffs

Historically, national governments have wielded tariffs on certain agricultural products as points of negotiation to influence international trade. The movement toward free trade is an incremental process. Both the supply of certain food products and the demand for other products are much broader in global markets than they were

100 years ago, when diversified farms tended to serve local markets exclusively. Large modern farms, which often specialize in just a few commodities, routinely sell products for shipment to far-flung markets when that sort of trade is economically feasible.

Environmental Conservation

The Natural Resources Conservation Service (NRCS) operates many programs that foster responsible environmental stewardship of rural land. For example, the NRCS provides conservation funding (cost share) for land improvements and administers easement programs, such as the easements provided in the Wetlands Reserve Program (WRP).

The 2014 Farm Bill consolidated the conservation programs run by the FSA and NRCS for greater efficiency at the local level. It also tied the governmental subsidy of crop insurance premiums to basic conservation practices for highly erodible lands and wetlands.

Chapter 5

Data Collection

The expression "Quality in, quality out" is most often associated with the computer world, but it is also applicable to an appraiser's collection, analysis, and reporting of data relevant to the appraisal problem. The information on social, economic, governmental, and environmental forces that appraisers gather is fundamental to the determination of property value. Reliable and meaningful data is vital to every appraisal. The selected data provides

1. The background against which the specific property is appraised
2. Information from which to infer possible trends affecting land values
3. Numbers used in the calculations performed within the three approaches to value
4. A basis for judgments about the highest and best use and reconciliation of the value indications derived in the approaches into a final opinion of defined value

In essence, the quality of an appraiser's data is directly related to the quality of the final appraisal report. Therefore, the sources from which an appraiser secures data are crucial to the appraisal process.

Market data is studied and analyzed to give an appraiser a broad overview of the market and to help determine the subject property's highest and best use. Data on sales of comparable properties is required to apply the sales comparison approach, and solid income data (rentals, leases, cropshare information, profit and loss data, etc.) is also needed to develop the income capitalization approach. Reliable cost and depreciation information from sales of improved properties and unimproved land sales are essential to the cost approach.

The most important part of any appraisal is the analysis of available data from the market. It is critical that the reader of an appraisal report be given the full benefit of an appraiser's knowledge of the market and analysis of data from that market. Appraisers study and analyze comparable properties to derive a market capitalization rate to use in the income capitalization approach, to estimate depreciation in the cost approach, and to make adjustments that are applied in the sales comparison approach.

Data Types

Different types of data are useful at different stages in the valuation process. First, an appraiser may collect macro-level demographic data on the market area as well as competitive supply and demand data on competitive properties to perform market analysis. Next, the appraiser may collect micro-level data such as physical characteristics of the subject property, zoning restrictions, and the income anticipated from alternative uses for highest and best use analysis. The analysis undertaken in the application of the three approaches to value generally requires micro-level data on the subject property, such as cost data for the property improvements to be used in the cost approach, and micro-level data on competitive properties, such as information on comparable sales transactions to be used in the sales comparison approach.

Appraisers must be aware of the patterns, customs, and practices that characterize the communities in which they work. One invaluable tool in making accurate value opinions is an understanding that people within and outside of a community have a strong influence on property value through their opinions and actions.

An appraiser who works in a given rural community develops a knowledge of that community and its peculiarities. An understanding of the motivations for buying and selling rural land and the extent to which external forces influence property value is often acquired through many years of experience in the community. If an appraiser does not have this type of familiarity with a particular area, extensive efforts must be made to obtain valid information from long-standing community members, particularly those involved in real estate transactions.

Rural communities have their own demographics, which generally differ from their urban counterparts. Like any other neighborhood, a rural community is characterized as a group of similar or complementary

macro-level data. Data that relates to the social, economic, governmental, and environmental forces that affect real property values. This type of data is usually not specific to any particular property but is applicable in many assignments of similar types of properties.

micro-level data. Details about the property being appraised, comparable sale and rental properties, and relevant local market characteristics.

land uses. When analyzing properties whose basic function is the production of agricultural commodities, the definition of *community* becomes even more detailed for the appraiser. Farm communities are characterized by operation sizes, types of ownership, soils, crops grown, and land uses. The boundaries of farm communities may extend over several counties or larger areas.

Macro-Level Data

An appraiser working in a specific market on a daily basis will often have general data on file. Macro-level data consists of information about the social, economic, governmental, and environmental forces that affect property value. This information is part of the background knowledge that appraisers bring to their assignments. All macro-level data is ultimately understood in terms of how it affects the economic climate in which real property transactions occur.

Economic Trends

The economic stability of a rural community is perhaps the single largest determinant of that community's future. Advances in technology, transportation, and communications have changed the complexion of many small rural towns and communities. Some small rural communities have adapted to these changes and remain vibrant and viable.

New construction, consumer activity at local businesses, property maintenance, vacant buildings, and the number of properties for sale can often provide an indication of the current status of a community's prosperity. Discussions with local farmers and ranchers, lenders, community leaders, chamber of commerce officials, business owners, farm equipment dealers, and school officials, among others, can confirm or refute an appraiser's initial impressions. The community's economic direction or stability can be verified with information from a variety of sources, such as (1) building permits issued by the local or county authority, (2) real estate transactions documented in the county recorder's office, and (3) the increase or decrease in real estate sales activity as shown by multiple listing services or online listings of real estate companies specializing in rural property. Any rumors concerning the relocation of businesses or industries to or from the community must be verified to accurately assess the economy of a rural area.

Demographics

The quality of life considerations for members of rural communites are essentially the same as those in other communties. The degree to which these needs are met affects community property values. The characteristics of the population often dictate the principal concerns of that population. If many young families live in an area, the number and quality of schools definitely influence property purchases.

Table 5.1 Economic Trends and Indicators

Economic Trends	Economic Indicators
International Trends 	Changes in: · balance of foreign trade · rates of foreign exchange · commodity price levels · wage levels · interest rates · industrial production levels · volume of retail sales
National and Regional Trends 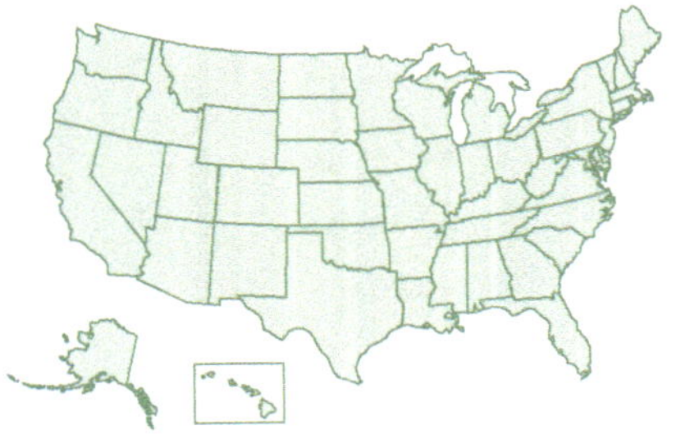	Changes in: · gross national product · gross domestic product · national income · the balance of payments to other nations · price level indexes · interest rates · aggregate employment and unemployment statistics · the number of housing starts and building permits issued · the dollar volume of construction · other macro-level data Note: A time series of economic indicators, which describes and measures changes or movements over a period of time, may reveal fluctuations in long-term trends and help put current statistics in perspective—i.e., help determine the current position of the economic cycle.
Local Trends 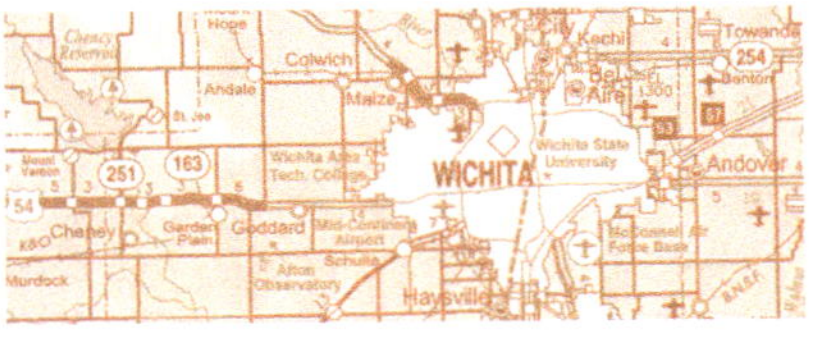	Changes in: · population · net household formation · the diversity of the community's economic base · the level and stability of employment · wage rates · household or family income
Trends Affecting Rural Land	Changes in: · size and complexity of business operations in farming, ranching, timber harvesting, drilling, or mining · level of mechanization or labor-intensiveness · degree of dependence on government subsidies or government-leased lands · prospective competition from imports

Accessible health care facilities may increase property values in a given area. Proximity to places of worship, recreational facilities, cultural attractions, and shopping facilities may influence the decision to purchase a property in a rural area. The condition and quality of the roads to and from these facilities also affect the value of all types of properties in the community.

In many rural areas, the availability of off-farm employment has historically been significant. However, the recent trends of farm consolidation and decreases in small farm operations have reduced the need for off-farm employment for large farm operators and increased the need for other sources of income for small farm operators who are struggling with profitability. Many farm operators must have income from other owned or rented land or other sources to meet the debt service on their farms. An established non-farm economy in the area can have an effect on property values. If the industry creates housing needs, the development of small parcels could create an increase in both land value and property taxes, which could, in turn, force the sale of agricultural land for the higher values commanded by residential lots.

Figure 5.1 provides a sample of demographic data obtained from the US Department of Agriculture (USDA) Census of Agriculture on a county in Wisconsin.

Demographic Data Sources

There are numerous online sources for social and demographic information pertaining to rural communities. In some areas, primary sources or local agencies may be the only sources of more localized information on ethnic composition, schools and enrollment, churches, health care providers and facilities, local government structure, major employers, unemployment figures, recreational facilities, and shopping facilities.

One of the best sources of social and demographic data for a community is the local chamber of commerce. Some state agencies publish profiles of individual communities that include demographic data. Other sources include commercial data providers (e.g., The Site To Do Business or STDB at www.stdb.com) and some federal agencies such as the following:

- FRED Economic Data (https://research.stlouisfed.org/fred2)
- FSA (www.fsa.usda.gov)
- USDA Census of Agriculture (www.agcensus.usda.gov)

Governmental Regulations

As mentioned in Chapter 1, the legal right to own land is subject to the powers of government. Part of an appraiser's research must focus on the effects of local, state, and federal government regulations. The sources of data on governmental regulation are many and varied. Some of the required data is available at the local level from regulatory agencies or local deed records, while other data must be obtained from state regulatory agencies. Still other data must be obtained at the federal level. Regulatory requirement data is available on the Internet at little or no cost. Appraisers must be aware of any current government regulations that could influence the value of the subject property.

Figure 5.1 USDA Census of Agriculture Fact Sheet

Waushara County – Wisconsin

Ranked items among the 72 state counties and 3,079 U.S. counties, 2012

Item	Quantity	State Rank	Universe [1]	U.S. Rank	Universe [1]
MARKET VALUE OF AGRICULTURAL PRODUCTS SOLD ($1,000)					
Total value of agricultural products sold	134,149	38	72	886	3,077
Value of crops including nursery and greenhouse	104,150	12	71	619	3,072
Value of livestock, poultry, and their products	30,000	53	72	1,298	3,076
VALUE OF SALES BY COMMODITY GROUP ($1,000)					
Grains, oilseeds, dry beans, and dry peas	29,817	46	71	1,081	2,926
Tobacco	-	-	10	-	436
Cotton and cottonseed	-	-	-	-	635
Vegetables, melons, potatoes, and sweet potatoes	67,134	2	70	44	2,802
Fruits, tree nuts, and berries	(D)	22	70	(D)	2,724
Nursery, greenhouse, floriculture, and sod	2,328	24	71	710	2,678
Cut Christmas trees and short rotation woody crops	(D)	3	67	37	1,530
Other crops and hay	3,034	37	71	964	3,049
Poultry and eggs	51	52	72	1,621	3,013
Cattle and calves	5,616	53	72	1,729	3,056
Milk from cows	18,480	53	68	356	2,038
Hogs and pigs	(D)	(D)	70	(D)	2,827
Sheep, goats, wool, mohair, and milk	947	8	68	176	2,988
Horses, ponies, mules, burros, and donkeys	275	14	69	981	3,011
Aquaculture	(D)	7	53	(D)	1,366
Other animals and other animal products	(D)	6	70	(D)	2,924
TOP CROP ITEMS (acres)					
Corn for grain	30,545	42	69	796	2,638
Vegetables harvested, all	30,140	3	70	29	2,801
Forage-land used for all hay and haylage, grass silage, and greenchop	16,222	54	72	1,174	3,057
Soybeans for beans	14,918	45	68	1,015	2,162
Potatoes	8,372	4	69	33	2,354
TOP LIVESTOCK INVENTORY ITEMS (number)					
Cattle and calves	12,315	54	72	1,775	3,063
Layers	2,779	38	72	1,083	3,040
Goats, all	2,253	7	70	181	2,996
Horses and ponies	1,129	40	72	1,041	3,072
Broilers and other meat-type chickens	1,082	34	71	931	2,723

Other County Highlights, 2012

Economic Characteristics	Quantity
Farms by value of sales:	
Less than $1,000	118
$1,000 to $2,499	60
$2,500 to $4,999	49
$5,000 to $9,999	65
$10,000 to $19,999	60
$20,000 to $24,999	13
$25,000 to $39,999	36
$40,000 to $49,999	14
$50,000 to $99,999	62
$100,000 to $249,999	50
$250,000 to $499,999	17
$500,000 or more	48
Total farm production expenses ($1,000)	100,705
Average per farm ($)	170,110
Net cash farm income of operation ($1,000)	40,118
Average per farm ($)	67,766

Operator Characteristics	Quantity
Principal operators by primary occupation:	
Farming	264
Other	328
Principal operators by sex:	
Male	543
Female	49
Average age of principal operator (years)	57.4
All operators by race [2]:	
American Indian or Alaska Native	2
Asian	8
Black or African American	-
Native Hawaiian or Other Pacific Islander	-
White	932
More than one race	-
All operators of Spanish, Hispanic, or Latino Origin [2]	8

See "Census of Agriculture, Volume 1, Geographic Area Series" for complete footnotes, explanations, definitions, and methodology.
- Represents zero. (D) Withheld to avoid disclosing data for individual operations.
[1] Universe is number of counties in state or U.S. with item. [2] Data were collected for a maximum of three operators per farm.

An appraiser may look at taxation, including assessment rates, to discern one basic effect of government regulation in a particular area. Eminent domain takings in proximity to agricultural lands can increase or decrease a community's property values. Pollution and erosion control regulations can increase the cost of running an operation and decrease income and land value.

In some areas, agricultural lands are subject to county or parish zoning regulations. The government can change these zoning and land use regulations as demand dictates. Changes in the highest and best use can make it economically infeasible to keep land in agricultural production. Although land values generally increase, the change in zoning to a commercial, residential, or industrial use can increase real estate taxes. Income tax laws and FSA programs can also influence land values.

> **Governmental Regulation Data Sources**
>
> Property assessments, millage rates, and zoning information can typically be obtained at county or parish courthouses. In many states, city or town planning and zoning authorities control zoning within village limits. State environmental agencies enforce pollution and hazardous waste control laws and can provide information on these regulations. The FSA office and the USDA's National Agricultural Statistics Service (NASS) (www.nass.usda.gov) are sources for agricultural data. (Due to confidentiality laws, it is impossible to get individual farm-level data from FSA offices.) The Bureau of Land Management (BLM) (www.blm.gov), Forest Service (www.fs.fed.us), and state land departments can provide information on land leasing rules.

Environmental Data

As in most cases, all environmental forces that affect land value must be considered. These can be issues that are specific to a property or considered external to a property. Detriments and hazards are present to some extent on nearly all properties. An appraiser should determine the types of nuisances and detrimental factors present on the property–such as odors, undesirable business establishments, dumps, landfills, potential hazardous wastes, and noxious plants–as well as neighborhood hazards such as annual variations in yields and the incidence of hail, flood, and drought. In addition, appraisers must consider the presence of wetlands, highly erodible lands, and endangered species.

Physical observation of the property will reveal which environmental concerns warrant further investigation.

> **Environmental Data Sources**
>
> The local Natural Resources Conservation Services (NRCS) office, planning and zoning commission, state environmental agency, and chamber of commerce can generally provide environmental information to satisfy most of an appraiser's needs.

Climatic Data

Of the many characteristics that determine the agricultural use of land, climate is the most unpredictable. Long-term weather records help farmers establish norms and make predictions, but the climate in any future year cannot be ensured. Climatic conditions govern the suitability or unsuitability of an area for producing specific commodities. Agricultural technology has been able to alter some climatic conditions. The installation of irrigation systems in low rainfall areas

Climatic Data Sources

Climatic information can be obtained from the NRCS (www.nrcs.usda.gov), county extension services, and local libraries and chambers of commerce.

can ensure adequate water resources for plant growth, and heated greenhouses can protect plants from freezing temperatures. However, even a property that has been modified to overcome one extreme may be subject to other climatic influences that can affect production. For example, drought can exhaust irrigation water supplies and hail can destroy greenhouses.

Appraisers are not expected to be meteorologists, but they should be well enough acquainted with climatic norms to be able to make sound judgments about crop suitability, yields, and the supply and dependability of water resources in an area, particularly in situations when the feasibility of a site's proposed use is being analyzed.

Soil Data

Soil properties and topography combine with climatic conditions to make an area suitable for the production of certain commodities and unsuitable for the production of others. However, advances in land-forming and improvements in irrigation have lessened some of the restrictions for marginal cropland. The vegetation of rangeland is dictated by both soils and climate. Soils play a very important role in the appraisal of agricultural properties, and they will be discussed in detail in later chapters.

Crop and Yield Data

Data on crops and yields is important in the income capitalization approach and can often assist appraisers in making comparisons in the sales comparison approach. Data for every major commodity produced in the nation is available at national and state levels through the National Agricultural Statistics Service (NASS). However, published data sources diminish as the defined area narrows from the state to the county to the local level. Local data sources may provide information such as actual crop share, lease, and cash rental arrangements.

Crop and Yield Data Sources

The NASS compiles extensive data for every commodity produced within the state's jurisdiction at the state and often county levels. Statistics by state are accessible through the NASS national website. Crop and yield data for each county and for the state is obtained by surveying the state's agricultural producers. Other sources include seed dealers, county extension services, farm managers, and marketing companies.

Livestock Data

In the appraisal of livestock operations, assembling data relating to the specific operation is crucial. The areas that must be researched

Soil Data Sources

One extremely valuable tool for rural appraisers is the soil surveys published by the NRCS (http://websoilsurvey.sc.egov.usda.gov/App/HomePage.htm) for individual counties and areas. In cooperation with state agricultural experiment stations and other federal and state agencies, the USDA has been publishing soil surveys since 1899. Soil surveys contain detailed maps and descriptions of soil types in the county or area surveyed. They describe soils and soil characteristics (horizons, consistency, structure, and texture), site characteristics (drainage, topography, climate, and fertility), and effects on soil productivity. The soil surveys also describe soil properties that provide a scientific basis for comparing one parcel of land with another.

The contents of and uses for soil surveys have been expanded over the years. Since 1957, soil surveys have included interpretations for each of the soils found in an area and cited on the area maps. The kinds of interpretations included in recent surveys vary with the needs of the area, but the following interpretations are included in most surveys:

- estimated yields of the common agricultural crops under defined levels of management
- land capability interpretations
- soil woodland interpretations
- rangeland interpretations
- engineering uses of soils
- interpretations for community planning
- suitability of the soil for drainage and irrigation
- suitability of the soil for wildlife and recreation

Information derived from these interpretations is useful in the valuation process. For example, estimated yields can be used in conjunction with expected or average commodity prices to provide a basis for estimating a property's gross annual income. State Department of Agriculture statistical summaries are also excellent sources of this type of information. Rangeland studies indicate the number of animal units that can be maintained on a range without damaging its natural vegetation. Some states publish productivity bulletins or guides that index soils with a relative productivity rating. Engineering studies of soil uses provide information that is essential to the development of rural land for urban or recreational use. Studies that indicate the suitability of land for irrigation and drainage help appraisers determine the highest and best use of land.

include the livestock breeds, commodity prices, market stability, marketing facility locations, the operation's prominence in the neighborhood, potential death losses from disease and predators, feed requirements, and government support programs. As with many segments of agriculture, livestock operations have generally grown larger and more complex with corporate ownership and vertical integration.

Livestock Data Sources

Local FSA, NRCS, and county extension service offices are excellent sources for livestock data. Information for a specific enterprise can also be obtained from sources such as marketing cooperatives, major poultry companies, creameries, livestock auction yards, meatpacking companies, the Agricultural Marketing Service, state agriculture departments, universities, and state offices of the NASS.

Data on Timber and Minerals (Natural Resources)

Sources of data on natural resources are limited, and the information obtained from these sources must be scrutinized. Appraisers must have an understanding of the timber or mining operation as well as

Timber and Mineral Data Sources

Although data sources on timber and minerals are few, a wealth of information is available through the US Forest Service and the Bureau of Land Management (BLM). These two federal agencies should be the focus of an appraiser's data collection efforts. State agencies associated with these federal agencies–such as the state department of forestry, the state bureau of mines, or the state land department–can also provide meaningful information. While timber and mining companies possess data, most are reluctant to divulge information. Timber cruisers are excellent suppliers of information on the quantity and quality of standing timber, while geologists or assayers can provide information relating to the supply, price, and concentrations of mineral deposits.

Water Resource Data Sources

Besides individual water districts and companies, the main source of project information on water resources is the Bureau of Reclamation (www.usbr.gov), the federal agency that oversees all operations of the dams and reservoirs storing irrigation water. For pump well data, individual well drillers and pump companies can be a good source of information. The NRCS and Extension Service often compile information on the cost of irrigation, which can be invaluable to an appraiser's income capitalization approach calculations.

information about supply and demand, competitive operations, governmental regulations, and income and expenses. A good understanding of all relevant factors is difficult to develop with so few sources available to appraisers.

Water Resource Data

A valuable commodity in the arid regions of the United States, water and the lack of it was the cause of many feuds and deaths in the nineteenth century. (In the twenty-first century, the feuding has largely moved to the courtroom.) Irrigation projects and wells are the lifeblood of agriculture in the arid western states. Many western rivers and streams have been dammed to store water for irrigation. Thousands of miles of canals and pipelines distribute that water to farms in the fertile, but arid, valleys. Irrigation projects, districts, companies, and cooperatives govern the use and distribution of the water to their members or subscribers. Most often, water rights are appurtenant to the land and regulated by a state agency, which can provide the necessary data.

Micro-Level Data

Micro-level data includes details about the property being appraised and comparable properties that have been sold or rented, as well as local market characteristics relevant to the analysis of the subject property. In the valuation process, micro-level data is used to determine highest and best use and to make the specific comparisons and analyses required to develop an opinion of value. The micro-level data about a subject property provided in land and building descriptions helps appraisers select other micro-level data pertaining to sales and rentals of comparable properties, construction costs, and specific characteristics of the local market that influence the value of the subject and comparable properties.

Lease Data

Collecting lease data is vital to the income capitalization approach to value and requires tact and diplomacy because lessees and lessors are usually the only source of lease information. The financial information in a lease is often considered to be of a private nature, so appraisers need to earn the trust of the lessee or lessor to secure the necessary data.

When a cropshare lease is being analyzed, details concerning variables such as fertilizer and seeding rates, costs, and crop yields must be obtained to project the landlord's net income accurately. Allowing an appraiser to inspect property records will take time away from the landlord or tenant's schedule. Considerate appraisers will be aware of this demand and arrange interviews accordingly.

> **Lease Data Sources**
>
> Lease data can sometimes be obtained from studies conducted by universities or the American Society of Farm Managers and Rural Appraisers or from private lending institutions such as banks, Farm Credit System lending entities, and life insurance companies. Farm managers can also often provide lease information. Published surveys by land grant universities are also an excellent source of lease data.

Cost Data

Improvement costs for typical buildings on agricultural properties are readily available, but extensive research and investigation may be needed to acquire accurate information. It may be difficult to estimate the reproduction costs of many older buildings with outmoded designs and structural features without enlisting the assistance of contractors and specialty suppliers.

> **Cost Data Sources**
>
> Most appraisers use cost guide publications such as the *Marshall Valuation Service* (www.marshallswift.com) for building cost information. However, local contractors can provide more accurate cost information in cases when distance may be a factor. For example, a cost guide may not provide the most accurate estimate of the cost to build a dwelling on a ranch located 75 miles from the nearest town. A local contractor's estimate will be more accurate because it will include the added travel and transportation costs incurred along with the construction costs.
>
> When atypical improvements are encountered, appraisers should look to building supply sources to obtain cost estimates. Appraisers should contact equipment manufacturers to obtain current cost data on specialized equipment (such as that used in vegetable canneries or meatpacking houses).

Market Data

Market data is created by the actions of buyers and sellers in an open and competitive market, thus sources of market data may include market-support businesses and market-related businesses as well as the market itself. The reliability of the sales comparison approach to value depends on an appraiser's ability to gather, assemble, analyze, and interpret market data. Poor market data can render the entire approach useless and indefensible, while good market data can provide meaningful and supportable conclusions. It is vital that the market be thor-

Market Data Sources

Where available, appraisers can make use of various real estate multiple listing services and sales reporting companies, which periodically provide sales information extracted from county courthouse records. This type of service can be purchased for a reasonable fee, but the data is not necessarily conclusive or accurate. Appraisers must investigate the sales data and rely on the reporting service only in the initial stages of investigation. Closing agents such as title insurance companies and attorneys can provide sales information, as can real estate brokers and real estate auctioneers. Lending institutions are excellent sources of information on sales of their acquired properties and sales in which they provided financial assistance to the purchaser. Exchanging sales data with other appraisers is, of course, a popular way to add sales information to an appraiser's data bank. Searching individual county courthouse records remains a very sound method of collecting sales information.

Once a good base of sales data has been compiled, an appraiser's analysis can reveal trends in the marketplace and determine if a more thorough investigation of the market is warranted. Discussions with real estate brokers, lenders, and individual buyers and sellers can provide valuable information that an appraiser can use to support trends observed in analyzing the market data.

oughly investigated to collect all possible data and that sales data be confirmed with the buyer, the seller, or another knowledgeable party.

Collecting Data

Every rural property is unique, and appraisers must concentrate their information collection, confirmation, and analysis efforts on the data that is most pertinent to the particular appraisal problem. Most appraisers will develop their own techniques for analyzing and gathering market data.

Collecting Macro-Level Data

The sources of macro-level data generally provide information on relatively distinct geographical areas:

1. The national level
2. The regional and state level
3. The neighborhood and market area level

National and USDA Data

General national data is seldom a part of the final written appraisal report, so its importance to an appraiser's data collection effort is often underestimated. Information is compiled by government agencies that report and monitor national and world economic trends, and these trends affect rural properties in many ways. The state of the national economy influences not only the amount of food products purchased, but also the types of products in demand. High demand for one product over another will filter back to the value of the land that produces the preferred product. Despite government programs that attempt to stabilize prices for agricultural commodities, any sustained downtrend in the economy may have decided effects on rural property value. The Federal Crop Insurance program, administered by the Risk Management Agency of the USDA, is the primary price and yield support mechanism for agricultural crops.

Agricultural land and facility values tend to be indirectly influenced by interest rates and directly influenced by commodity, livestock, poultry, egg, and milk prices. Because agricultural production requires extensive capital expenditures, the debt service on property acquisition may become prohibitively high. If the total return on investment (including anticipated property appreciation) is not equal to or greater than the return available from more liquid investment opportunities, owners will tend to sell rural property for other purposes rather that retain it in agricultural production, which may increase the supply of land and lower its value.

The cost of energy is also an important consideration. When world petroleum prices increase, the cost of agricultural production increases. The costs to transport and process agricultural commodities are also influenced by increases in energy costs.

World population and the demand for and supply of food and fiber products have an effect on rural land values. Increased demand for agricultural commodities for export or the curtailment of existing demand may be important influences. Increased agricultural production in areas outside the United States can affect commodity prices and lower land values for agricultural properties, as can any economic downswing that affects the purchasing ability of foreign markets. Shifts in monetary exchange rates that make it more favorable to purchase from other markets can also affect rural land values.

Of course, the effects of national and world economic forces on property value are not immediately evident. They are not directly translatable into adjustments in the sales comparison approach or changes in the gross income estimates for agricultural properties. The effects of economic forces are cumulative, and many years of sustained activity may pass before noticeable results occur. A change in government policy that creates favorable export conditions for agricultural commodities will not cause immediate changes in land values. If these conditions are sustained, however, the gradual increase in agricultural productivity and income will eventually result in higher land values. For example, a federal mandate for the increased production of ethanol created a new demand for corn, which in turn affected grain prices and then eventually land values in the Corn Belt between 2007 and 2013.

Because government actions indirectly change agricultural land values, these factors are not used by appraisers in specific value calculations. Rather, knowledge of national and world economic trends alerts appraisers to specific changes (e.g., an increase in gross income averages or an increase in transportation or operational charges due to higher energy costs) that will be useful in explaining, and even in anticipating, changes in land value.

Although the effect of national and international economic trends on rural property value cannot always be quantified, knowledge of and sensitivity to these broad influences can help appraisers understand and respond to changes in the attitudes and behavior of buyers who are responding to these same trends. Since the anticipation of profit is the essential motivation for purchasing property, any economic trend that affects the buyer's anticipation of profit will ultimately affect property value.

National and USDA Data Sources

Various agencies within the USDA (www.usda.gov) compile general national data that appraisers can use. The Forest Service stores a wealth of timber and minerals information, as well as vegetation and grazing fee data. The FSA can provide historical data on national prices, production, and reserves for the country's major agricultural commodities. The NRCS is an excellent source of information on soils, water, conservation measures, environmental improvements, and agricultural pollution control.

Current information on the prices, supply and demand, location, quality, and condition of cotton, tobacco, milk, livestock, grains, fruits, vegetables, and other commodities can be obtained from the Agricultural Marketing Service (AMS) (www.ams.usda.gov).

The NASS, the statistical and data collection arm of the USDA, compiles information on acres planted, acres harvested, crop yields and production, and production values for every agricultural commodity produced in the United States.

Regional and State Data

When appraisers begin area analysis, quality regional and state data become very important. Agricultural properties depend on complementary services for economic viability, and the presence of these services within a reasonable distance is vital to the success or failure of the agricultural enterprise. Access to markets, transportation, support businesses, and financial institutions is essential to rural properties and must be considered on a regional or state basis.

The income from agricultural production depends on markets for the commodity produced. The location of markets (including terminal livestock markets, terminal grain markets, ethanol plants, livestock and milk processing plants, and loading, sorting, and buying facilities) is important, especially to the value of large operations. For example, the presence of large commercial confinement feeding operations near rangeland in the West can enhance the value of the rangeland itself.

Costs of transportation from the farm to the market can influence the profit margin of agricultural enterprises insofar as these costs must be absorbed in the prices obtained for agricultural goods. Good facilities for transporting the agricultural commodity from the producing property to a processing or handling property and then on to its final market quickly and effectively are fundamental concerns of agricultural producers. Therefore, purchasers consider transportation in their property selection. In the appraisal of rural properties, an analysis of the area to determine the influence of transportation on values is essential.

Support businesses within the region or state greatly influence the desirability, salability, and rentability of rural properties. The importance of seed, feed, fertilizer, machinery, and equipment dealers in rural communities is obvious. However, the importance of businesses such as canneries, vegetable processors, meatpacking plants, livestock commission yards, and grain drying, loading, and storing facilities must not be overlooked. In agricultural communities, the suppliers of production, processing, and distribution needs surround the area where the raw product is produced. The accessibility of suppliers is often correlated to the desirability of the area, which ultimately affects land values.

Another factor to be considered is the availability of financing and the type of financing most common in the area. This is the basis of the financing terms adjustment in the sales comparison approach. Current interest rates and terms for both production-type loans and long-term real estate loans must be known because the cost of debt service can figure prominently in the analysis of expenses and property income.

Regional and State Data Sources

Regional and state data can be obtained online from state agencies or from the state or regional offices of federal agencies. Most of the previously mentioned federal agencies have state offices that can provide local information. The BLM and the Forest Service have excellent timber, minerals, and range data for every state where there are BLM or Forest Service lands to be managed.

Universities, particularly land grant institutions through the Agricultural Extension Services (AES), can provide a wide array of data relating to most aspects of rural property. Although their exact titles vary by state, state agencies of the federal departments of commerce, agriculture, forestry, lands, and environmental services can be excellent data sources.

Neighborhood and Local Data

Although general data helps appraisers gain a broad knowledge of rural properties, rural appraisers must rely on local data to approach the appraisal problem directly. The forces at work within each community define which data will be most useful. An awareness of local forces helps appraisers understand current market activity and anticipate and identify trends that affect rural properties.

Collecting Micro-Level Data

Figure 5.2 lists many sources of information pertaining to sales and rentals of agricultural properties. Sales of comparable properties should be classified as unimproved or improved. Sales of unimproved properties of only one land class will lend support for land class ratio

Figure 5.2 Suggested Data Sources

Sales of Comparable Properties

- Real estate professionals (such as brokers)
- Title and abstract companies
- Attorneys who are familiar with real estate transactions
- Local tax appraisal districts or assessors
- County deed records
- Declarations of value in disclosure states
- Local agricultural lenders
- Local farmers and ranchers who are knowledgeable about the subject area
- Sales data published on agricultural real estate companies' websites
- Online subscription services (such as www.landsalesbulletin.com)
- Published news

Rental Information

- Local county agents
- Local agricultural lenders (such as the Farm Service Agency, Farm Credit Services, and commercial banks)
- Local agricultural suppliers or marketing outlets
- Local farmers and ranchers who are known to lease or rent agricultural lands
- Local trust departments and farm management firms
- Published surveys by land grant universities
- Local newspapers

analysis. The valuation of the land component of improved property sales is facilitated by analyzing information gathered from the unimproved land sales.

Listings and Offerings

Whenever possible, appraisers should gather information about properties offered for sale. For example, an appraiser can request to be added to the mailing lists of banks, brokers, and others who offer properties for sale. Many regional brokers have websites that provide listing information as well as e-mail subscriptions for regular updates about new listings or price changes. Working relationships with brokers, Farm Credit offices, and local bankers will provide a great deal of information.

Advertisements of listed properties suggest the strength or weakness of the local market for a particular type of property and the sales activity in a particular area. Information on purchase offers may also be obtained from brokers or managers. Listings usually reflect the upper limit of value, while offers commonly set the lower limit of value.

Collecting Competitive Supply and Demand Data

By its very nature, rural property is not driven by the same supply and demand factors typically seen in more urban settings. While competitive properties for a small commercial building may be found in the neighboring community, competitive properties for a large farm or ranch may be many miles away in an adjoining county. Therefore, gathering data on rural competitive supply may be very time-consuming without a network of individuals in the market.

A working knowledge of relevant market activity is essential with rural properties, as most competing properties will not be in close proximity to the subject property. Surveying supply is usually done by researching available properties through brokers, agents, and other market participants. Appraisers can then interview owners, managers, and brokers of competitive properties in the region. Field observations and interviews are especially important because rural properties can have so many variations, and it is important to understand the motivations of the buyers and sellers of these properties.

Verifying Data

The analysis of market data is only as credible as the data that is collected and presented to the client in a report. A primary purpose of verifying a real property sale is to make sure that the sale occurred under conditions that meet the definition of value used in the appraisal, such as "at arm's length" in an appraisal of market value. The verification process also provides appraisers with an opportunity to obtain additional and sometimes more accurate information about the property and to gain fresh insight into the attitudes and motivations of the buyer and seller, which is of particular importance for agricultural properties with operating businesses having many moving parts.

An example of important data that is often uncovered through the confirmation process would be information that is not part of the public records, such as how many cultivated acres were on a property that recently sold, whether the land was irrigated, and how much of the irrigated land was irrigated using either precision leveling or center-pivot irrigation. Although aerial imagery can shed some light on these issues, the imagery may be dated and may not reflect the actual conditions on the date of the sale. Similarly, the presence of field tile drainage on a row crop farm may not be evident in the information that is publicly available. In the verification process, an appraiser can inquire about how much of the farm was tiled and whether the drainage was working. In the verification process, appraisers can ask about the values the seller and buyer assigned to the buildings and improvements to assist in allocating contributory values of those buildings and improvements. When information is scarce, any opportunity to learn more about a property or transaction can reap benefits for the analyst.

Chapter 6

Legal Descriptions

Land boundaries differentiate separate ownerships, and the land within one set of boundaries may be referred to as a *parcel*, *lot*, *plot*, or *tract*. These terms may be applied to all types of improved and unimproved land, and they are often used interchangeably by market participants. Appraisers, however, should use the terms consistently in their appraisal reports to avoid confusing their clients. A *parcel* of land generally refers to a piece of land that can be identified by a common description and is held in one ownership. Every parcel of real estate is unique. To identify individual parcels, appraisers rely on legal descriptions, surveys, or other descriptive information typically provided by the client or found in public records. A legal description identifies a property in such a way that it cannot be confused with any other property. A legal description, though not required, is usually included or referenced in an appraisal report. In addition to the specific description of the real estate, the legal description may indicate easements, restrictions, mineral rights, water rights, or partial interests.

History has recorded many court battles relating to the legal description of property. Clients often instruct appraisers to obtain legal descriptions from sources such as the local assessor, the register of deeds, or a will. However, portions of the original property may have been sold, traded, or altered since the original legal description was created. Ownership of a farm or ranch property is often held by more than one entity.

It is incumbent on appraisers to demand the exact legal description of the property to be appraised from the client. Obviously, many clients require an appraiser's assistance in obtaining this document. In the United States, most states require that deeds, mortgages, easements, and other documents relating to real estate be publicly recorded, and most of these records are cross-indexed. This system allows appraisers to search for a property through the grantor-grantee records or through the land index by legal description. Appraisers should be sure to work with exact legal descriptions because offices such as the assessor often work with abbreviated or laymen's legal descriptions.

Obtaining the legal description often leads to the difficult task of precisely identifying the physical property. A large volume of quasipublic information is available to appraisers, depending on the political subdivision in which the property is located. Information from independent surveyors or the mapping departments of county tax assessors may assist the client or appraiser in identifying or correcting imperfections in the written legal description. Cadastral maps, US Geological Survey maps, FSA/NRCS/RD photos, and numerous other sources of information may also be consulted to help appraisers identify the exact physical property from the written legal description. Some web-based services have searchable section, township, and range maps with aerial imagery (e.g., Agridata, USDA Web Soil Survey). In extreme situations, the client may need to obtain an abstract of title or title insurance commitment to create a correct legal description. A combination of documented official information, support data from unofficial sources, and careful field work can ensure a good legal description of the property for appraisal purposes.[1] Federal and state legislation requires that appraisals comply with appropriate professional standards, and a legal description is one way to identify the property and satisfy the requirement found in professional standards.

A property may be difficult to visualize if the field boundaries or land uses are inconsistent with the exact legal description. Federal, state, or private leased lands may be fenced jointly with lands in private ownership. The exact locations of irrigation wells, livestock watering facilities, and building improvements are extremely important in determining the legal boundaries of a property. There are often discrepancies between what the owner or client believes are the correct boundaries of the physical property or location of portions of that property and the true legal description. A precise survey may be required prior to an appraiser's completion of the appraisal assignment. The use of personal GPS devices or GPS-enabled smartphones allows for a relatively accurate location determination with respects to property features. Using aerial imagery in combination with a GPS device or smartphone can take much of the guesswork out of site visits.

1. For a more extensive discussion of the sources for, and construction of, legal descriptions, see Stephen Estopinal and Wendy Lathrop, *Professional Surveyors and Real Property Descriptions: Composition, Construction, and Comprehension* (Hoboken, NJ: John Wiley & Sons Inc., 2012).

Some of the systems currently in use to identify real estate include

- the street address system
- the block and parcel number system
- the US Rectangular Survey System (and the analogous Dominion Lands Survey System in Canada)
- map-based systems
- federal surveys based on the coordinate system
- the State Plane Coordinate System (SPSC)

Efforts have been made to standardize these systems to accommodate researchers and make use of modern recordkeeping technology.

In the United States, three methods are commonly used in legal descriptions of real property:

- metes and bounds system
- rectangular survey system
- lot and block system

An appraiser should be familiar with these forms of legal description and know which form or forms are common in the area where the appraisal is being conducted.

Metes and Bounds System

The metes and bounds system is the primary method of describing real estate in the 13 original American colonies, Texas, Hawaii, and a few other states east of the Mississippi River. This system of legal description is centuries old and was brought to America by the early settlers. This system was used for legal descriptions of real estate in the United States prior to the development of the rectangular survey system. Under the metes and bounds system, real estate is described using natural or man-made objects as reference points. The starting point for the description is usually an arbitrary point, typically called the *point of beginning* (POB). Original metes and bounds descriptions often did not relate to the reference points used to describe other properties in the immediate area. Modern metes and bounds descriptions, however, have reference points that are tied to adjoining surveys.

The metes and bounds system is based on distances (metes) and bearings (bounds) from succeeding reference points around the perimeter of the real estate. Distances and directions are thus described relative to naturally occurring or man-made objects, which identify the corners or sides of the parcel. Unfortunately, oak trees, stone heaps, roads, and rivers change over time. Originally there were no consistent general reference points in the area, county, or state from which to describe individual parcels of land.

The following is a portion of a metes and bounds description used in the Hartford, Connecticut probate court in 1812:

> Commencing at a heap of stone about a stone's throw from a certain small clump of alders, near a brook running down off from a rather high part of said ridge; thence by a straight line to a certain marked white birch tree, about two or three times as far from a jog in a fence going around a ledge nearby; thence by another straight line in a different direction, around said ledge and the Great Swamp, so called; thence in line of said lot in part and in part by another piece of fence which joins onto said line, and by extension of the general run of said fence to a heap of stone near a surface rock; thence as aforesaid, to the "Horn," so called, and passing around the same as aforesaid, as far as the "Great Bend" so called, and from thence to a squarish sort of jog in another fence, and so on to a marked black oak tree with some stones piled around it; thence by another straight line in about a contrary direction and somewhere about parallel with the line around by the ledge and the Great Swamp, to a stake and stone bounds not far off from the old Indian trail; thence . . .

This example points out the potential ambiguity of older metes and bounds descriptions.

The modern metes and bounds system employs permanent man-made monuments and precise descriptions of distance (linear measurement) and bearing (angle measurement). The surveyor installs permanent markers, commonly called *benchmarks*, at some or all reference points around the described real estate. The corner or marker where the survey or legal description actually begins is the point of beginning. A POB can be used in association with a rectangular survey, recorded plat, or metes and bounds description.

In a metes and bounds description, distances are typically measured in feet, to the nearest tenth or hundredth of a foot. Distances can also be measured in rods, chains, or miles. Bearings or directions are measured in degrees (°) of a 360-degree circle, with further subdivisions of 60 minutes (') per degree and 60 seconds (") per minute. Thus, 29°14'52" would be read as 29 degrees, 14 minutes, and 52 seconds. It is important to note that the 360 degrees of a metes and bounds description do not relate to navigational headings.

The 360-degree circle of the metes and bounds system is divided with north-south and east-west lines into four quadrants of 90 degrees each, as shown in Figure 6.1. A course is followed to connect monuments or points within the description. Directions for the course are quoted relative to degrees of variation east, west, north, or south from the crossing of the horizontal and vertical lines. New directions are given for each succeeding reference point within the description. Bearings or directions first indicate the direction north or south of the horizontal line, and then the degrees, minutes, and seconds east or west of the vertical line. For example, north 50 degrees east describes a line that runs 50 degrees east of north, or in an easterly direction.

Likewise, south 30 degrees west describes a line that runs 30 degrees west of true south. In the diagram in Figure 6.1, minutes and seconds have been omitted for clarity. Note that 90 degrees west can either be north 90 degrees west or south 90 degrees west. Each change in direction within the legal description uses the same quadrants of the 360-degree circle, but the crossing lines are moved to the new point of reference.

To describe the property in Figure 6.2, start at the POB, which is determined by measuring from the permanent reference mark–N80°E, and go a distance of 152 feet. Imagine looking down on the POB from a helicopter with the nose pointed north. Then proceed clockwise along the parcel's perimeter. The next point is reached by following the bearing of South 80 degrees East (S80°E) a distance of 180 feet (metes). To get to the next point, travel along the direction (bounds) South 15 degrees West for 160 feet. The next point is reached by traveling from the preceding point South 85 degrees West for 151 feet. To return to the POB, head North 4 degrees, 11 minutes, and 19 seconds East a total of 199.50 feet from the preceding point in the description.

The written legal description for this parcel would first identify the county and state within which it lies and then the permanent marker: "Beginning at the permanent reference mark, thence N80°00'00"E, 152.00 feet to the point of beginning; thence S80°00'00"E, 180.00 feet, thence S15°00'00"W, 160.00 feet; thence S85°00'00"W, 151.00 feet; thence N4°11'19"E, 199.50 feet to the point of beginning."

Metes and bounds descriptions are used extensively within the rectangular survey system. Because physical features such as rivers, lakes, oceans, highways, railroads, and parcels of excluded land do not conform to the rectangular survey system (which is described in detail below), a metes and bounds description is required. The description is referenced from the permanent reference mark, which is within the rectangular survey.

Figure 6.1 Quadrants of a Circle

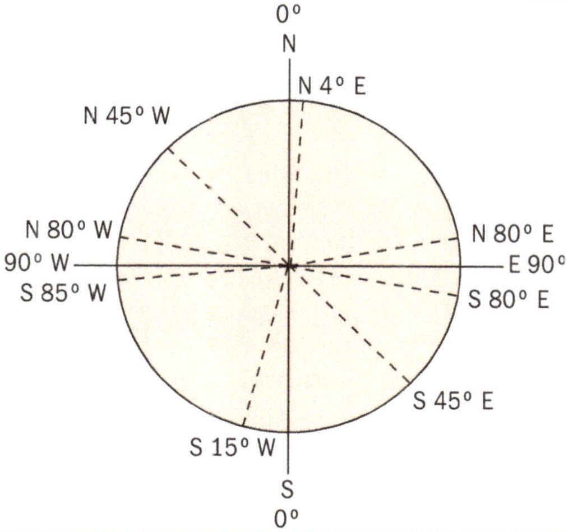

Figure 6.2 Metes and Bounds Description

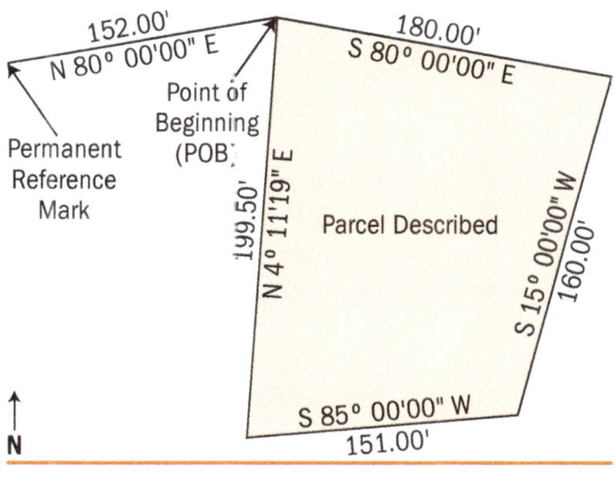

Rectangular Survey System

The rectangular survey system of legal description was created by the US Congress in the late 1700s, after the United States acquired the lands of the Louisiana Purchase, Mexican Territory, and Oregon Territory. The rectangular survey system works well on the open plains, but natural phenomena such as shorelines and rivers or man-made objects such as highways often preclude a complete legal description under the rectangular survey system. Augmentation with the metes and bounds system is then required to complete the legal description.

The original ordinance passed by the Continental Congress contained the basic elements of the present system, which is sometimes referred to as the *governmental survey*, and included this statement:

> The surveyors shall proceed to divide the said territory into townships of 6 miles square by lines running due north and south and others crossing them at right angles. The plat of the township shall be marked by subdivision into lots 1 mile square of 640 acres in the same direction as the external lines and numbered from 1 to 36.

Although the rectangular survey system appears complicated, it was designed to be mathematically pure, concise, and consistent. The system is based on sets of intersecting lines: principal meridians, which run north and south, and base lines, which run east and west. The intersection of these lines is used as a reference point to locate and describe land. Both meridians and base lines are precisely located in reference to degrees of longitude and latitude. For example, the sixth principal meridian is a north-south line arbitrarily established by the government surveyors west of Lincoln, Nebraska. It intersects an east-west base line, which is the state boundary between Kansas and Nebraska (see Figure 6.3).

Base Lines

Base lines were established by the original government surveys on true parallels of latitude. Several base lines serve as the boundaries between states. Figure 6.3 shows the system of base lines and principal meridians established for the United States.

Principal Meridians

The second axis for each reference point is a principal meridian, running north and south. Individual meridians were arbitrarily located along the respective base lines, without reference to specific points. Most meridians were originally named for physical or political features within the immediate area. More than 25 meridians are used in the governmental survey of the United States. Some are confined to one state or a portion of a state, while others are used in several states. For example, the Fifth Principal Meridian includes lands in Ar-

Figure 6.3 Principal Meridians and Base Lines

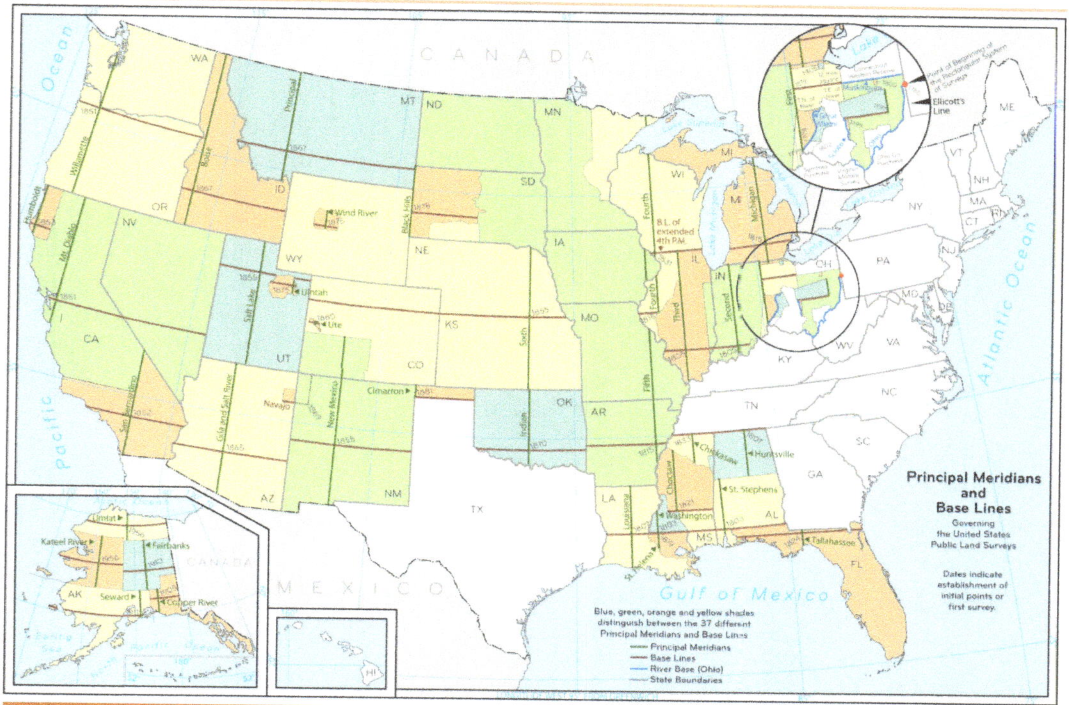

Source: BLM, www.blm.gov/wo/st/en/prog/more/cadastralsurvey/meridians.html

kansas, Missouri, Iowa, Minnesota, North Dakota, and eastern South Dakota. Regardless of the specific location, the basis for the rectangular survey is the intersection of two axes, the base line and the meridian. From this central point, the rectangular survey lays out blocks of land six miles square.

Ranges

The next step in the rectangular survey was to draw parallel north-south lines at six-mile intervals east and west of the each principal meridian. The resulting six-mile-wide vertical columns are called *ranges* and are numbered consecutively east and west from the respective principal meridian. The first range west is thus referred to as *Range One West*, and is typically written as R1W. Similarly, the eighth range east of the meridian is called *Range Eight East*, written R8E. It begins 42 miles east and ends 48 miles east of the principal meridian.

Townships

In a similar fashion, parallel east-west lines were drawn every six miles north and south of the base lines. These six-mile wide horizontal spaces are called *townships.* They are numbered north and south from the base line in the same fashion, such as T4N for Township Four North and T5S for Township Five South. Township Five South would occupy a strip of land starting 24 miles south of the base line and extending to 30 miles south of the base line.

Thus, series of parallel lines create north-to-south columns (ranges) and east-to-west horizontal spaces (townships). The bisecting of ranges and townships creates blocks of real estate that are six miles by six miles, which can be referred to with clear numerical definition. The shaded portion in the upper right quadrant of Figure 6.4 identifies T4N-R5E, which begins 18 miles north of the base line and 24 miles east of the meridian. Similarly, T5S-R8W begins 24 miles south of the base line and 42 miles west of the meridian. Under the rectangular survey system, each township is numbered relative to the juncture of a specific base line and meridian (see Figure 6.4).

Figure 6.4 Numbering of Townships

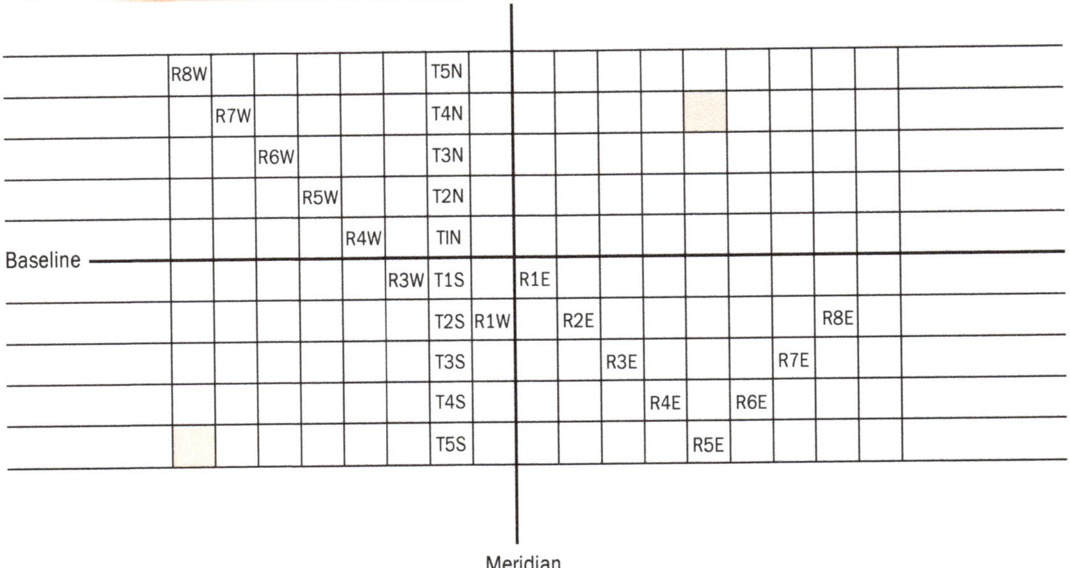

Sections

Figure 6.5 illustrates the subdivision of each township into 36 parts, called *sections.* Sections within a township are created by parallel lines running both north-south and east-west one mile apart, creating the 36 sections. Within the rectangular survey system, townships are six miles square, or 36 square miles, while sections are one mile square or one square mile. A standard section therefore includes 640 acres. Sections within the township are numbered from 1 to 36, beginning in the northeast corner and proceeding west and east alternately in a southerly manner throughout the township. Figure 6.5 shows the 36 sections of Township 5 South - Range 8 West (T5S-R8W), together with all adjoining sections located in adjoining townships. On a map, the section number is written in the center of the section.

The rectangular survey system further divides each section into quarters containing 160 acres (e.g., NE¼ of Section 3); quarter quarters containing 40 acres (e.g., NE¼NE¼); and quarter quarter quarters containing 10 acres (e.g., NE¼NE¼NE¼). The system can also subdivide sections and portions thereof by one-half (e.g., S½, or N½NE¼, or S½N½ of Section 17), as illustrated in Figure 6.6.

Since a standard section contains 640 acres, an acre contains 43,560 square feet, regardless of its configuration. Due to the curvature of the earth and human error, however, any section within the rectangular survey might contain more or less acres depending on the original survey. The curvature of the earth creates problems for the square design of the system. An integral part of the system is the process of correction at township lines. The northern and western sides of each township are used for correction of the survey and often contain more or less than 640 acres. Thus, Sections 1 through 6 plus Sections 7, 18, 19, 30, and 31 have special considerations within each township. In general, the S½ of Sections 1 through 5 (see Figure 6.7) are surveyed as 320 acres. Additionally, the S½ of both the NW1/4 and the NE1/4 are surveyed as 80 acres. Finally, the N½ of the N½ of Sections 1 through 5 are surveyed as lots, typically numbered 1 through 4 from east to west.

Figure 6.5 Subdivision of a Township

36	31	32	33	34	35	36	31
1	6	5	4	3	2	1	6
12	7	8	9	10	11	12	7
13	18	17	16	15	14	13	18
24	19	20	21	22	23	24	19
25	30	29	28	27	26	25	30
36	31	32	33	34	35	36	31
1	6	5	4	3	2	1	6

Figure 6.6 Division of Sections

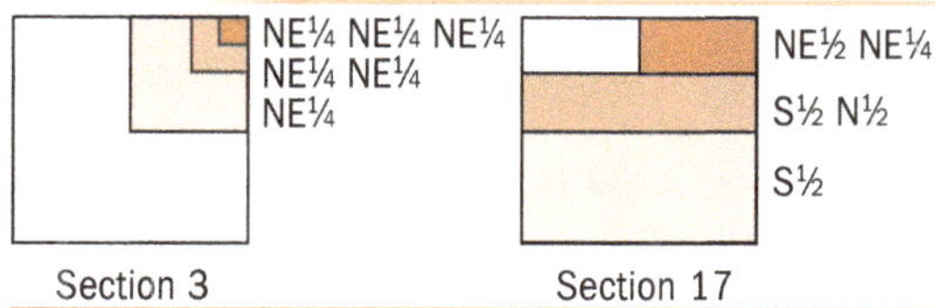

Figure 6.7 Correction at Township Lines

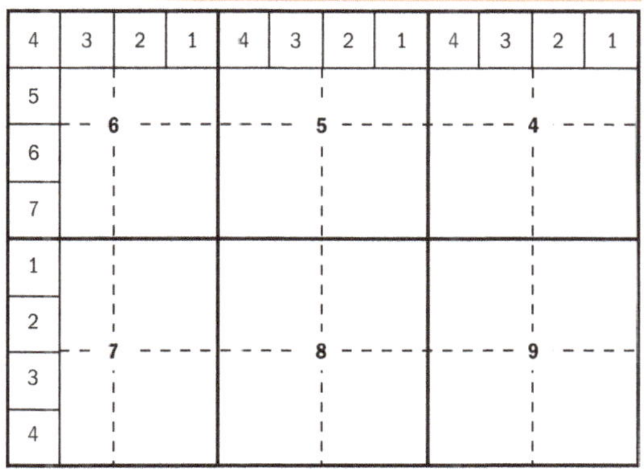

4 Indicates Lot Number

6 Indicates Center of Section

The surveyor thus created lots along the northern boundary of the township containing whatever acres were necessary to correct the survey for the curvature of the earth.

Similarly, the E½ of Sections 7, 18, 19, 30, and 31 were surveyed to contain 320 acres (in two quarters), and the E½ of both the NW¼ and the SW¼ to contain 80 acres each. To finalize these sections, the surveyor used Lots 1 through 4, numbered from north to south along the west side of each section. The required number of acres was included by the surveyor for each section to correct the township line. Section 6 of the rectangular survey is unique in that it requires correction on both the north and west boundaries. The survey is completed as described above, starting with the SE¼ as the fixed portion. The S½NE¼, the E½SW¼, and the SE¼NW¼ were then surveyed with 80 or 40 acres, respectively. Finally, Section 6 typically includes seven lots numbered from northeast to west and then to south along the north and west boundaries.

Agencies such as the US Geological Survey (USGS) or US Coast and Geodetic Survey (USCGS) have set permanent markers at locations throughout townships within the rectangular survey system. These markers can be located at section corners, half-mile lines, or other locations. Even though the US Rectangular Survey System is relatively precise, any section within the township may contain far more or less than 640 acres. Many irregular sections in which the total acreage in a section may be much larger or smaller than 640 acres are located along the Mississippi River.

Written Descriptions

Several different methods may be used to write a legal description under the US Rectangular Survey System. Although each method is correct, some may be more or less difficult to read or more or less acceptable in the local area. In most cases, the reader can best decipher the description by reading it backwards. To find the exact section and township, the reader must first identify the specific principal meridian. A typical legal description might read:

> S½SW¼NE¼ Section 15-T17N-R21W of the 6th PM, Custer County, Nebraska.

The reader can most easily understand this legal description by first determining the reference point of the sixth principal meridian (PM). (Note that the state and county are not needed in this type of legal description.) Next, appraisers identify the township and range, the section, and finally the specific portions of the section. The description is read backwards, first NE¼, then SW¼ of NE¼, and finally S½ of SW¼ of NE¼. The same legal description might also be written:

> T17N-R21W of the 6th PM Custer County, Nebraska:
> S½SW¼NE¼ Section 15 (or Section 15: S½SW¼NE¼)

One important factor in writing and understanding legal descriptions is the proper use of punctuation and the words *of* and *and.* A comma, semicolon, or conjunction (*and*) denotes a break in the subportions of a section. The legal description above refers to 20 acres total. If the description is written as S½, SW¼NE¼ or as S½ and SW¼NE¼, the land described contains 360 acres. The word *of* denotes a continuation, not a break, in the legal description. Thus, S½ of SW¼ of NE¼ is equivalent to the original tract of 20 acres and is the proper "reading" of the legal description. Extreme care is required in the writing or reading of legal descriptions prepared according to the rectangular survey system of land description.

Other Methods of Description

Real estate can also be described by referring to a document previously recorded in public records. The document referenced must contain a full legal description of the parcel in question or the parcel to be excluded from the description in question. A previously recorded legal description could be included in a deed, mortgage, or similar document. For example, suppose that several years ago Robert Baker received a deed from Lorelei Adams that contained a long and complicated metes and bounds description. Baker recorded the deed in the public records office, where it became Page 456 of Book 1089. If Baker later wants to deed the land to Angelo Cooper and Cooper wishes to obtain financing from the bank, the following land description can be used in both the new deed to Cooper and the mortgage to the bank:

> All the land described in the deed from Adams to Baker recorded in Book 1089, Page 456, County of ABC, State of XYZ, at the public recorder's office for said county and state.

Because the recorder's books are open to the public, anyone could go to Book 1089, Page 456, and find the detailed description of the parcel. Again, this method of legal description can be used to identify the desired parcel or a parcel to be excluded from the description in question.

A method of vertical description is required to describe or transfer air rights or subsurface rights. The starting point or surface from which to measure a vertical height or depth is called a *datum.* The most commonly used datum in the United States is mean sea level. A number of political subdivisions have established their own datum surfaces (points) for use in local surveys. From the datum point, benchmarks relative to that point are set by government surveyors. A surveyor need not travel to the datum point to determine an elevation. Rather, the benchmarks, which include the local elevation, are used as metes and bounds reference points. In selling or leasing subsurface drilling (mineral) rights or in restricting air rights, the datum chosen is often the actual surface of the parcel. For example,

Figure 6.8 Contour Lines Show Vertical Distances

Contour map showing elevation changes

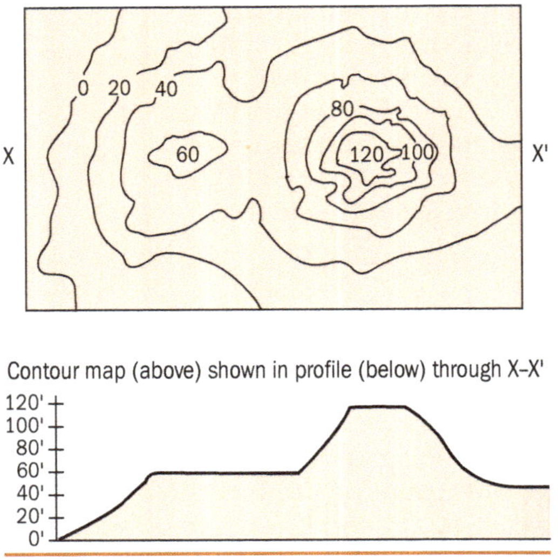

Contour map (above) shown in profile (below) through X–X'

an oil lease may permit the extraction of oil and gas from a depth greater than 500 feet, but not more than 5,000 feet beneath the surface of the parcel of land.

Contour maps, commonly referred to as *topographical maps*, can also be used to describe land. These maps show contour lines, which connect all points having the same elevation. The purpose of contour maps is to show hills, valleys, slopes, water runoff, and other features. A legal description might identify an easement that includes all lands on the subject parcel below the 20-foot contour line. Figure 6.8 illustrates how vertical distances are shown using contour lines.

Appraisers should be familiar with these forms of legal description and know which form or forms are common in the area where their appraisals are being conducted.

Chapter 7

Soils and Rangeland Capability

Soil is the basic component of value in a rural property used for agricultural commodity production because it is the medium of productive capability. Of course, soil alone will not make a profitable agricultural enterprise. Use and management of the soil determine the profit that can be realized from agricultural production. The analysis of soils and their productivity is generally most important in the appraisal of permanent plantings, cropland, and hayland. In livestock ranching, properties are usually larger and the soil is analyzed less intensively. Rangeland analysis is typically focused on what are known as "ecological sites" on the ranch.

This chapter is divided into two parts. The first explores the information to be considered in the soil analysis, and the second concerns the analysis of rangeland's productive capability.

Soil Characteristics

Every appraisal of a rural property that produces agricultural products should include a full description of its soils or rangeland characteristics and an analysis of their potential uses and productivity. Soil must be evaluated in terms of its potential productivity and desirability, which affect the agricultural property value. Value may or may not be directly proportionate to the kinds of soil and their relative productivity, but knowledge of this basic component is essential for all rural property appraisers. Although rural appraisers are not usually experts in soil science or agronomy, they do have access to published digital soil surveys and productivity guides through the USDA's Natural Resources Conservation Service (such as the Web Soil Survey) and other subscription services. Appraisers gather this expert knowledge and make economic judgments by relating soil information to the earning potential of rural property. The discussion in this chapter explains what soil is and emphasizes the soil characteristics that have the greatest effect on property value.[1]

Soils are natural bodies capable of supporting plants. Soil properties in a specific place result from the combined effects of climate and living matter on the parent material (such as rock). The soil development process is influenced by changes in the contours of the land over time. The characteristics of soils and the surrounding cultural environment are important determinants of the uses to which soils are suited. Among the most important soil characteristics are texture, structure, consistence, thickness, and arrangement of the soil horizons, or layers. These characteristics and their interrelationships, together with soil and site features such as depth to bedrock, depth to root-inhibiting layers, available water capacity, drainage, and flooding, give an initial indication of the soil's potential productivity and the type of management that may be required to obtain the greatest land productivity.

Soil Horizons

Knowing the composition of the soil layers can help appraisers determine how well the soil can support certain plant life. Soil layers, or *horizons*, are three-dimensional areas with certain characteristics produced by soil-forming processes. The horizons are roughly parallel to the soil surface, but they vary in thickness and may have irregular boundaries. The top six feet of soil are the most significant for agricultural appraisals. The horizons in these top six feet are commonly known as the *surface soil*, *subsoil*, and *substratum*.

The term *topsoil* has several meanings. It may simply describe fertile soil material that is rich in organic matter and often used to improve plant growth on road banks, gardens, and lawns. But for an appraiser's purposes, this is referred to as part of the *surface soil*. The *subsoil* is a layer generally below the surface soil. The surface soil and the subsoil layers together make up the *solum*, the part of the soil where soil formation is most active and most of the soil's plants and animals live. The *substratum* is the part of the soil below the solum. Figure 7.1 illustrates a generalized soil profile.

1. The discussion of soil, soil horizons, and texture in this chapter is based on information from the Soil Conservation Service, *USDA Soil Survey Manual, Soil Survey Staff Handbook No. 18* (Washington, DC: 1993) and USDA Natural Resources Conservation Service, *Soil Taxonomy: A Basic System of Soil Classification for Making and Interpreting Soil Surveys*, 2nd ed. (Washington, DC: 1999).

Texture

An appraisal of agricultural property includes a description of the texture of the surface and subsoil layers. The *texture* of a mineral soil refers to the relative proportions (by weight) of three basic soil particles–sand, silt, and clay. Texture is determined by the weight of the three particles after all coarse fragments (greater than 2 millimeters in diameter) have been removed. Soil textural class names describe the various kinds of soil textures based on the proportions of clay, silt, and sand. There are 12 basic classes of soil texture:

- sand
- loamy sand
- sandy loam
- loam
- silt
- silt loam
- sandy clay loam
- clay loam
- silty clay loam
- sandy clay
- silty clay
- clay

Each class has a defined range of clay, silt, and sand content, as shown in the soil triangle in Figure 7.2. For example, loam is a soil material that consists of 7%–27% clay, 28%–50% silt, and less than 52% sand. The presence of coarse particles smaller than 10 inches is indicated by modifiers such as *gravelly* sandy loam or *cobbly* loam. Cobbly loam is loam soil that contains fragments 3-10 inches in diameter. Larger particles, such as stones or rocks, are defined in terms of the influence they have on soil use. Even if soil is naturally productive, or can be made so through management, stoniness plays a role in determining whether land is used as cropland, rangeland, pasture, or forest.

Figure 7.1 Generalized Soil Profile

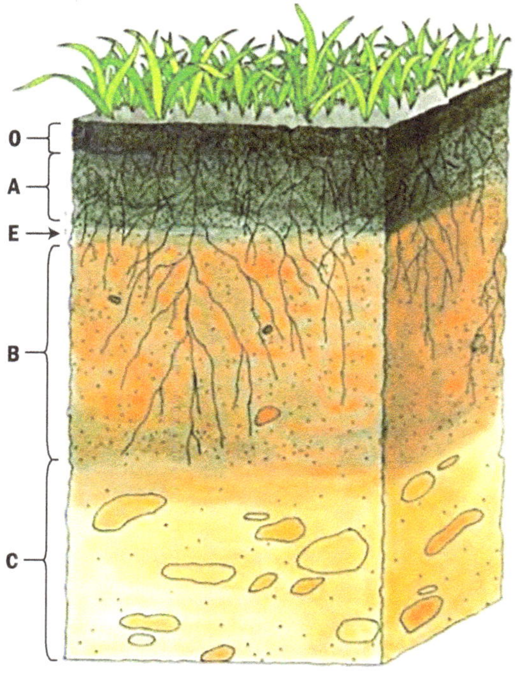

In a soil profile, the three major horizons (i.e., layers of soil with characteristics produced by soil-forming processes) are

- **A:** the surface horizon
- **B:** the subsoil, i.e., the part of the soil that lies below plow depth
- **C:** the substratum

Additional layers may include

- **O:** an organic horizon on the surface
- **E:** a master horizon, usually between the A and B horizons, for soil with a significant loss of minerals

Source: NRCS Soil Profile Gallery, www.nrcs.usda.gov/wps/portal/nrcs/detail/soils/survey/office/ssr7/profile/?cid=nrcs142p2_047970

Figure 7.2 Soil Triangle

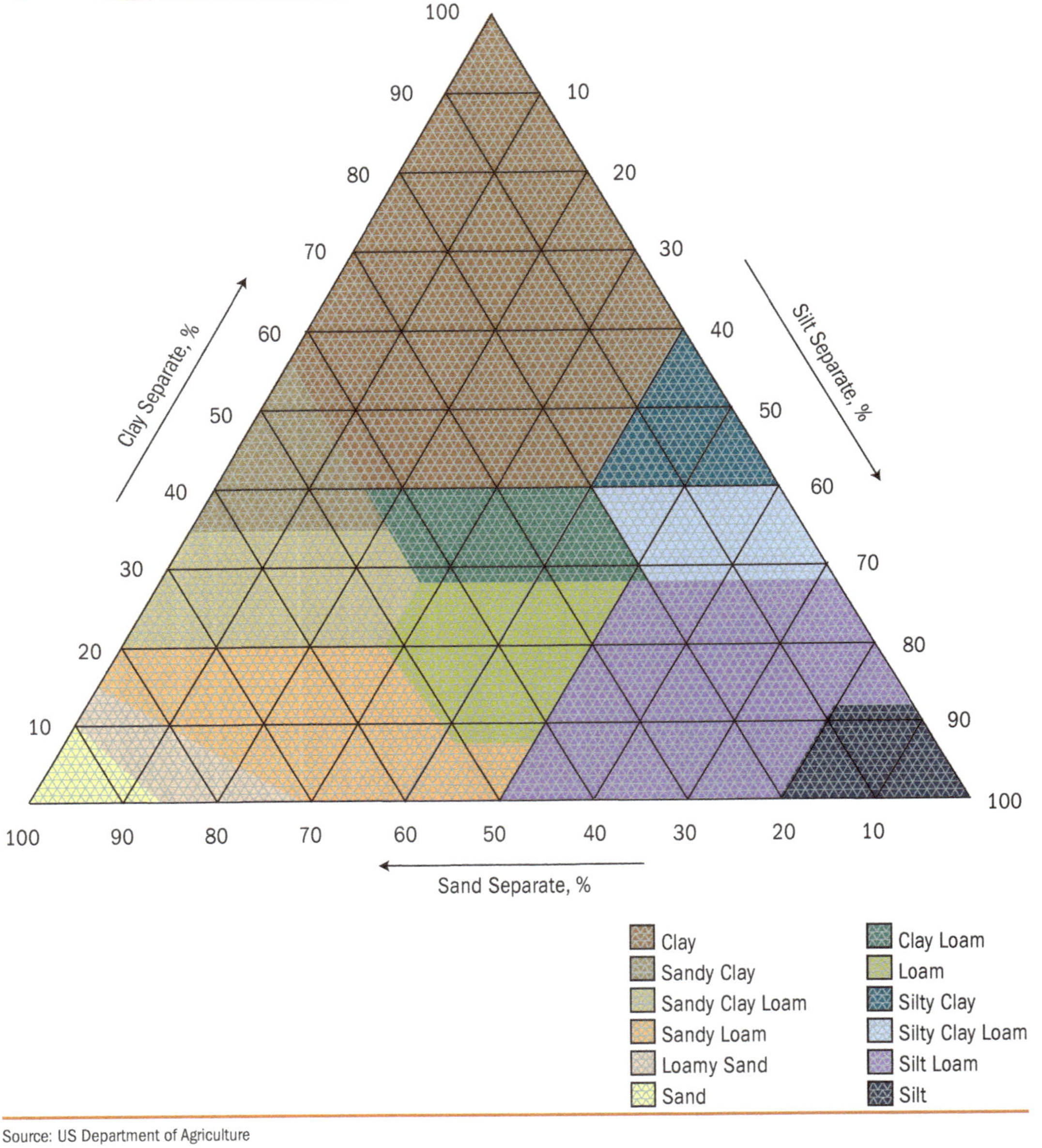

Source: US Department of Agriculture

A soil layer's texture is probably its most stable quality over a long period of time, but there are ways in which texture can be altered. For example, heavy and continuous soil blowing during a drought can remove the finer particles from the soil and leave the surface soil more coarse. Erosion by wind or water can also change the depth and fertility of the soil. A soil's texture, structure, and consistence, as well as each horizon's physical and chemical characteristics, must be considered to determine that soil's present and potential fertility.

Fertility

The inherent fertility of soil depends on the amount of organic matter and plant nutrients that are naturally present in the soil in a form that plants can use. Fertility can be altered by past use. For example, a property may have been planted with the same row crop for years with little or no return of plant nutrients and organic matter to the soil, while another property may be virgin soil. Still another property may have been farmed for years, but because farm management has properly rotated crops and carried out other measures, adequate levels of organic matter, nitrogen, phosphorus, potash, and other plant nutrients have been maintained. Both inherent soil characteristics and past land management must be investigated to determine soil's current fertility level.

Color

Soil color is another characteristic that appraisers should not overlook. Typically, the darker the surface soil, the more productive the soil is likely to be. Dark soil generally indicates a high level of organic matter. Care must be exercised, however, because in a few soils, soil color simply reflects the color of the parent soil material. While color is an important consideration, conclusions about soil capability cannot be drawn from soil color alone. This single characteristic may suggest a variety of conditions.

pH

The pH of the soil must also be considered. Strongly alkaline soils with pH levels of 8.5 or higher support limited vegetation. In extreme cases, alkaline can render soil virtually sterile. At the opposite extreme, high acidity tends to lock up soil nutrients, making them unavailable to plants. Furthermore, some plants are sensitive to high acidity and cannot be successfully grown on acid soils. In most cases, soil alkalinity and acidity can be modified or corrected with the addition of lime (to correct acidity), sulphur (to correct alkalinity), or other soil modifiers. While these soil modification techniques are not typically a direct concern of appraisers, local practices affect soil productivity and land value.

Topography, Drainage, and Climate

In addition to the major soil characteristics previously discussed, appraisers examine several other aspects of the soils and the environment:

- topography
- drainage
- climate

Topography is an important physical characteristic of the landscape in that it can limit the soil's productive capability or potential uses. Soils that are level, in depressions, or at the base of slopes may require surface drainage and tiling or surface land forming. Sloping areas may be susceptible to erosion, especially if used for row crops, and may require conservation tillage, terracing, a crop rotation program, or other practices (singly or in some combination) to control water erosion and maintain the soil's productivity.

The topography of a property must be seriously considered. In some cases, topography can help identify a soil type. Slope–an aspect of topography–is a soil characteristic. More importantly, the topography of a property often determines the type of agriculture that the property will accommodate. For example, soils on relatively steep topography in areas of moderate or high rainfall commonly have severe erosion problems. In areas where the soils are used for cultivated crops because of favorable air flow, however, the same soils might accommodate a horticultural enterprise. Sometimes soils on nearly level valley floors have seasonal limitations because of slow air movement or the storage of cold air. Finally, topography suggests the degree of hazard from wind or water erosion. Topography can be described in terms of "percentage of slope," but appraisers more frequently use terms such as *gently rolling*, *rolling*, *steep*, and *level*.

Drainage can refer to both the drainage patterns on the landscape and the rate at which water moves through the soil. The relationship between topography and drainage is obvious, but in many areas drainage problems are not revealed by topography alone. Strongly compacted soil layers, which result from natural soil structure, land management, or both, can adversely affect drainage. For example, soil particles can pack together tightly, leaving little room for air or moisture to pass through. This type of subsoil condition, known as *hardpan*, is commonly accompanied by internal drainage limitations that have a serious impact on soil productivity. In areas with no previous history of poor drainage, irrigation combined with certain geologic features can raise the water table to the soil surface and cause drainage problems. In some instances, poor drainage can be corrected through major investments in tiling or land forming. Throughout some parts of the country, drainage districts (much like irrigation districts) have been formed to reclaim or improve land that has poor natural drainage.

Drainage can be linked to water quality in a region. For example, the City of Des Moines sued farmers in three counties in northwest Iowa over nitrate levels that the water treatment department believed were elevated due to subsurface tilling of agricultural land.

Because internal soil drainage can be difficult to measure, appraisers must be sure to investigate this aspect of the soil very carefully. Soil surveys (which are discussed later in this chapter), site visits to the property, and information obtained from the present owner are the best sources of information about the drainage characteristics of soil.

Climate and agricultural enterprises are inextricably related. Most successful agricultural enterprises are compatible with the local climate. If they are not, climatic conditions can be mitigated with adaptations such as frost protection, drainage, irrigation, or the growing of tolerant crops. The most important climatic characteristics are rainfall, length of the growing season or frost-free season, temperature, evaporation, sunshine, and winds. The area's susceptibility to weather hazards such as unseasonable frost, hail, early or late snow, hurricane, or drought must also be considered. While total annual precipitation may indicate the desirability of a climate for agriculture, an appraiser should have some knowledge of the rainfall pattern. For example, many agricultural regions in the West experience a warm-season dry spell that necessitates irrigation or the cultivation of early-maturing or drought-resistant crops. Drought conditions, such as the historically low groundwater supplies in California from 2012 to 2015, can be a major factor in land value and an expense of operating an agricultural enterprise.

The effects of irrigation cannot be overlooked. For example, drought is generally more serious on sandy soils than on fine-textured soils, but sandy soils often respond well to irrigation. If irrigation water is available or the area receives abundant rainfall, sandy soil will likely have a much higher productive capability than it would in an area without irrigation potential or adequate rainfall. Regardless of climate, the rate of internal drainage and the water-holding capacity of soil have a direct bearing on its productivity.

Soil Inventory for Appraisal

A soil inventory can provide much of the information used in applying the valuation approaches. For example, the soil characteristics of property are considered by appraisers in making adjustments in the sales comparison approach. One kind of soil is judged to be more desirable than another. Consequently, the value the local market ascribes to a given soil can be used in deriving an opinion of value by the sales comparison approach. In the income capitalization approach, the farm's soil must be considered in relation to potential gross farm production and the operating expenses and level of management required to achieve that potential. The farm's soil is the principal determinant of its income and expense.

Appraisers make use of a variety of resources in investigating and inventorying the soils of a rural property. Aerial photographs of the property and the surrounding area are valuable tools, as are topographical maps, soil maps, and interpretive maps, such as those made to highlight soil productivity. Many kinds of interpretive maps can be prepared to display the suitability or potential of the soils for various uses.

Soil Mapping

Soil scientists have identified and mapped soils throughout broad areas of the United States. Modern, detailed soil surveys are available for more than 95% of the United States. The surveys are produced by the NRCS in cooperation with the National Cooperative Soil Survey, a joint effort of the USDA and other federal agencies, state agencies including the agricultural experiment stations, and local agencies. The NRCS also provides an interactive online Web Soil Survey (http://websoilsurvey.sc.egov.usda.gov).

Soil surveys consist of a set of soil maps and a narrative that describes the soils and their suitability for selected uses, including urban and agricultural uses. The text describes soil characteristics such as color, consistence, structure, texture, and thickness of horizons; soil and site characteristics such as drainage, topography, and climate; and the effects of land use and nature on the soils' productivity. Because they describe the properties of major soils found in the area, the soil surveys provide a scientific basis for comparing one property with another. The maps are site-specific and provide adequate information. The aerial map in Figure 7.3 depicts the manner in which fields are typically mapped.

Soil Series

Soils that have the same origin and are similar in all major respects except for surface layer texture are given a name that designates the series to which they belong. The soil series name generally refers to a geographic location or feature and usually indicates the location where the soil was originally classified. Willamette soils, for example, were first identified in the Willamette Valley of Oregon.

An appraiser who is valuing a property that has Willamette silt loam will find the following in a soil survey:

- a description of the color, structure, consistence, and texture of the horizons
- the type of root growth
- the pH
- the degree of slope

Figure 7.3 Soil Map

Source: NRCS Web Soil Survey, http://websoilsurvey.sc.egov.usda.gov

The survey will also identify what that the particular soil series is used for: Willamette soils are used mainly for vegetable crops, berries, orchards, small grain, hay, and pasture, and sometimes for wildlife habitat, recreation, and homesites.

Productivity Ratings

Productivity ratings are essential to rural property appraisal. Compiled through various methods, productivity ratings summarize the effect of surface texture, subsoil characteristics, topography, climate, drainage, organic matter, and fertility on the productive capacity of soils. Productivity ratings are used in analyzing the most suitable land use and cropping patterns for a property, and thus are part of highest and best use analysis. Productivity ratings provide a basis for estimating yields and potential income. They are also used as a means for comparing properties and making adjustments for varying productive capabilities.

Soil capability ratings and yield indexes are used to estimate the productivity of all types of rural properties. Yields per acre indicate the productive capabilities of farms, orchards, and vineyards. The productive capability of a livestock range is often measured in terms of the animal units (AUs) or animal unit months (AUMs) of forage that each acre can support. Range condition and the corresponding stocking rate can vary considerably, even on the same soil. A range survey showing composition by species and production is needed. The productive capability of timberland may be measured in thousands of board feet (MBF) per acre of usable timber.

Several soil productivity ratings are used in the United States:

- the NRCS Land Capability Classification System
- state soil productivity ratings
- Bureau of Reclamation classifications
- the Storie Index
- corn suitability rating (CSR) and productivity index (PI) rating

Suitability ratings provide a relative ranking of all soils mapped in a county based on their potential for crop production. The CSR and PI indexes rate the relative productivity of individual soil types, which can be weighted according to the land mix of a particular property to determine that property's overall relative productivity.

NRCS Land Capability Classification System

The most widely accepted land capability classification system is the one developed by the NRCS. This classification is based on the limitations of the soil rather than its potential yield, meaning that it indicates the general suitability of soils for most kinds of field crops. This classification also indicates a soil's needs, response to management, and risk of damage.

The system was originally designed to help the NRCS advise farmers about proper soil management techniques. Because the system is recognized throughout the United States, an appraiser can refer to this system when communicating with a client who is not familiar with the soils in the subject property's area.

Land capability class data is included in map unit description reports available in the Web Soil Survey and the Soil Data Mart. The NRCS system includes land capability classes, subclasses, and capability units (see Table 7.1). All soils in one land capability class have similar limitations and restrictions in use, but they may have different productive capabilities. A subclass designation is used to identify the most important kind of limitation, such as wetness or climate. Finally, the capability unit identifies soils within a subclass that require similar management and have similar risks and productivity.

State Soil Productivity Ratings

Several state agricultural colleges have developed soil ratings used in their respective states. Some rating systems attempt to relate soil productivity to specific crop yields, while others assign soils a numerical value, or fertility index.

Table 7.1 NRCS Soil Hierarchy

Land Capability Classes	
Class I soils	Few limitations that restrict use
Class II soils	Moderate limitations, which reduce the choice of plants that can be grown or indicate the need for moderate conservation practices
Class III soils	Severe limitations, which reduce the choice of plants, indicate the need for special conservation practices, or both
Class IV soils	Very severe limitations, which restrict the choice of plants, indicate the need for very careful management, or both
Class V soils	• generally not considered suitable for cultivation • not likely to erode but have other limitations that are impractical to remove and limit their use
Class VI soils	Severe limitations that make them generally unsuited for cultivation
Class VII soils	Very severe limitations that make them unsuited for cultivation
Class VIII soils and land forms	Limitations that nearly preclude their use for commercial crop production
Capability Subclasses	Soil groups within one class
Subclass *e*	The main limitation is the risk of erosion
Subclass *w*	The main limitation indicated is excess water in or on the soil due to poor soil drainage, wetness, a high water table, or overflow
Subclass *s*	The soil is limited mainly because it is shallow, droughty, or stony or has another problem in the rooting zone, such as a low moisture-holding capacity, salinity, or alkalinity
Subclass *c*	The chief limitation is a climate that is too cold or dry
Capability Units	In some areas, soil capability units are used to identify soils that are similar in terms of management requirements, risks, and productive capability. For example, two soils in subclass IIe may be classed IIe-1 and IIe-2 to distinguish the degree of management needed to control the erosion problem.

Bureau of Reclamation Classification

The Bureau of Reclamation land classification system is used to determine the suitability of lands for sustained irrigation farming. It is a systematic classification that divides land into categories based on its characteristics. Suitability is determined by correlating the land's potential productive capacity with the anticipated cost of production and the land's adaptability to a wide range of crops.

The Storie Index

The Storie Index is a numerical soil rating system that has been used widely in California. The index expresses the soil's relative suitability for general intensive agriculture at the time of evaluation. The rating is based on soil characteristics only and is derived by evaluating factors such as soil depth, surface texture, subsoil characteristics, drainage, salt and alkali content, and relief. Not considered are factors such as the availability of water for irrigation, climate, and distance to markets, which might determine the desirability of growing certain crops in a given locality. By itself, the index is not a direct indicator of land value. When economic factors are known to the user, however, the Storie Index provides additional, objective information for comparing the land tract values.

Four general factors are considered in the Storie Index rating:

A. The characteristics of the soil profile and the depth of the soil

B. The texture of the surface

C. The dominant slope of the map unit

X. Other, less permanent factors that are more readily subject to management or modification, such as drainage, flooding, and nutrient level

For some soils, more than one X factor is used. Each of the four general factors is evaluated in relation to a basis of 100%, which expresses the most favorable or ideal condition for general crop production. Lower percentage ratings are assigned for conditions that are less favorable. Factor ratings are selected from tables prepared originally from data and observations that relate soil properties, plant growth, and crop yield.

The Storie Index tends to be somewhat rigid and assumes that each of the factors contributes equally to the productivity of the soil, which is seldom the case. Nevertheless, the index can be a useful guide in estimating soil productivity.

In the Web Soil Survey, Storie Index ratings are available under the Suitabilities and Limitations tab in the Land Classifications section. Other soil databases may also include Storie Index ratings.

Corn Suitability Rating or Productivity Index Rating

In some Midwestern states, the corn suitability rating (CSR) and productivity index rating (PI) are used to estimate a soil's capacity to produce primary crops. This system rates each soil type on a property and converts the individual ratings into a single index. This system provides a method for making qualitative comparisons between farms. For example, the price of tillable land divided by the index can be used as a unit of comparison and tracked over time as evidence of changing market conditions.

Characteristics of an "Ecological Site"

The carrying capacity for a rural property composed partly or wholly of native rangeland is estimated by analyzing the *ecological sites* that exist on the property and then verifying the historical stocking rates and grazing practices. Appraisers are charged with estimating the carrying capacity of the subject and comparable properties that form the basis for valuation. A thorough understanding of ecological sites is essential to properly estimating the productivity of rangeland properties. *The National Ecological Site Handbook* defines a rangeland *ecological site* as a conceptual division of the landscape. It is defined as "a distinctive kind of land based on recurring soil, landform, geological, and climate characteristics that differs from other kinds of land in its ability to produce distinctive kinds and amounts of vegetation and in its ability to respond similarly to management actions and natural disturbances."[2]

This concept of a rangeland *ecological site* has two parts that are critical to understand:

- An ecological site is a distinctive kind of rangeland that differs from other kinds of rangeland in its ability to produce a characteristic natural plant community.
- An ecological site supports a native plant community typified by an association of species that differs from that of other ecological sites in the kind or proportion of species or in total production.

The other factors detailed in the definition contribute to the makeup of an ecological site. From the appraiser's perspective, the key is understanding that an ecological site is unique unto itself and that its uniqueness is due not only to the site's soils but also its climate and topography.

With a basic understanding of the concept of ecological sites, an appraiser can proceed to develop the data necessary to analyze a property and estimate its carrying capacity in the appraisal process. Approaching the process from a methodology standpoint seems to be the most practical way to demonstrate the analysis of rangeland and the development of estimated carrying capacity. In addition to understanding the concept of ecological sites, appraisers must possess a general familiarity with the plant community in the region to be analyzed and some understanding of the effect of wildlife on a property's livestock carrying capacity. Appraisers must be able to identify the major forage species of herbs, legumes, and woody plants to evaluate range condition with sufficient accuracy to estimate the carrying capacities of both the subject property and comparable properties.

In many areas in the western United States, the estimation of carrying capacity on public lands is determined by the public entity that controls the land asset. However, for non-government-owned land, appraisers should assess the range condition and estimate the carrying capacity of the property based on an analysis of the rangeland. The accuracy of the estimate should be verified with historical operating data, if available.

2. US Department of Agriculture, *National Ecological Site Handbook* (2014).

Rangeland Analysis

Ecological sites are identified in most jurisdictions using the county soil surveys that have been prepared by the USDA since 1960. Most of the soil surveys for states with rangeland assets include extensive data on the ecological sites found in the area, including the following:

- physical descriptions
- forage composition by percentage of climax vegetation species (the species of plants that would colonize land given the opportunity to pass through a series of interim stages to a state of equilibrium, i.e., the "climax" of the process)
- condition indicators
- minimum, average, and maximum forage production in pounds of air-dried forage to be produced annually by a given site

For counties that do not have published soil surveys, the local NRCS office usually makes various ecological site descriptions applicable to the local area. The information found in the local office's technical guide provides an extensive description of each ecological site in the area, including the forage species, annual vegetation produced, recommended initial stocking rates, and recommended rates for stocking on improved rangeland as a result of mechanical or other improvement practices.

A critical part of the property analysis and, ultimately, the estimation of carrying capacity is the ability of an appraiser to properly assess range condition. Range condition is a function of management, not weather. Even if a site is in excellent condition–i.e., with 100% of the climax vegetation present listed as appropriate for the site–its condition can be reduced drastically by poor or inappropriate management.

A layman's assessment of range condition is usually based on whether the range forage produced looks lush due to normal or above-normal rainfall, or minimal due to drought or low rainfall. However, rangeland that is appropriately managed can be in excellent condition even in the middle of a drought because the ratio of climax vegetation to optimum vegetation is the measure of condition. Certainly, in periods of drought or other adverse climatic conditions, the total annual yield of air-dry vegetation on a site may be very small. Nonetheless, as long as the proper ratios of plant species are maintained, the condition may remain excellent or good. If the ratios are not maintained, the condition is likely to be fair or poor.

Livestock and wildlife tend to graze or browse selectively, repeatedly grazing or browsing on those plant species they like most. Continuous, heavy grazing of potential plant communities for extended periods reduces the vigor and ability of the plant communities to reproduce. If these practices are permitted to continue, retrogression occurs and many, if not all, of the most desirable species will be eliminated from the site. The plants eliminated first are usually the tall and midgrasses, the most desirable forbs, and the best legumes and browse. These most desirable plants are referred to as *decreasers*, since they tend to decrease as grazing pressure continues. As the population of decreasers begins to decline, the percentage of secondary species that are less palatable tend to increase in the plant community. These less desirable species are called *increasers*, since they tend to increase as the more desirable species decrease.

The terms *decreasers* and *increasers* describe the site's initial response to continuous heavy use. The type of use must be analyzed to identify the source of the grazing pressure and institute appropriate management actions to stabilize the site. For example, a midgrass or a tall grass that has been decreased from cattle grazing may increase when the site is grazed by sheep and goats.

When improper management techniques are permitted to continue, the population of both increasers and decreasers is eventually eliminated. When this occurs, the plants that replace the native climax plants are not native to the site and would not normally be found on the site. These plants are called *invaders*. Invaders usually have special adaptations, such as spines, burrs, toxins, low palatability, or very slow growth, which prevent overgrazing by animals. Common examples of invaders are mesquite, juniper cedar, and cactus.

The process of retrogression described above may occur on a given site or sites and is purely a function of management (grazing practices). All rangeland responds to good management. Some sites respond faster than others, and all respond faster in favorable years. However, the bottom line is that no favorable climatic condition can prevent deterioration in range condition if management practices are inappropriate.

Figure 7.4 shows the area of a tract of land being appraised from the Web Soil Survey for Jim Hogg County, Texas. The subject property (i.e., the "area of interest" in the survey interface) encompasses approximately 5,100 acres. The soils found on the subject property are indicated in Table 7.2.

Using the NRSC's Web Soil Survey, the soils of the subject property were measured and are displayed on the soil map in Figure 7.5, which indicates the location and dispersion of the soil types on the subject property. The soil survey categorizes the soils on the subject property into ecological sites as indicated in Table 7.3. The data generated for an area of interest includes a brief description of the terrain and topography, a forage inventory guide by percentage, and the annual air-dried forage production when in climax condition.

Figure 7.4 Subject Property Layout

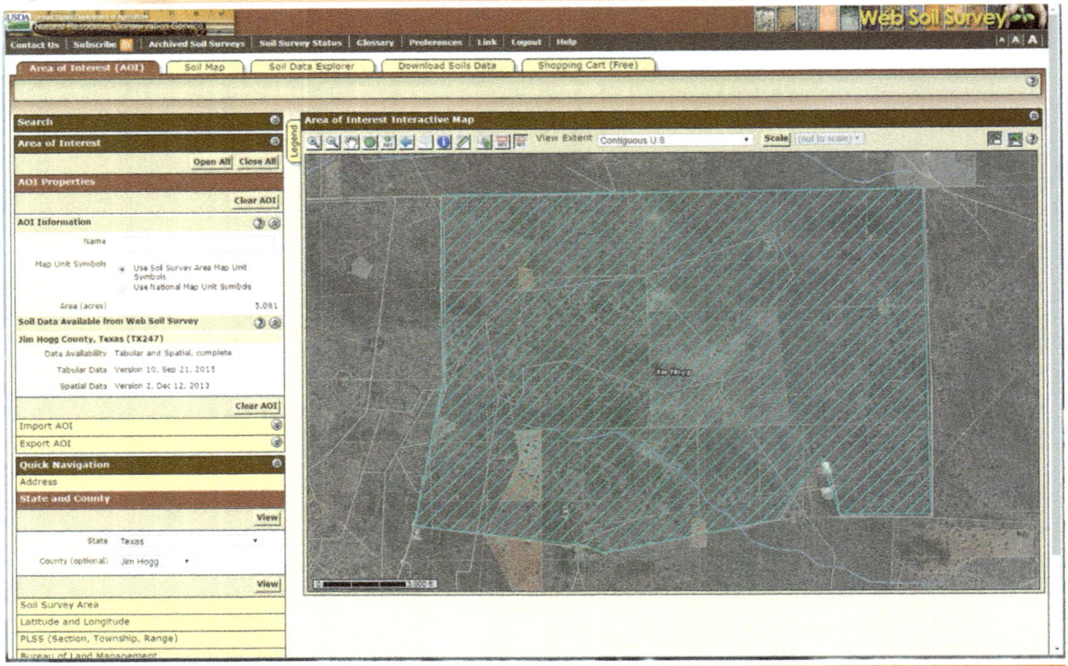

Source: NRCS Web Soil Survey

Table 7.2 Soil Types of Jim Hogg County, Texas (TX247)

Map Unit Symbol	Map Unit Name	Acres in Area of Interest (AOI)	Percentage of Area of Interest
Cm	Comitas loamy fine sand, 0%-3% slopes	649.9	12.8%
Cu	Cuevitas-Randado association	229.8	4.5%
Dl	Delmita loamy fine sand, 0%-3% slopes	2,017.3	39.7%
Dn	Delmita fine sandy loam, 0%-3% slopes	1,202.8	23.7%
Ns	Nueces-Sarita association, 0%-3% slopes	981.4	19.3%
Totals for Area of Interest		**5,081.2**	**100.0%**

Source: NRCS Web Soil Survey

Figure 7.5 Soil Map of Area of Interest

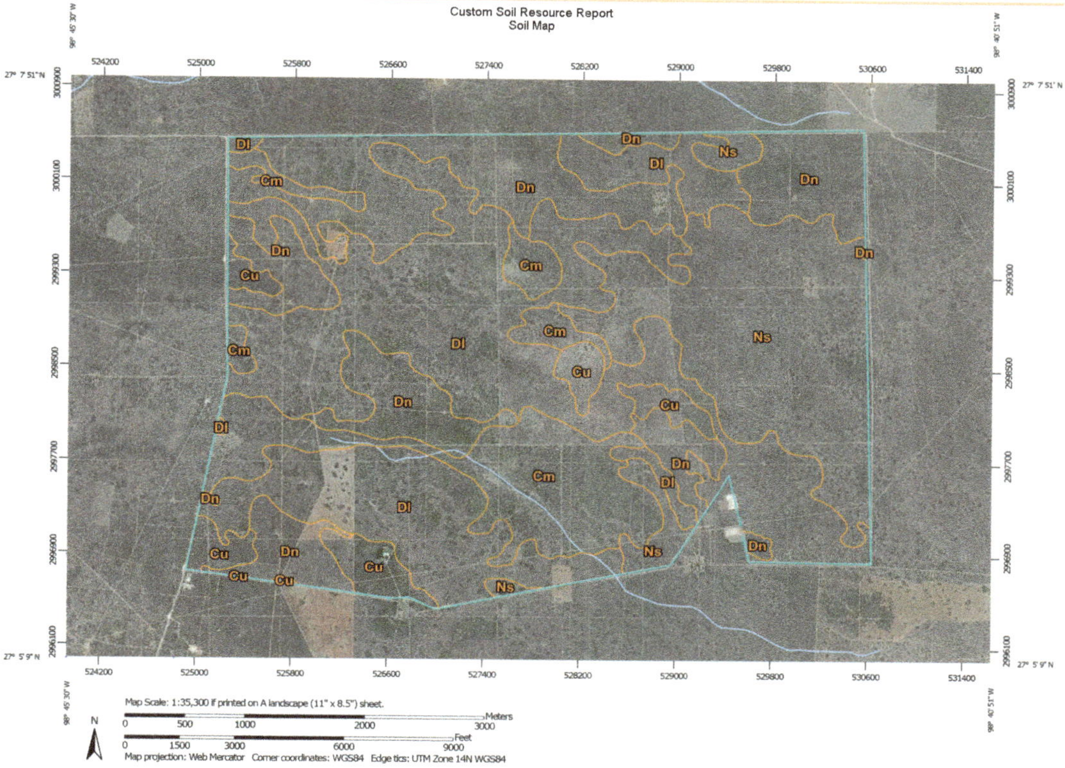

MAP LEGEND

Area of Interest (AOI)
- Area of Interest (AOI)

Soils
- Soil Map Unit Polygons
- Soil Map Unit Lines
- Soil Map Unit Points

Special Point Features
- Blowout
- Borrow Pit
- Clay Spot
- Closed Depression
- Gravel Pit
- Gravelly Spot
- Landfill
- Lava Flow
- Marsh or swamp
- Mine or Quarry
- Miscellaneous Water
- Perennial Water
- Rock Outcrop
- Saline Spot
- Sandy Spot
- Severely Eroded Spot
- Sinkhole
- Slide or Slip
- Sodic Spot
- Spoil Area
- Stony Spot
- Very Stony Spot
- Wet Spot
- Other
- Special Line Features

Water Features
- Streams and Canals

Transportation
- Rails
- Interstate Highways
- US Routes
- Major Roads
- Local Roads

Background
- Aerial Photography

MAP INFORMATION

The soil surveys that comprise your AOI were mapped at 1:31,700.

Please rely on the bar scale on each map sheet for map measurements.

Source of Map: Natural Resources Conservation Service
Web Soil Survey URL: http://websoilsurvey.nrcs.usda.gov
Coordinate System: Web Mercator (EPSG:3857)

Maps from the Web Soil Survey are based on the Web Mercator projection, which preserves direction and shape but distorts distance and area. A projection that preserves area, such as the Albers equal-area conic projection, should be used if more accurate calculations of distance or area are required.

This product is generated from the USDA-NRCS certified data as of the version date(s) listed below.

Soil Survey Area: Jim Hogg County, Texas
Survey Area Data: Version 10, Sep 21, 2015

Soil map units are labeled (as space allows) for map scales 1:50,000 or larger.

Date(s) aerial images were photographed: Dec 12, 2010—Jun 17, 2011

The orthophoto or other base map on which the soil lines were compiled and digitized probably differs from the background imagery displayed on these maps. As a result, some minor shifting of map unit boundaries may be evident.

Source: NRCS Web Soil Survey

Table 7.3 Soil Types of Area of Interest

Map Symbol	Soil Map Unit Name	Total Acres	Ecological Site Name
Cm	Comitas loamy fine sand, 0%-3% slopes	649.9	Loamy Sand 25-30" PZ (R083EY705TX)
Cu	Cuevitas-Randado association	229.8	Shallow Sandy Loam 20-30" PZ (R083CY487TX)
Dl	Delmita loamy fine sand, 0%-3% slopes	2,017.3	Loamy Sand 20-25" PZ (R083CY463TX)
Dn	Delmita fine sandy loam, 0%-3% slopes	1,202.8	Red Sandy Loam 20-25" PZ (R083CY467TX)
Ns	Nueces-Sarita association, 0%-3% slopes	981.4	Sandy 20-28" PZ (R083EY706TX)

Source: NRCS Web Soil Survey

Analysis of the data available from the Web Soil Survey reveals that the forage produced on the Comitas loamy sand ecological site should consist of the following:

- 20% little bluestem
- 10% each of Arizona cottontop, brownseed paspalum, and plains bristlegrass
- 5% each of fall witchgrass, hooded windmill grass, one-sided crinkleawn, pink pappusgrass, switchgrass, and tanglehead
- 5% other shrubs and perennial grasses
- 3% other perennial forbs
- 2% other annual forbs

The total annual dry-weight production forage on this ecological site is 4,500 pounds per acre in a favorable year, 3,500 pounds per acre in a normal year, and 2,000 pounds per acre in an unfavorable year.

With this data, an appraiser can begin to estimate the indicated carrying capacity of the subject property or a comparable property.

Estimation of Carrying Capacity

Most agricultural extension offices and online resources suggest that the best way to determine carrying capacity is to look at the property's history and historical trends to see if the soil is improving, stable, or declining. The second-best way is by direct comparison–i.e., analyzing the historical figures for neighboring properties if the information is not available for the subject property. If historical stocking rate information is not available or the local extension service or Natural Resources Conservation Service offices cannot provide this sort of information, these resources may still be useful to appraisers in measuring annual forage production on the land in question and calculating a carrying capacity estimate.

The calculations in a carrying capacity estimation involve certain basic measurements of agricultural productivity:

- the annual potential of dry vegetation given the soil types on the site
- management techniques for grazing of the site
- how much of the available forage is actually consumed by the livestock
- the corresponding portion of the available forage that is "walked down" by the livestock

Table 7.4 illustrates the relevant data culled from the Web Soil Survey about the major soil type on the area of interest in Jim Hogg County, Texas, that was shown earlier in this chapter.

Table 7.4 Carrying Capacity Data for Area of Interest

Variable	Sandy Loam Ecological Site—Jim Hogg County
Annual potential air-dry vegetation on site	Total annual forage 3,500-2,000 pounds yearly, average (say) 2,750 pounds
Grazing management practice	50%—take half of the forage and leave half for seed stock
Percentage of vegetation used by livestock	95% used by livestock
Vegetation walked down by livestock in the grazing process	20% of vegetation walked down by livestock or otherwise not usable—i.e., 80% usable after walk down accounted for on site

A common practice of grazing management is to "take half and leave half" on a site. That is, the "take" half of the amount of forage produced on the site is calculated as 25% to the livestock and 25% to trampling, insects, and decay. The "leave" half is assumed to be left alone to protect the soil and increase plant vigor and protection. The base (to get to the 25% carrying capacity) is determined by the annual potential production of the soil. This is a product of the grazeable acres and the distance to water resources, topography, forage preferences, and brush cover and density. The local agricultural extension offices typically have this information available or can provide some guidance.[3]

In addition to information about the productive capability of the rangeland (i.e., the amount of forage that can be grown), carrying capacity estimation also requires a working knowledge of the rate of consumption of most common forage species found on a given site or

3. Other resources in the professional literature include Jason K. Ahola, "Estimating Carrying Capacity: How Many Cows Can I Graze?" *Progressive Cattleman* (Published on February 24, 2013, available online at www.progressivecattle.com/topics/range-pasture/5357-estimating-carrying-capacity-how-many-cows-can-i-graze); Melvin George and David Lile, "Stocking Rate and Carrying Capacity," *Ecology and Management of Grazing: An Online Course* (Available online at http://studylib.net/doc/7488692/stocking-rate-and-carrying-capacity); and Amanda Hancock, NDSU Central Grasslands Research Extension Center, "Doing the Math: Calculating a Sustainable Stocking Rate" (Available online at www.ag.ndsu.edu/archive/streeter/2006report/aums/Doing%20the%20Math.htm).

sites by the type of livestock that will use the land. The calculation of carrying capacity is expressed in terms of animal units per month or animal units, as described in Chapter 16.

Rural appraisers who engage in the appraisal of livestock ranches need at least a working knowledge of the processes related to rangeland analysis discussed in this chapter. The various practices described help appraisers develop or verify realistic stocking rates for use in the valuation process.

Chapter 8

Site Improvements

A parcel of land becomes a site when it is improved and ready to be used for a specific purpose. Both on-site and off-site improvements can make a site suitable for its intended use or development. Like buildings and other structures, site improvements can be subject to physical, functional, and external obsolescence.

On-site improvements on rural properties are typically referred to as *land improvements.* Land improvements commonly include perimeter and interior fencing, interior roads, cell/radio towers, livestock handling facilities, high-tension power lines, above-grade utility pipelines, water and septic systems, irrigation wells, and associated irrigation improvements such as underground piping, reservoirs, tailwater ditches, and land forming. Off-site improvements include water supply companies, irrigation and drainage districts, and utilities including electricity, gas, and telephone.

Typically, the market does not allocate a separate value for site improvements on rural lands. However, the value of these improvements is reflected in the overall market value, appeal, and acceptability of the property. Land improvements typically reflect a general level of neighborhood acceptance and use. For example, in a wheat-producing region where cattle provide a secondary income source, perimeter and interior fencing may be important. However, a similar region without a secondary livestock use may have no fences at all. Specific types of rural property may require appropriate land improvements, which vary from region to region based on the uses in the area. A cattle ranch needs fences and corrals, a hop farm needs hop trellises, and irrigation and drainage facilities are common land improvements in certain regions.

When inventorying land improvements, appraisers consider their age, condition, estimated depreciation, and any annual maintenance expense. Appraisers must assess the adequacy of the improvement and its contribution to the operation and value of the property. Appraisers also note whether the improvements are located advantageously to provide benefits such as protection from the elements or access to public roads.

Definitions

Site improvements are land improvements that are necessary to make a property suitable for its intended use or development. Care must be taken to differentiate between site improvements, personal property, fixtures, and chattels.

Personal property includes moveable items of the property that are not permanently affixed to, or part of, the real estate. Examples of personal property are portable irrigation pipes, livestock squeeze chutes, and portable buildings. A *fixture* is an article that was once personal property, but has since been installed or attached to the land or building in a rather permanent manner. It is legally regarded as part of the real estate. Examples of fixtures include windmills and water troughs

Chattels or *trade fixtures* are articles that are owned and attached to a rented space or building by a tenant and used in conducting a business. A trade fixture is to be removed by the tenant when the lease expires, unless this right has been surrendered in the lease. Examples of chattels include fuel tanks, gearheads, and irrigation motors. The courts consider the manner in which the item is affixed, the character of the item, and the intention of the party who attached the item in determining whether that item is a trade fixture, and therefore personal property, or part of the real estate.

The distinction between fixtures (real property) and personal property (chattels) varies significantly from region to region. Appraisers must know how the courts and market participants in the specific jurisdiction or area define these items. Fixtures are part of the realty and must be included in the property value. If personal property is to be included, it should be identified and described as such in the appraisal report.

Irrigation

Some agricultural land must be irrigated to obtain its greatest productive capability. When irrigation is economically feasible and brings the land to its highest and best use, the value of the land is increased. Generally defined, *irrigation* is "the artificial application of water to plants to sustain or enhance plant growth."[1]

The value of irrigated property is increased by the smooth integration of all the factors that affect irrigation. These factors include basic aspects of the land such as climate, which establishes the need for irrigation, and the soil and topography, which determine if the land can be irrigated. Soil and topography may also dictate the requirements for an appropriate irrigation system. The effectiveness of irrigation on a particular property depends on the essential element of water–its physical source, the rights governing its use, its quality

1. Megan Stubbs, *Irrigation in US Agriculture: On-Farm Technologies and Best Management Practices* (Washington, DC: Congressional Research Service, 2015), 1.

and conveyance, the methods and rates of applying it, and its cost. In a well-designed and balanced irrigation system, all aspects work together and contribute to increased property value. For example, a better irrigation system produces greater yields and greater net income, thus adding value to the property to the point of diminishing returns. Beyond the maximum point at which the crop can use the water provided, the investment in irrigation does not produce a commensurate return.

Typical on-site irrigation improvements include land leveling or shaping, wells and casings, underground transmission lines, concrete ditches, storage reservoirs, tailwater recovery ditches, and permanent set sprinklers and drip irrigation systems. Other necessary irrigation equipment such as pumps, gearheads, motors, and side-roll or center-pivot sprinkler systems are typically considered personal property subject to local custom and legal interpretation. In some regions, center-pivot sprinkler systems that are owned by the tenant are treated as an improvement that conveys with the real property and as a trade fixture.

Center Pivot Irrigation Sprinkler System

Photo by Howard Johnson, USDA Natural Resources Conservation Service

Aerial View of Cropland with Center Pivot Irrigation

Source: US Geological Survey, http://water.usgs.gov/edu/pictures/full-size/wuir-centerpivot-aerial-large.jpg

Irrigation and drainage districts service large areas and fund their activities through property taxes and user charges. These service companies have significant improvements, which provide important off-site improvements to rural properties within their jurisdictions. Irrigation districts often hold title to water rights as trustees for the beneficial use of their members. The quality of water rights, the financial structures, and the charges imposed by individual districts vary dramatically and directly impact the income-producing capability of the properties served. Adjacent properties in different irrigation or drainage districts that are identical in all other respects may have significantly different market values due to these factors. Not all regions have irrigation districts, and the use of irrigation water is handled by the state authorities through a permitting process.

Irrigation systems are typically tailored for each situation, based on the soil type, topography, and intended crop production. For example, sandy, undulating soils are better suited to sprinkler irrigation than level soils with a higher silt or clay content, which may require gravity or surface irrigation. The use of larger equipment and better technology have allowed many landowners to land-form their cropland to maximize irrigation benefits over most soil types and topography. Along with remote monitoring stations, these irrigation improvements reduce water usage and overall costs.

Fences

The fencing of agricultural properties serves many purposes, including controlling livestock, managing game animals, controlling property access, and providing windbreaks (see Figure 8.1). Fence condition and construction on a specific property should be measured against area or neighborhood norms for properties with the same highest and best use. As is typical for most land improvements, fences that meet area norms are often not assigned a separate value but are treated as part of the land. Common deficiencies requiring attention or adjustment by appraisers include the absence or poor condition of necessary fencing on a livestock ranch, superadequate construction such as fencing on a row crop farm in a non-livestock area, and inadequacies such as the lack of needed interior or pasture fences.

Fence construction and costs vary with design, materials, and intended use. Typical ranch fences may be four or five strands of barbed wire on wood or metal posts. Wood rail snow fences are required in certain intermountain areas. Fences on sheep and goat ranches generally are constructed of net wire on posts. In ranching areas that have experienced a change in highest and best use from livestock grazing to recreational hunting, barbed wire fences are commonly replaced with 8-ft. net wire fences designed to control the movement of deer and other wild game.

Ranches located in areas with rough, mountainous terrain can be extremely difficult and expensive to fence. Therefore, it is not uncommon to find exterior or perimeter fences that follow natural geographic boundaries or divisions but do not coincide with the property's legal boundary. State laws vary, but in some areas title to lands outside boundary fences may be claimed through adverse possession statutes by the adjacent landowner unless a formal boundary agreement exists.

Figure 8.1 Agricultural Fencing

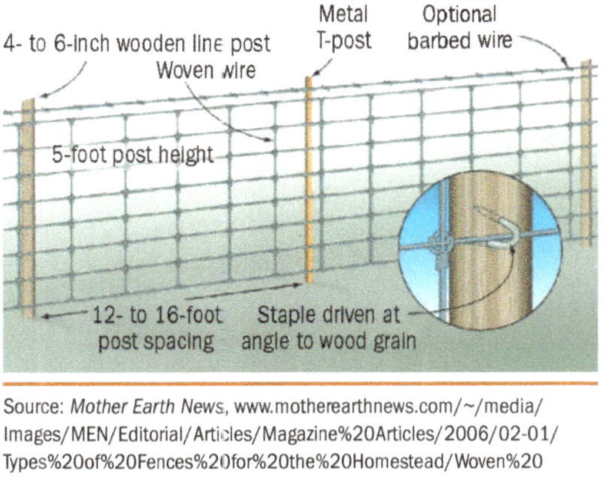

Source: *Mother Earth News*, www.motherearthnews.com/~/media/Images/MEN/Editorial/Articles/Magazine%20Articles/2006/02-01/Types%20of%20Fences%20for%20the%20Homestead/Woven%20Wire%20Fence.jpg

Livestock Handling Facilities

Livestock handling facilities, which include corrals, pens, and loading chutes, are important improvements on ranches, stock farms, and farm feedlots. Most cattle handling gates and fencing used today can be moved as needed and would be classified as personal property. When site improvements related to livestock handling are essential for making a property suitable for its intended use, the market often views these improvements as a part of the land and accords value consideration only for deficiencies–that is, when the improvements are absent or deteriorated. Similar improvements on highly improved special-use properties such as commercial feedlots, dairies, and horse farms represent a much more substantial investment and warrant greater attention and detail in site description and value estimation.

Properties used for ranching or livestock grazing must have sufficient livestock water. The amount of water needed varies with the temperature, humidity, and amount of moisture in the feed and with the size, condition, and species of the animals. The availability and distribution of livestock water sources directly affect how efficiently the range is used because animals tend to overgraze the forage that is closest to water sources.

Typical livestock water sources include creeks, streams, wells equipped with windmills or electric pumps, and earthen ponds that impound runoff water. Water quality for livestock is less critical than it is for domestic use, but animals tend to congregate to the freshest or least brackish water, resulting in excessive grazing pressure in these areas. Pasture arrangement is obviously affected by water sources, and all pastures or livestock enclosures should have a water source. A common rule of thumb is for livestock ranches to have sufficient watering so that animals are not required to walk more than one mile to water. When water sources such as wells or ponds are located at greater distances, water distribution via pipelines and water troughs is necessary.

Domestic Wells

The availability, quality, quantity, and safety of domestic water sources are important to rural properties. These water sources provide water for rural residences, but they are also important for standard farming and ranching practices, including the on-site preparation of chemical sprays and mixtures. The location of domestic wells in relation to the residence, other improvements on the property, and septic systems should be noted. Many rural areas now have space, distance, and design regulations affecting the proximity of domestic wells and septic systems. Appraisers should be familiar with these ordinances and analyze properties accordingly.

Other factors of importance are the depth and volume of wells and water pressure, storage, and distribution systems. Local well drilling or service companies can be helpful in determining the adequacy of existing systems, including necessary well volume and storage requirements. Domestic wells may become less important due to the increasing availability of public water sources in rural areas.

Figure 8.2 Domestic Well

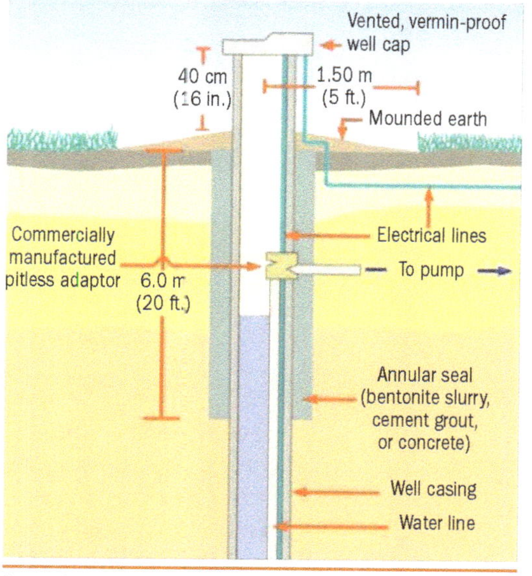

Source: Ontario Ministry of Agriculture, Food, and Rural Affairs, www.omafra.gov.on.ca/english/environment/facts/06-117f2.jpg

Sewage

Most rural properties do not have sewer service and must rely on private septic systems. The location and adequacy of the septic tank, absorption fields, or sewage lagoons are important criteria in determining and describing the adequacy of the sewage system. The effectiveness of the system is directly related to soil properties. The suitability of the site would likely need to be cleared with the health department or similar local state authorities.

Septic tank absorption fields are subsurface systems of tile or perforated pipe that distribute effluent from a septic tank into the natural soil. Soil permeability, depth to bedrock, and susceptibility to flooding determine a system's effectiveness and the potential effect on groundwater. Sewage lagoons are shallow ponds constructed to hold sewage while aerobic bacteria decompose the solid and liquid wastes. The lagoon floor and sides must be of nearly impervious soil material to minimize seepage and groundwater contamination.

Figure 8.3 Typical On-Site Wastewater Treatment System

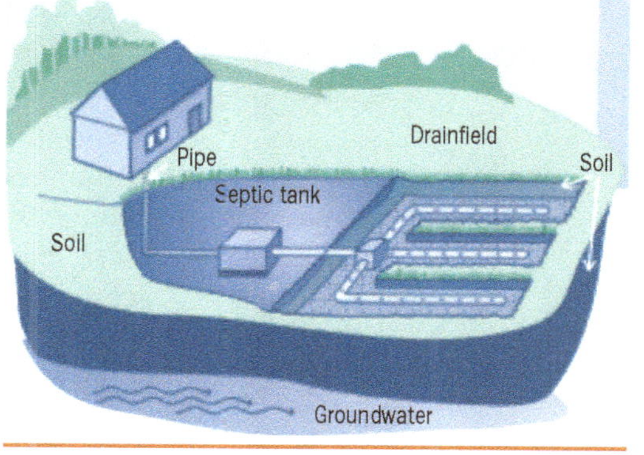

Many older septic systems do not have enough capacity to service modern appliances. Moreover, a rural residence originally designed for intermittent recreational use may need septic renovation prior to conversion to a permanent residence. Most areas have a minimal acreage requirement to allow for the use of a private septic system.

Figure 8.4 Typical Septic Tanks with Ground-Level Inspection Risers and Screen

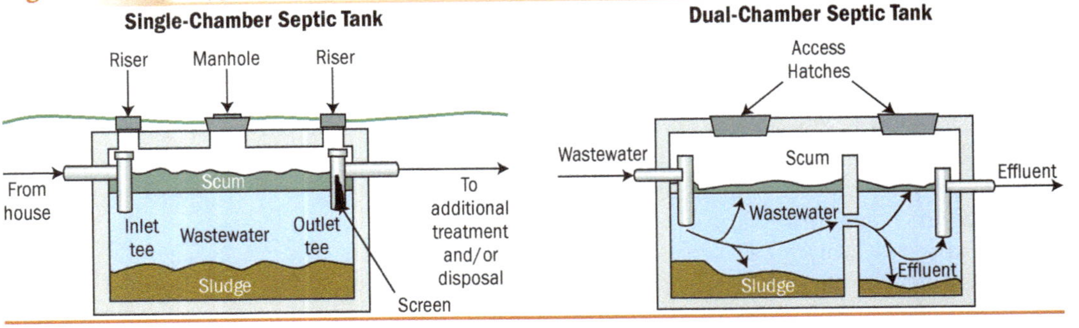

Interior Roads

Access or public roads serving rural properties are a location factor. Although they constitute an important off-site improvement, they are not typically discussed with on-site improvements. On the other hand, interior roads are an important site improvement and should be addressed. The suitability or adequacy of interior roads, like many land improvements, must be assessed against area norms for similar properties with the same highest and best use. Construction, quality, location, anticipated maintenance costs, and ability to serve the intended purpose are important review criteria.

Paved interior roads on most rural properties may be an over-improvement. Similarly, an inadequate access road may represent a deficiency.

Drainage

Agricultural drainage may be defined as the removal and disposal of excess water from land (see Figure 8.5). The sources of excess water may be precipitation, snowmelt, irrigation water, overland flow or underground seepage from adjacent areas, artesian flow from deep aquifers, floodwater from channels, or water applied for irrigation or other uses. Drainage systems are needed to supplement natural drainage in many areas. The amount of water to be removed by these systems depends on the relative effectiveness of any natural and constructed drainage.

Agricultural drainage is divided into two broad classes:

- surface drainage
- subsurface drainage

Some installations serve the dual purpose of surface and subsurface drainage. Adequate drainage can (1) facilitate early plowing and plant-

Figure 8.5 Rural Drainage System

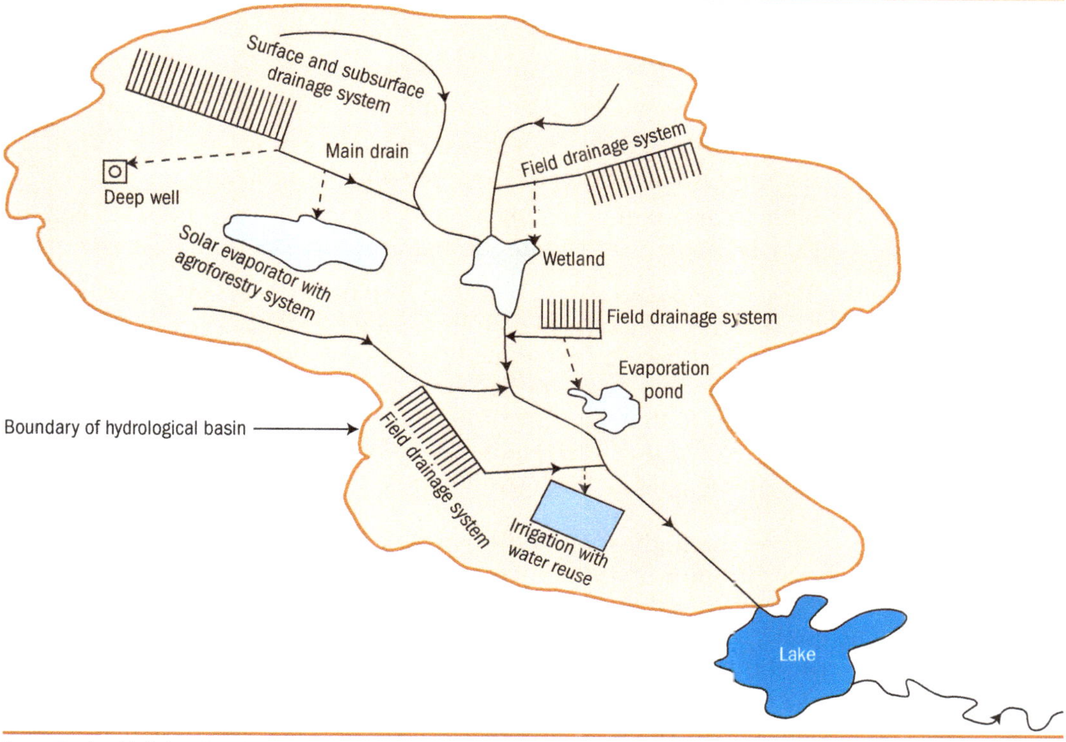

Source: Food and Agriculture Organization of the United Nations, www.fao.org/docrep/w7224e/w7224e01.gif

ing, (2) lengthen the effective growing season, (3) increase the depth of the root zone, (4) aerate the soil, (5) minimize erosion, (6) leach excess salts from the root zone, and (7) promote higher soil temperatures.

Surface drainage systems depend on the mechanical forming of land into appropriate terraces, diversion ditches, waterways, open ditches, and dams. Subsurface drainage systems employ drain tiles, which should be strong, dense, and resistant to freezing, thawing, acids, and sulfates. The drainage system should be designed to meet the needs of a particular soil, topography, and irrigation network. For example, soils with a high clay content require closer spacing of underground drainage tile lines than soils with higher sand content. Some drainage systems are dual purpose in that they facilitate drainage of the cropland and store the runoff to be used later for irrigation. Drainage water, like irrigation water, can be moved by gravity or by pumping, and the cost of moving it must be considered. Well-maintained drainage systems with clear, unclogged drains have the greatest utility and value.

Utility Development

The effect of utility development on rural properties is threefold. Appraisers should address

1. The utility services that are available on site on the valuation date
2. Those utilities that affect the intended use or alternative future uses but are not currently extended to the property
3. Utility easements, such as pipelines and electrical lines, which may be physically present on the property but do not service it

The presence or absence of standard utility services such as water, sewer, electricity, telephone, and natural gas may or may not influence the property's intended use. Water, sewer, and gas service are commonly not available on rural properties, necessitating private development of these services via wells, septic systems, and propane or butane systems. The development of broad-based cellular telephone systems has made telephone service more generally available in rural areas.

When the highest and best use or intended use of the property requires utilities that are not present, appraisers should estimate the cost to cure this deficiency. For example, division of a large rural property may result in the severing of essential utility services to some of the tracts, which may affect the highest and best or intended use. Depending on the highest and best use determination, appraisers may find it necessary to estimate the cost of providing utility services to the parcels.

The presence of utility easements for electrical transmission lines or oil and gas pipelines may limit alternative uses of the property by preventing the installation of center-pivot irrigation or tile drainage systems.

Chapter 9

Building Description

Typically, buildings on rural properties vary widely in terms of importance with respect to intended uses. While some improvements contribute little to no value, other highly specialized buildings can make major contributions to property value. In rural areas, single-family residences range from small frame houses to elaborate estates. Appraisals of rural residences may be documented on the same forms used for urban homes and are often submitted to the same lenders. Different underwriting problems arise when the home is located in a rural area, but the description portion of the appraisal report would remain very similar to those for urban homes.

In ranch areas, buildings may account for an insignificant portion of property value, especially if a large acreage is included in the property. On the other hand, properties can have significant improvements that do little to support the current uses but contribute to the overall property value. These improvements would typically suffer from obsolescence.

Farm buildings may make a small overall contribution to the land value but remain integral to the operation of the farm. These farms may include outbuildings for livestock and machinery, shop facilities, and many other structures for the storage of grain or other crops. For building-intensive properties such as grain elevators, cotton gins, feedlots, swine facilities, poultry facilities, and dairies, a substantial contribution may be made by the buildings while the land may be an insignificant part of the overall value.

Some rural properties with specialized agricultural operations include equipment or other fixtures that are classified as personal property rather than real property, which complicates the valuation process. In those situations, the property interests being valued must be clearly identified by the appraiser. This means that a clear allocation of value between business assets, such as equipment, and real property, such as permanent buildings housing that equipment, is critical. Allocating value between real property, personal property, and intangible business assets is a complex appraisal assignment requiring specific expertise and training.

Whatever the type of rural property, appraisers must describe all of the buildings included on the property and consider the building type or types present. The construction, use, size, and floor plan of the buildings must be considered, as these factors relate to the highest and best use of the property.

Zoning and Building Codes

Rural areas typically have no or very liberal zoning regulations. Although the zoning may be nonexistent, construction and sewage disposal permits would likely still be required. For rural buildings used for livestock or crops, permits are typically not required. For more building-intensive properties, especially those associated with confinement livestock, there are numerous state laws concerning waste disposal and the quality of air and water. These laws can influence the location and operation of the facilities even in areas without zoning. Most states have a permitting process for livestock facilities that exceed a minimum size based on animal unit capacity.

The US Environmental Protection Agency (EPA) requires that an animal feeding operation (AFO) above a certain size in animal units have a National Pollutant Discharge Elimination System (NPDES) permit to discharge waste. An AFO that meets certain criteria will be classified by the EPA as a *concentrated animal feeding operation (CAFO)*.

In areas that have zoning, appraisers must consult with the zoning offices to ascertain if the proposed improvements are legal and if the zoning has any effect on existing improvements. Because environmental, water quality, air quality, floodplain, coastal, and health codes can all affect rural property, appraisers must be aware of these laws or regulations and understand their influence on existing and proposed structures. Building codes can also affect rural real estate, and appraisers must be aware of any codes that exist in the area.

Measuring Building Size

Gross building area is the standard by which nonresidential agricultural buildings are measured. *Gross building area* is the total floor area of the building, excluding unenclosed areas, measured from the exterior of the walls. Sheds attached to the sides of a barn, for instance, are also measured and may be considered separately or as a part of gross building area depending on local custom.

Consistency is the key for using measurements to compare buildings. Obviously, when accepting building measurements obtained from another appraiser or from a public source, appraisers must know the measurement standards used.

Accurate building measurements are very important in the completion of an appraisal. Building measurements are used in the cost approach to estimate replacement cost, in the sales comparison approach for comparative purposes, and in the income capitalization approach to estimate market rent.

Table 9.1 EPA Classifications of Animal Feeding Operations

Animal feeding operation (AFO)	A facility where
	· animals have been, are, or will be stabled or confined and fed or maintained for a total of 45 days or more in any 12-month period, and
	· crops, vegetation, forage growth, or post-harvest residues are not sustained in the normal growing season over any portion of the lot or facility
Large CAFO	An AFO with at least
	· 700 mature dairy cows
	· 1,000 beef cattle or heifers
	· 2,500 swine (each 55 pounds or more)
	· 10,000 swine (each under 55 pounds)
	· 30,000 ducks (other than liquid manure handling systems)
	· 5,000 ducks (liquid manure handling systems)
	· 30,000 laying hens or broilers (liquid manure handling systems)
	· 125,000 chickens except laying hens (other than liquid manure handling systems)
	· 82,000 laying hens (other than liquid handling systems)
	· 1,000 veal calves
	· 500 horses
	· 10,000 sheep or lambs
	· 55,000 turkeys
Medium CAFO	An AFO with at least
	· 200 mature dairy cows
	· 300 beef cattle or heifers
	· 750 swine (each 55 pounds or more)
	· 3,000 swine (each under 55 pounds)
	· 10,000 ducks (other than liquid manure handling systems)
	· 1,500 ducks (liquid manure handling systems)
	· 9,000 chickens (liquid manure handling systems)
	· 37,500 chickens except laying hens (other than liquid manure handling systems)
	· 25,000 laying hens (other than liquid manure handling systems)
	· 300 veal calves
	· 150 horses
	· 3,000 sheep or lambs
	· 16,500 turkeys
	where either
	· pollutants are discharged into waters of the United States through a man-made ditch, flushing system, or other similar man-made device
	or
	· pollutants are discharged directly into waters of the United States which originate outside of and pass over, across, or through the facility or otherwise come into direct contact with the confined animals (for example, if animals come into contact with surface water running through the area where they are confined)

Source: EPA, *Compiled CAFO Final Rule*, www.epa.gov/sites/production/files/2015-08/documents/cafo_final_rule2008_comp.pdf.

Building Description Considerations

With the variety of agricultural buildings found in the market, many different factors must be considered depending on the type of building being appraised. Appraisers must consider the complete building description, which includes an exterior description, an interior description, the identification of equipment, and an analysis of the total facility.

The exterior building description should provide information about the type, quality, and design of the materials that were used or will be used in the construction of the building being appraised (see Figure 9.1). Details concerning the type of foundation, floor, exterior walls, roof structure, roof cover, and insulation should be included.

The description of the interior of an agricultural building should include discussion of the floor, wall finish and supports, partitioning, manure disposal construction, interior height, heating systems, ventilation systems, and lighting (see Figure 9.2). Because agricultural buildings vary tremendously, appraisers must be familiar with the

Figure 9.1 Exterior Description

A. Substructure
  1. Pole
  2. Concrete
  3. Footings
  4. Piers
  5. Beams
  6. Manure pits

B. Exterior walls
  1. Solid wall (load-bearing)
  2. Curtain wall (non-load-bearing)
  3. Material
  4. Insulation

C. Roof construction
  1. Structure
  2. Cover
  3. Drains
  4. Insulation

D. Utilities
  1. Electrical source
  2. Water source
  3. Gas source
  4. Interior distribution
  5. Multiple sources (required or not required)

Figure 9.2 Interior Description

A. Floor structure
  1. None (dirt)
  2. Concrete
  3. Drains
  4. Type (dairy, poultry, swine)

B. Interior construction
  1. Clearspan
  2. Walls (type, position)
  3. Insulation

C. Equipment
  1. Dairy
  2. Poultry
  3. Swine
  4. Cattle
  5. Electrical service
  6. Plumbing

D. Heating systems
  1. Type and fuel

current interior construction standards of the type of building being appraised. Again, the housing of large machinery is an important consideration in describing the building interior. Today's buildings are larger and have higher ceilings than the buildings of 35 years ago. Columns can be disadvantageous if they cause difficulty in the internal movement of machinery or hamper manure removal, and the utility of a building would suffer due to these interior supports. Heating can be an important factor, especially if a building houses very young animals. Ventilation is also important in the removal of dust and odors that can inhibit animal growth.

To analyze an agricultural facility as a whole, an appraiser should make a determination of the quality and condition of the structure and, more importantly, its conformity to current construction standards. Typically, integrators supply the animals that will be housed in specialized agricultural buildings. The integrator's standards are often the measurement of the functional utility of the structure. In addition, the large machinery used in agricultural enterprises today requires larger doors, higher eaves, and less interior partitioning than is typical in older buildings. When a group of buildings makes up a total operation, the functional utility of the group of buildings must be described. Do the buildings provide for an efficient flow of operation so that all buildings are used for maximum efficiency?

Equipment can be an integral part of a building, and a building's capacity for housing modern equipment is often the basis for assessing its functional utility. Dairy, swine, and poultry producers must use the most modern equipment to maximize the overall utility of the structure. Integrators often require the most modern equipment, or they will not place animals in the buildings. An integrator's refusal to place the necessary livestock in the building could indicate obsolescence and totally eliminate the value of the building.

Certain equipment may be considered personal property rather than real property. For example, much of the equipment used for irrigation, such as water pumps and motors, is moveable and thus not real property. The sprinkler span on a pivot irrigation system is detachable and is also considered personal property.

Specialized Agricultural Improvements

The remainder of this chapter provides suggestions for writing building descriptions for various types of agricultural buildings. Appraisers should be familiar with the standards for the area in which the appraised buildings are located and adapt these general guidelines to the specific area. It is important to understand the dynamics of agricultural product prices and how they can affect the value of agricultural land, plants, and buildings. Because technology changes quickly, external and functional obsolescence can be very significant for many agricultural operations and can also happen quickly. Many specialized agricultural facilities such as those described below (e.g., swine or poultry facilities) are large-scale, highly mechanized operations. A good understanding of the business operation is essential for the accurate valuation of agricultural properties, just as it is for industrial properties. These types of facilities are generally valued by appraisers who have specialty practices.

Swine Facilities

Commercial swine facilities are located in several parts of the United States. The majority of hogs raised in the United States are owned by integrators who contract with building owners for all aspects of the production process. Each integrator has its own set of building configurations and designs, and appraisers must be familiar with the operation of swine facilities and know what drives the market. In describing a swine facility, the factors to be addressed concern the construction of the facility and the buildings and equipment necessary for a profitable operation. A major factor is the ability of a facility to meet the needs of local integrators who may want to offer contracts to place their pigs in existing facilities. Most new swine facility construction is built to meet the needs of a specific integrator using that company's plans and specifications.

Dimensions

The size and scope of most swine facilities is based on the needs of the integrators that constructed them. That is, the integrator must be able to coordinate all components of the livestock operation, which may be located on multiple properties that are significant distances apart. Appraisers must consider the typical size and current standards prevalent in the area and judge the subject property in light of these standards. Accurate measurements of the facility are needed to ascertain the number of animals that can be housed.

Swine Facility

Source: Peter Menzel Photography, http://cdn.c.photoshelter.com/img-get/I0000PVe9_.TXToQ/s/750/750/USA-AG-PIG-04-xs.jpg

Foundation

For most swine facilities, the foundation is usually concrete with concrete slats over a 4- to 8-foot manure pit.

Exterior Walls

Corrugated, prefinished steel panels were the standard for several years, but roll-up curtains are the current standard in many areas because they provide more efficient ventilation. Appraisers must consider the standard for the local market and determine which type of exterior walls are preferred in that market.

Ceilings

Ceiling types range from corrugated, prefinished steel with blown-in insulation to sloping ceilings with roof vents for additional ventilation. Some barns have aluminum, glass board, or PVC interior ceilings and walls that resist corrosion.

Heating

Heating must be sufficient for the climate in which the property is located.

Lighting

Lighting and wiring vary with the quality of the facility's construction.

Swine Facility Foundation

Source: Gannett CDN, www.gannett-cdn.com/-mm-/6d449287f98d12f855f02891b744362790642ae2/c=0-0-748-562&r=x513&c=680x510/local/-/media/DesMoines/DesMoines/2014/06/06//1402089711000-hog-confinement-01.jpg

Ventilation

Depending on the design of the facility, building ventilation can take the form of fans, curtain walls, or moveable panels to assist normal air flow. Most ventilation systems include pit fans to provide manure pit ventilation. Many newer barns use outdoor lagoons instead of interior manure pits, and thus have no pit ventilation fans.

Feed and Water Systems

Feed and water systems are often considered part of the building equipment and typically have a shorter life than the building structure due to wear and deterioration from use.

Pens and Crates

Different types of pens and crates are often found in the various buildings that make up a swine facility. These pens and crates should meet current standards. They will have a shorter life than the building structure.

Manure Storage

Manure storage can take the form of a lagoon, a storage tank, or a manure pit within the building. Appraisers must observe and consider the size of the manure storage area as it relates to the number of animals and the amount of manure that will be produced. Manure facilities are emptied and injected into the ground for fertilization purposes at various times of the year, depending on local climate and cropping patterns. The facility should have either sufficient acreage or contracts with other producers for manure disposal. Local and state regulations concerning manure disposal should be reviewed by appraisers to confirm that the facility is in compliance. If the facility is not compliant, obsolescence may exist.

The largest expense of a CAFO providing services to an integrator on a contract basis is often the cost of manure removal. If manure is not removed from the site, it has to be stored, which can require the use of lagoons or other storage areas that are considerably larger than what are usually found on smaller animal feeding operations. For example, large hog facilities can have lagoons that hold millions of gallons of waste.

CAFO operators may face the risk of litigation from neighbors who claim that odors from the waste storage is a nuisance or from governmental authorities who have identified a facility as a source of contamination in the local water supply. In addition, closing down a CAFO can leave the operator responsible for disposing of the stored waste, so reuse of a CAFO for some other agricultural purpose is often complicated by the environmental liabilities resulting from the facility's prior use.

Table 9.2 Manure Removal from Livestock Operations (2004-2006)

	Percentage of Total Manure Produced		
	Dairy	**Hogs**	**Broilers**
Manure removed from operation	19	26	61
–Sold by operation	4	5	22
–Operation paid to haul it away	7	3	3
–Operation gave it away	8	18	36
	$ per CWT of Production		
Prices			
–Revenue from manure sales	0.28	0.22	0.20
–Expenses to haul manure away	0.39	0.34	0.31

Data from authors' calculations, based on data from the Agricultural Resource Management Survey, version 4, 2004 (hogs), 2005 (dairy), and 2006 (broilers).

Source: James M. MacDonald and William D. McBride, *The Transformation of US Livestock Agriculture: Scale, Efficiency, and Risks/EIB-43* (Economic Research Service/USDA, 2009), 32.

Essentials of the Building Description: Swine Facilities

The building description for a swine facility must include discussion of all physical and functional curable items, the age and condition of short-lived items, the age and condition of the long-lived portion of the facility, and other items of functional and external obsolescence. Also, the overall integrity of the facility must be addressed. For example, can the farrowing space be used efficiently? The overall facility must be considered in order to ascertain functional obsolescence due to either the lack or excess of a particular capacity. External obsolescence in swine facilities is typically the result of lower hog prices.

Figure 9.3 Subfloor Design for Deep-Pit Manure Tanks

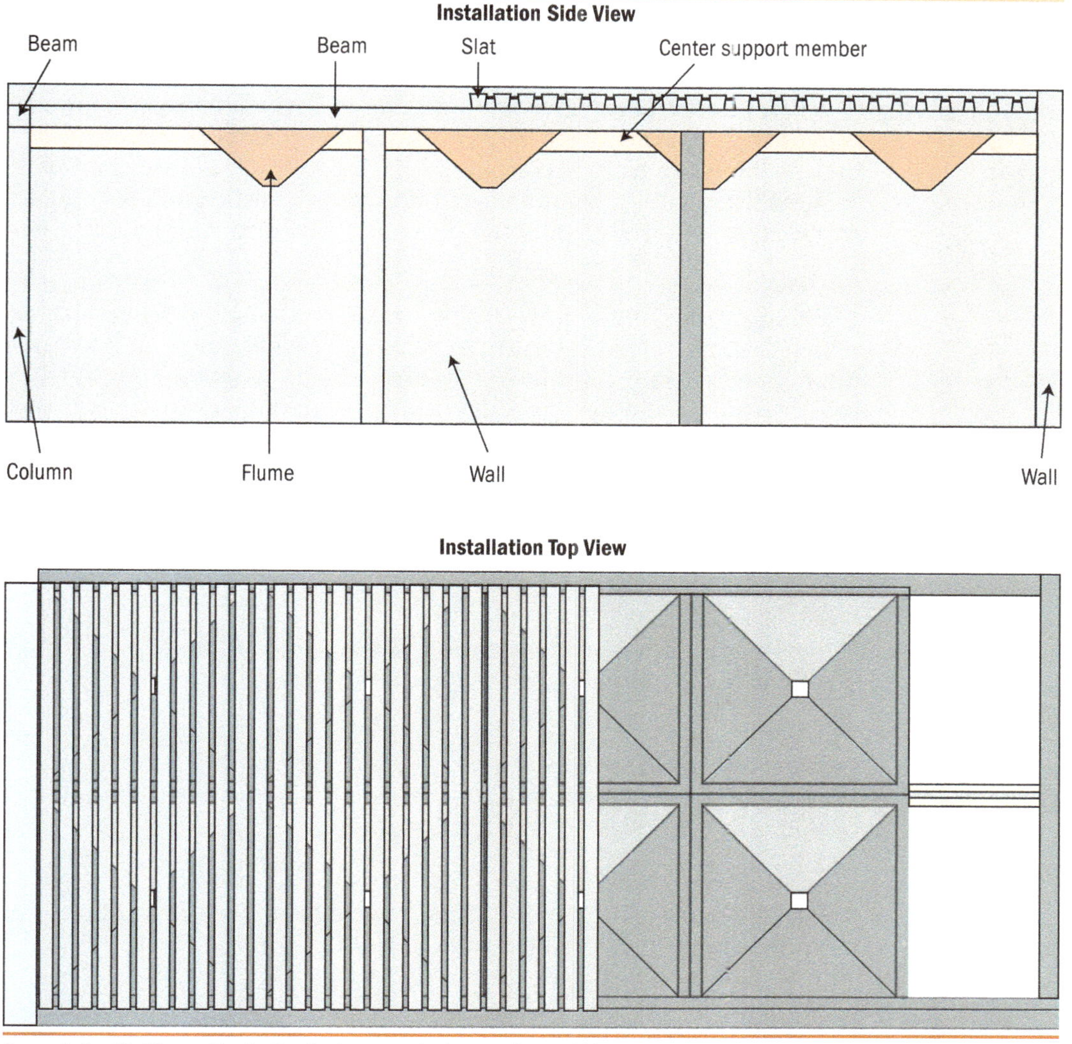

Source: *National Hog Farmer*, http://nationalhogfarmer.com/environmental-stewardship/regulations/environmental-energy-building-designs-0426

Utilities

Many farms need dual sources of water and electricity, and the lack of these dual sources increases the risk of operations. Most facilities have backup generator systems, and these systems must be large enough to supply the farm until normal electrical service is restored. Well water and a community water system can provide dual water sources for a farm.

Poultry Facilities

Commercial poultry facilities differ considerably depending on the type of operation and the area in which they are located. Poultry facilities are constructed to house broilers, caged laying hens, breeder stock, and pullets. Different building construction and equipment is needed to operate these facilities. For example, most new broiler houses today are fully enclosed and incorporate tunnel ventilation, evaporative cooling technologies, improved lighting, and automated controls to manage internal temperatures, airflows, and lighting.[1]

Dimensions

The dimensions of a poultry facility relate to the facility's overall size, which is more relevant. Again, the facility must be large enough to be financially feasible, but not so large as to be unprofitable. Appraisers must consider whether the subject property conforms to the typical size and current standards prevalent in the area. Some states have enacted regulations on cage and space requirements. It is important to understand if a property complies and how value might be affected. An accurate measurement of the facility is needed to ascertain the number of birds that can be housed.

Foundation

Depending on the type of operation, the foundation may vary from a dirt floor to a concrete floor with a manure pit.

Exterior Walls

Corrugated, prefinished steel panels are used in cold climates, and roll-up curtains are common in warmer climates. Appraisers must consider the standard for the local market.

Interior Construction

Clearspan construction, which facilitates manure removal, is predominant in newer poultry facility buildings.

Heating

Heating must be sufficient for the climate.

Lighting

Lighting and wiring vary with the quality of the facility's construction.

Ventilation

Depending on the facility's design, building ventilation may be provided by fans, curtain walls, or moveable panels that assist normal air flow.

---

1. James M. MacDonald, *Technology, Organization, and Financial Performance in US Broiler Production* (Washington, DC: USDA Economic Research Service, 2014).

Figure 9.4 Components of a Poultry Facility

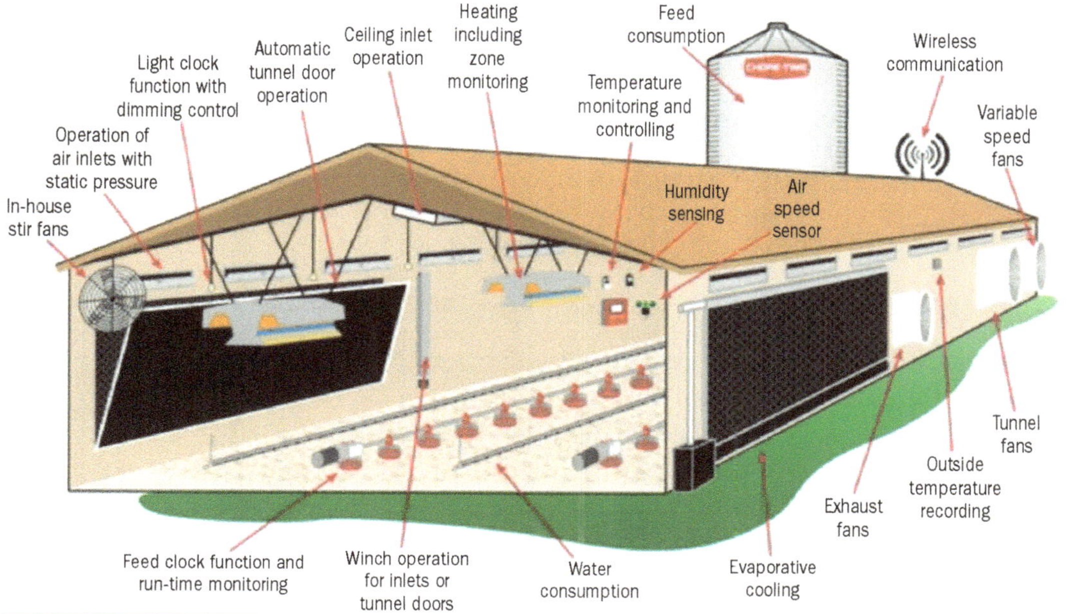

Source: WealthResult.com, www.wealthresult.com/wp-content/uploads/2013/08/Chicken-House-Diagram.jpg

Feed and Water Systems

Feed and water systems are considered part of the equipment and vary greatly with the type of operation. Feed and water systems have a shorter life than the building structure due to wear and deterioration.

Cages

Cages are generally part of the equipment found in buildings that house poultry. However, cages are not typically used to house broilers. If cages are present, they should meet current standards. They will have a shorter life than the building structure.

Poultry Cages

Source: http://broilerhouses.co.za/wp-content/uploads/chickens-in-layer-cages.jpg

Ground storage is used in many poultry operations. The manure is removed by loader or tractor in batches to manure pits within the building. Appraisers must observe and consider the size of the manure storage as it relates to the number of birds and amount of manure to be produced. Local and state regulations concerning manure disposal must be reviewed by an appraiser, and the facility must be in compliance. If the facility is not compliant, obsolescence may result.

Utilities

Many farms need dual sources of water and electricity. The lack of dual sources increases the risk of operations. Most facilities have a backup generator system, which must be able to supply the farm until normal electrical service is restored. Well water and community water systems can provide dual water sources.

Essentials of the Building Description: Poultry Facilities

The building description for a poultry facility must include discussion of the following factors:

- physical and functional curable items
- short-lived items, including their age and condition
- long-lived components, including their age and condition
- functional obsolescence
- external obsolescence

Feeding Trough

Source: Agriculture Proud, https://agricultureproud.files.wordpress.com/2012/10/cattle-feed-bunk.jpg

Cattle Feed Yards

Cattle feed yards are designed to accommodate large numbers of cattle being fed out prior to slaughter. These facilities are composed of improvements that fall into five general categories:

- feed pens, feed bunks, shades, working pens, loading and receiving chutes, hospital pens, and horse pens
- feed mill facilities
- grain storage and handling improvements
- office and building improvements
- site improvements

Pens

Pens in cattle feed yards are generally constructed of cable on pipe posts set in concrete, with a pipe top rail and center rub rails. The cables are tied to the end posts by coil springs. Bunk fencing consists of a pipe rail and an adjustable cable mounted on pipe posts. All pens are normally fitted with welded metal gates. Alleys run along the ends and rear of all pens and are of comparable construction. Each

Dairy Pens

Source: Connell Construction LLC, http://agriculturalbarns.com/wp-content/uploads/2014/05/opt-7.jpg

pen has water and possibly shade. The feed bunks are of either concrete or metal construction with a concrete apron, and they run the length of each pen. The feed alleys are all weather-surfaced and run along the feed bunk frontage. The hospital pens are of similar construction, but they have metal-sided working and squeeze chutes, all on concrete pads. The shipping and receiving pens are welded pipe with a crowding and working chute on a concrete slab with a sheet metal shade roof. A cattle scale is usually located in these facilities. The loading and unloading chute generally has two levels with concrete ramps and curved metal steel plate welded on pipe.

Feed Mill

A feed mill facility is comprised of a mill building and slab area used for batching the feed. The mill building consists of the mill office and storage area, which is at dock height on a slab foundation with a met-

Feed Mill Facility

Source: Dinklage Feed Yards, www.dinklagefeedyards.com/images/morgan_main.jpg

al or wood frame and walls. A mill truck scale (usually of 100,000-pound capacity) is located in the mill building under the truck load-out area. The milling equipment includes automatic steam-roller mills, which are fed grain from the cleaner or the hopper bins located above the roller mills. The welded steel hopper bins located in the mill building are mechanically operated. Concrete bulk material bins are also located within the commodity barn. Molasses, protein supplement, and fat tanks are located in the mill and supply product to the milling operation. The feed mill equipment includes numerous elevator legs, air lifts, centrifugal pumps, and air compressors. Grain is typically moved by augers. Processed grain and feed are usually moved by airlifts. Elevator legs, augers, airlifts, mixers, shakers or scalpers, steamrollers, and other components are engineered so that all capacities are compatible. Boiler rooms contain 125- to 200-hp boilers with water makeup systems, water softeners, and other equipment.

Grain Storage and Handling

Grain storage and handling improvements usually consist of corrugated, bolted metal bins with flat-bottom or concrete-slab sweep augers for unloading or hopper-bottom metal bins that unload into U-trough conveyors. The bins are connected by an elevator system that connects to a concrete dump pit, which receives grain and delivers it to the elevator for movement to the storage bins or directly to the mill.

Office and Building Improvements

The office and building improvements are typically of wood frame or brick veneer construction. The office portion is usually small. A truck platform scale with an electronic printer and a livestock scale with a manual printer are generally located in this building. A shop building of typical construction is also usually located on site, with heating depending on the climate. Other small shop buildings, sheds, or employee housing can be found on some sites.

Site Improvements

Site improvements on cattle feed yards generally include diesel and gasoline storage, liquefied petroleum gas (LPG) storage, water wells, water storage, and manure storage lagoons.

Essentials of the Building Description: Cattle Feed Yards

Again, appraisers should be familiar with standards for cattle feedlots in the subject property's area. Run-off lagoons usually have to be state licensed and permitted. Some states require a feed yard to include enough land to dispose of the run-off water on site. Appraisers should also consider the availability of grain, feeder cattle, and a slaughterhouse because excess shipping costs can affect the feasibility of the overall operation.

Grain Storage Facilities

Grain storage facilities typically found in grain production areas can vary from smaller on-farm setups to much larger elevators. On-farm storage facilities are designed to handle large quantities of grain and can be constructed of concrete or steel. Larger elevator facilities are typically located along railways, major rivers, and port facilities along the coasts of the United States. Newer facilities may have significant rail sidings or may be located near ethanol facilities.

The building description should include detailed discussion of any of the following components present in the facility:

- grain storage (upright or flat)
- elevator system
- warehouse storage
- driers
- seed equipment
- port facilities
- rail facilities
- office and scales

Grain Storage

The description of a grain storage facility should include a calculation of the total capacity of the facility, which is usually expressed in bushels. This number is normally based on owner records or actual measurements made by the appraiser. The number should be reconciled with the official licensed capacity. Any variances should be noted and addressed in the description.

The description needs to include the diameter and height of any upright storage and the exterior measurements and height of any flat storage. The tanks are either flat-bottom or hopper-bottom,

Grain Bins

Source: CTB Inc., www.brockmfg.com/uploads/photos/600/3_Grain_Bins_fix_copy.jpg

and the loading and unloading systems for the bins must be described. The bins are usually loaded by elevator legs from the top and unloaded either by U-trough augers or by a sweep in flat-bottom tanks. Flat storage can be loaded by an auger system in the roof or by portable augers. Unloading is generally accomplished with floor and portable augers.

Many newer facilities are equipped with a loop system that is used to both load grain into the top and remove it from the bottom. These are usually associated with a drive-over dump and load-out ramp.

The description needs to include details such as the size, age, and condition of the electric motors that power the loading and unloading equipment. Any driers or ventilation fans must be described, including the heat source, drying capability, and size of the electric motors in horsepower.

Elevator System

The elevator system includes the dump pit and the associated legs, which move the grain from the dump pit to the various storage and unloading areas. The length, width, and depth of the dump pit must be measured. The number and length of the elevator legs needs to be determined, along with the size of the distributor. Also, the size, age, and condition of all electric motors used in the elevator system needs to be

Figure 9.5 Grain Elevator

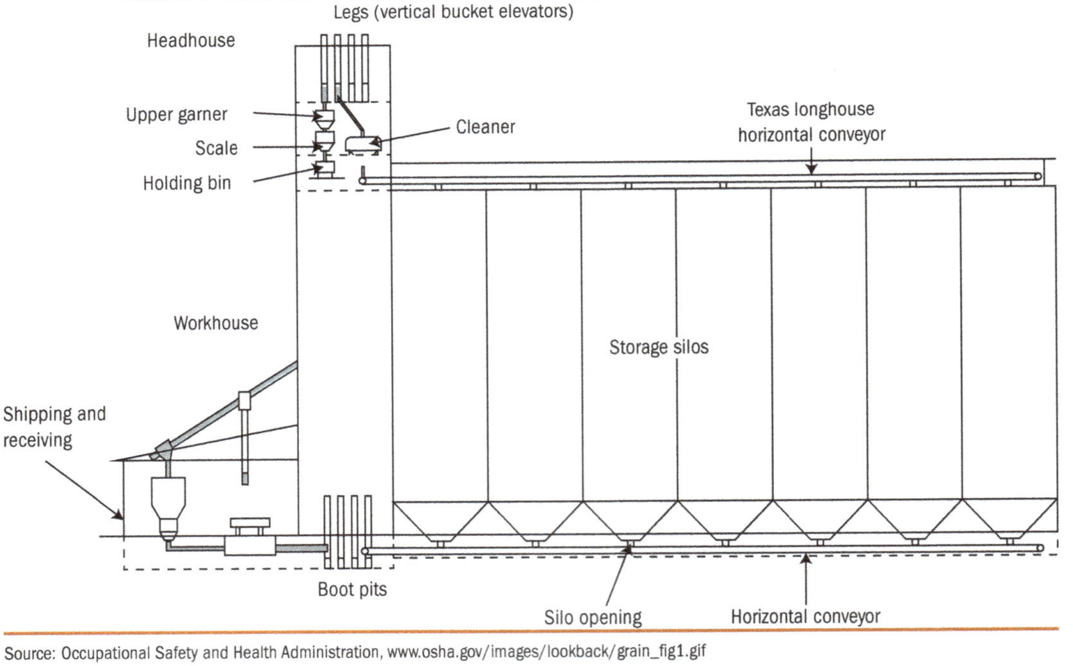

Source: Occupational Safety and Health Administration, www.osha.gov/images/lookback/grain_fig1.gif

recorded. The bushel capacity per hour of the dump pit and the elevator legs is also essential.

Warehouse Storage

The warehouse is used to store fertilizer, seed, or other products sold by the operation. The floor is typically concrete, and the walls and sides are metal. The amount of insulation and heating required is governed by the climate of the area where the complex is located.

Grain Warehouse

Source: www.ahrensagri.com.au/files/1681.jpg

Driers

The driers may be small, individual units attached to various bins or a large unit connected to the system with elevator legs that dry large quantities of grain. They may be continuous flow-through or batch driers.

Seed Equipment

When seed equipment is present, it adds another dimension to the elevator beyond handling local grain for shipping. Seed equipment includes equipment for cleaning, treating, weighing, sacking, and storing seed. All of this equipment should be itemized.

Port Facilities

Port facilities are located along major rivers to transport grain to the coasts for loading onto ships for export. Port facilities normally have docks for loading grain into barges or ships. The port facility should be described in detail, including the capacity and the methods for loading.

Rail Facilities

Many grain elevators have rail spurs and elevator legs for loading rail cars. The loading elevators and their capacity should be described in detail. The amount of rail spur is important because additional efficiency results if large numbers of rail cars can be loaded. Newer facilities often have high-speed rail loaders and loop track systems that accommodate unit trains of 100 to 110 grain cars. Train-loading facilities accommodating more than 100 cars have been constructed in response to freight rate trends established by large rail operators such as BNSF and UP. Older, smaller capacity spurs may have limited contributory value as a result of this trend.

Office and Scales

The office should be described to include its size, exterior finish, interior finish, heating and cooling system, and other features. The scales should be described in terms of size, capacity, age, and condition.

Essentials of the Building Description: Grain Storage

Grain storage facilities can vary widely from small on-farm storage consisting of only a few bins to large elevators with many bins. On-farm storage is typically sold as part of a larger tract of cropland, while elevators can sell as a business and include real property, FF&E, and intangible business assets. The intended use of these facilities is the handling and storage of grain, and each facility would be evaluated based on its capacity in bushels and ability to efficiently accomplish this. Any deficiencies in the ability to efficiently accomplish the process indicates obsolescence.

Cotton Gins

The general trend is to operate larger cotton gin plants very near the point of production in areas with substantial cotton production. Modern gin plants are being constructed with a 30- to 35-bale-per-hour output. Smaller gins are considered obsolete by current standards. The main components of a cotton gin are

- gin plant
- seed hopper
- trash bin
- warehouse

Gin Plant

A gin plant is typically constructed with a structural steel frame, a concrete floor with the appropriate recesses and tunnels, and a pre-engineered steel panel roof and sides. Side walls must be tall enough to house the gin equipment. Adequate windows, doors, and roof ventilators must be present due to dusty conditions. A covered load-out dock is usually attached to the gin building. The plant should include sufficient warehouse area to store ginned cotton until it is ready for shipment.

Seed Storage

Most cotton gins use large metal warehouse-type buildings to store cotton seeds. These are typically loaded with overhead ductwork. Older, smaller gins may have seed hoppers, which are multiple-hopper, welded-steel overhead bins with metal supports on a concrete foundation. The storage used must have sufficient capacity to store seed from the baled cotton.

Gin Hopper

Source: Cherokee Fabrication Co., Inc., www.cherokeefab.com/wp-content/uploads/2013/07/DCP_0531.jpg

Gin Plant

Source: www.farmflavor.com/wp-content/uploads/2012/12/1641012BM5376.jpg

Seed Storage Bin

Source: Cliff Granberry Corporation, www.cliffgranberrycorp.com/i/2/seed-house-2-2.jpg

Cotton Warehouse Storage

Source: Cherokee Fabrication Co. Inc., www.cherokeefab.com/wp-content/uploads/2013/06/Seed-Handling-Main2.jpg

Trash Bin

The trash bin or burr hopper is generally a metal frame building with a concrete floor and a corrugated metal roof and sides. An overhead conveyor system is used to move the cotton burrs from the gin plant to the trash bin. Some areas allow for the gin trash to be burned or stored on the ground, where it is exposed to the elements and can break down naturally.

Office

The description of the office should include the office size, exterior finish, interior finish, heating and cooling, and any other features present. The scales should be described in terms of size and capacity.

Location of Cotton Gins and Highest and Best Use

As previously mentioned, cotton gins are typically located in areas where a substantial amount of cotton is grown. Facilities located in marginal areas may suffer obsolescence if they cannot gin enough cotton to satisfy the test of financial feasibility in highest and best use analysis. Features that affect the feasibility of a gin are the uniform density (U-D) press and module feeder. A U-D press compresses cotton into a smaller bale unit than the conventional gin press does. Module feeders, which are used to feed cotton into the ginning process, are common in most areas, and the ability to handle the newer round modules or bales is considered essential as these newer cotton pickers gain popularity. Many cotton gins have closed and been salvaged or parceled out and moved to new locations due to the reduction in overall cotton production. Air quality regulations are stringent when gins are concerned, and the cost to comply with these regulations may influence the feasibility of a cotton gin.

Processing, Packing, and Cold Storage Facilities

Facilities for the processing or packing of fruits and vegetables typically consist of three major components:

- receiving area
- processing area
- packing and shipping area

Facilities can vary greatly, and the physical description of the assets should be consistent with the scale of the facility and its contribution to the total property value.

Legal restrictions and requirements differ from region to region. Some areas may have little or no restrictions or requirements, while others may impose a hierarchy of regulations and conditions. Environmental concerns are becoming increasingly onerous. The regulatory environment is dynamic, and appraisers must strive to keep informed of the latest changes in zoning and regulatory constraints and requirements. Zoning, permissions, and licensing may involve a hierarchy of government agencies on the local, state, and national levels. A large, modern processing facility may require a plethora of permits and licenses, and appraisers must adequately research and report on these. Environmental concerns regarding air and water quality, health considerations, traffic flows, noise, dust, and other issues may necessitate a permit or review process. In many cases, existing facilities that are not in compliance with current regulations may be allowed to continue to operate as legal nonconforming uses.

Receiving

Receiving areas receive the raw product when it arrives from the field. Depending on the commodity, the facilities may take various forms. The raw product may be delivered to the plant in bins, field boxes, gondolas, or any combination of these. Thus, the receiving area may consist of a simple loading dock or a complex of automated dumping stations that typically provides for the quantitative and qualitative measurement of the incoming product. Quantitative measurement is usually accomplished via truck scales, while qualitative analysis is performed at an inspection station, which may or may not be certified by a government or industry agency or commission. Receiving may also provide for precooling of the raw field product. Ancillary site improvements may be critical to the receiving process, and they should be described adequately. Incoming truck staging areas should also be assessed.

Processing

Main processing or packing buildings typically feature conventional industrial design and construction. Internally framed steel structures are probably the most prevalent, although many packing areas consist of open-walled, covered sheds. Again, appraisers should describe both the physical structures and the flow of the product through the process. Depending on highest and best use determination, some consideration may also be given to the possible conversion of the facilities to alternative uses. Many industries have witnessed structural changes in the packing process, with trends moving toward the packing of the product in the field. Large packing houses and sheds have experienced rapid depreciation in these industries, and the ability to convert the structures to alternative uses may be a major increment of value.

Utilities are often critical in the sorting and packing process, and their description should be consistent with their importance. Modern operations require large electrical inputs, which are typically controlled by centralized computer stations. In some operations, the disposal of culled product is a significant task, and the vehicle for disposal or transfer should be adequately described. Wastewater can be an undesirable by-product of packing and processing. Appraisers must explore and report on how wastewater generated by the plant's operation is disposed of or processed.

Packing and Shipping

The final stage in the operation, packing and shipping, typically demands storage areas for packing materials. The actual packing is generally accomplished at the end of the processing line. When wooden or paper packing materials are used, materials storage requires automatic fire sprinklers, alarms, and possible additional safety features. The finished, packed product may either be shipped directly or transferred to storage to await future shipment. With perishable commodities, cold storage is a key component.

Cold Storage

Building structures for the refrigerated storage of perishable products typically feature conventional industrial design and construction. The basic building shell may be wood framed, internally steel framed, or concrete tilt-up. In most areas, concrete tilt-up and metal sandwich panel construction are popular. The basic shell is improved with refrigeration equipment and fixtures, insulation, receiving and loading docks, and office and dispatch areas. Here again, appraisers should adequately describe both the physical structure and the flow of the product. The product flow should address temperature changes, as the process often involves rapid changes in product temperature.

Some specialized products demand more precise storage environments, such as the "controlled atmosphere" storage required for apples. Other products may require high or low humidity or other specific atmospheric conditions. The storage facility must be designed and constructed to provide the required environment. These types of specialized improvements may suffer significant, and perhaps rapid, obsolescence from changes in market preferences.

Receiving areas may include precooling rooms with high-volume, forced-air refrigeration, or hydrocoolers with recirculating refrigerated water cooling. Hydrocoolers may be portable units, and even permanent units may be considered personal property. Precoolers tend to be real property. Entrance doors are typically sliding or roll-up and should be large enough to handle modern, fully loaded forklifts. Temperature control devices at entrances may be forced-air doors or plastic curtains. Loading dock areas and hallways may also be insulated and refrigerated. They may serve as overflow storage when needed.

The design and layout of main storage areas are dictated by industry standards. Some products are stored most efficiently in large, open rooms, while others may demand smaller, isolated bays. Internal racking and shelving may be a desirable asset for some types of storage. Eave or ceiling heights should be measured and described because cold storage is a function of volume in three dimensions, not merely area in two dimensions. Thus, the appropriate unit of measure is often cubic feet rather than square feet, which is the more traditional unit of measurement.

Refrigeration equipment and insulation should be matched to the location and operation. Equipment and fixtures should be described commensurate with their importance to the total facility and should be identified as either real or personal property. Appraisers should be familiar with the rules for differentiating between real and personal property in the specific jurisdiction. The refrigeration vehicle (i.e., Freon or ammonia) should also be clearly identified, and any usage restrictions should be discussed. The total refrigeration capacity should be identified and reported, usually in tons of capacity.

Temperature alarms and back-up power systems may be highly sophisticated, involving significant cost. Facilities that store high-value products may employ multiple redundant systems.

Chapter 10

Market Analysis

The market value of a rural property is largely determined by its competitive position in its market. Familiarity with the characteristics and attributes of a specific property helps appraisers identify competitive properties and understand the comparative advantages and disadvantages that the subject property offers potential buyers or renters. An investigation of economic conditions, their effect on real estate markets, and the momentum of these markets helps reveal to appraisers how those externalities affect properties. In its broadest sense, market analysis provides vital information needed to determine the highest and best use of a property and to develop the three approaches to value.

In the past, appraisers have focused on regional and neighborhood analysis to provide background for their analyses. Unfortunately, many appraisers left out a critical step. They did not make a connection between fundamental market forces and the value implications for the subject property. Modern appraisal theory has

The Emergence of Market Analysis in Valuation Literature and Education

A survey of valuation literature reveals that market analysis has not been an explicit step in the valuation process as long as the other steps in that process, e.g., the application of the three traditional approaches to value. And market analysis has historically received less attention in the valuation of rural property. More recently, however, the analysis of market conditions has become a central element in appraisal practice and in valuation education, literature, and standards.

A six-step market analysis process was developed in the 1990s as a standardized but flexible model of the various data collection and analysis activities that appraisers traditionally perform. The six-step process has proven to be a useful tool in the valuation of urban and suburban real estate, but the process has been less successfully applied to certain rural property markets.

changed. A six-step market analysis process provides a framework to analyze market trends and demonstrate the connections of those trends to the value of the subject property.

The market analysis portion of an appraisal report provides an appraiser with the opportunity to describe and present a complete picture of the setting in which the subject property exists. Similarly, the discussion of the market in the appraisal report enables the client or reader of the report to visualize and understand the nature of the market. Many market analysis assignments can be performed using the six-step process to answer the following questions:

Step 1. Property Productivity Analysis:	What attributes does the subject property offer the market?
Step 2. Delineate the Market:	Who are the most likely users of the subject land and improvements?
	Where would they most likely look for similar properties?
Step 3. Demand Analysis:	Is the property use needed and desired by market participants?
Step 4. Supply Analysis:	What is the competition? How much is available?
Step 5. Market Condition Analysis:	What is the condition of the market? Is the market in equilibrium?
Step 6. Marketability Study:	How do the current market conditions affect the value and income of the subject property?

The market analysis process addresses fundamental questions about supply and demand in the marketplace.

In the analysis of a proposed property, a seventh step can be added to include (a) analysis of the financial feasibility of alternative uses and (b) threshold testing, often using the breakeven point of the investment as the threshold.

Levels of Market Analysis

Market and marketability studies can be accomplished with different levels of research and different analysis methods. Analysts need to identify the amount of market research needed, and the analysis methodology used, in appraisal analyses and general market analysis reports as part of the determination of the scope of work. The various levels of analysis that may be appropriate range from relatively general techniques to increasingly complex ones. Two broad categories of analysis are identified by the terms *inferred demand analysis* and *fundamental analysis.*

The term *inferred demand analysis* is used because the analysis supports an inference about demand for a specific property but does not directly quantify that demand. Fundamental analysis is a technique that determines a property's economic demand based on a direct study of the users of the space. In other words, the demand can be quantified.

Inferred analysis is used in most rural property assignments. There is typically no direct connection between demographic statistics and the amount of land used for a specific use. Land in transition often requires more intensive market analysis, and other publications from the Appraisal Institute can be consulted for a more detailed study of the specific problems that appraisers face.

Level A market analysis is typically performed using demographic data and inferred data gathered by the appraiser, such as comparable sales, listings, and rentals. Typically, a Level B analysis employs data from regularly published, area-wide market surveys prepared by private firms or public agencies. A fundamental Level C market analysis may be necessary if the land is near the edge of an urban area or is in transition to higher use.[1]

The value of rural land on the urban fringe can change rapidly as the demand for new housing rises. (In general, demand for commercial uses such as retail and office space tends to lag behind demand for residential use.) Market analysis helps identify the timing of a change in use from an agricultural or recreational use to a residential or commercial use (or to another more intensive use of the land). When urban influences are overtaking the rural character of an area being studied, appraisers will likely find that valuation assignments involving what were initially described as "rural" properties actually have more in common with valuation assignments for urban properties.

The traditional six-step process of market analysis is most applicable in the feasibility analysis of a new nonagricultural use of land in a rural area. In fact, the development of a new residential subdivision on what was previously farmland is an archetypal example of the application of market analysis.

Appraisers working in rural markets may be faced with highest and best use decisions where many different potential uses are competing for land near the edge of a growing town. Analysis of the fundamental forces of supply and demand in the marketplace can be very useful in determining which uses are financially feasible and which one use is maximally productive.

Step 1: Property Productivity

The typical rural appraisal report includes sections with descriptive details about the subject property's physical characteristics, legal

1. A full description of Level C fundamental analysis is outside the scope of this publication. For more information, see Stephen F. Fanning, *Market Analysis for Real Estate*, 2nd ed. (Chicago: Appraisal Institute, 2014) for a comprehensive discussion of the subject and examples of market analysis applications.

characteristics, and location. Property productivity analysis takes the analysis one step further. For example, consider the property productivity analysis of an agricultural property. The analysis should not be just descriptive. Rather, the analysis should investigate if the land is being used to its full capacity and if the improvements are best suited for the use.

Some pertinent questions related to the property productivity of an agricultural property are

- What kind of agricultural uses could the land support?
- Are the improvements best suited for the use?
- What are the legal issues that affect the potential uses?
- How are the nearby properties being used?
- Is the subject property located in the path of community growth?

The answers to questions like these form an integral part of productivity analysis. A major purpose of site and improvement analyses is to identify the attributes that motivate people to pay for the right to occupy and use a property.

Step 2: Delineating Rural Market Areas

In market area delineation, appraisers consider the potential uses of the subject property and identify a market for the defined use (or more than one market if the property has alternative uses) and the most likely user of that property type. Appraisers of rural land should understand the links between the local rural economy, the market area's economic base (agricultural, extractive, or recreational), and the national economy as well as the significance of the encroachment of suburban and urban land uses on rural land.

To analyze the influence of supply and demand factors on the value of a property, appraisers first delineate the boundaries for the area of study. Most of the boundaries of real property markets are physically observable, and areas of influence vary in size. For rural and agricultural properties, market areas are generally characterized by common crop or livestock uses or by similar demographic and economic characteristics. Properties within the immediate vicinity of the subject property should be identified and analyzed in the market analysis.

The general analysis usually begins with the subject property and extends outward, identifying all relevant actual and potential factors that may influence the value of the subject property. When the boundaries are properly established, conclusions regarding the effects of the market area on the value of the subject property are significant. Establishing boundaries allows an appraiser to identify the primary area from which other market data can be obtained so that appraisers and the intended users of the analysis have the best information available to explain present or future trends or changes in land use.

Some rural properties, such as timberland and larger cattle ranches, are not as subject to immediate market influences as other properties, such as agricultural properties on which the products that are grown are eventually bought in local markets rather than shipped to other areas. However, the vast number of rural properties must be considered in terms of the market that they are a part of.

For agricultural properties, market area boundaries are typically established by analyzing a combination of the following:

1. Property rights and user type
2. Land use
3. Terrain, topography, soils, and rangeland characteristics
4. Climatic and rainfall patterns
5. Governmental regulatory and legal attributes of the market area

The potential users of property and where they come from must be determined before the quantity of demand for the property can be measured. Identification of potential users is the first part of overall market delineation. Delineating the subject property's specific user provides a key link between the property productivity analysis and the demand factors for the property. The user type is helpful is determining the data that will be most useful in developing an opinion of value. The sales comparison approach will be most useful if most properties are sold to owner occupants. In some cropland areas, investors are the primary purchasers and owners, while tenants are the users. The income capitalization approach may be the most applicable approach in these markets.

Land use is a significant factor in establishing boundaries because market areas tend to exhibit similar characteristics. Land use, which is observable and easily understood, must be considered in establishing boundaries. Examples of land use would be dry cropland, irrigated cropland, permanent plantings, open improved pasture, native pasture, native rangeland, timberland, and Conservation Reserve Program (CRP) land. Land uses can be delineated further by referencing Farm Service Agency maps.

Market areas are usually characterized by only one or two major crops. Livestock areas are generally thought of as ranch properties, but livestock are integrated into the crop system in many areas. An example of a livestock area would be the dairy area in California. In ranch country, there may be sheep and goat areas and separate cattle areas. In some areas, multiple species are grazed together. Large confinement livestock and poultry operations are usually located in rural areas away from population centers.

The importance of different value influences in an agricultural market area depends on what is produced there. Agricultural production areas are served by highways and roads that lead to marketing centers where farm products are sold, and proximity to other forms of transportation such as rivers and railroad lines can be even more important in some areas. Like an urban neighborhood, the farm community depends on government services such as roads and schools and on the availability of power sources (e.g., electricity, natural gas).

Infrastructure to support the particular land use that is dominant in a district is important to all districts, but it is particularly important to agricultural districts. The infrastructure for agriculture includes

- equipment sales and repair
- livestock auctions
- veterinary services
- outlets for seed, feed, fertilizer, herbicides, and similar products
- processors or intermediaries to buy, sell, or transport farm products
- government offices such as the Farm Service Agency and Natural Resources Conservation Service
- colleges and universities with agricultural studies programs

Terrain, topography, soils, and rangeland characteristics are readily identifiable by visual inspection and map analysis. These physical characteristics are apparent even to the casual observer. A market area may have several types of terrain and topography or may be more homogeneous. Terrain and topography generally play a major role in determining market area boundaries. Soils and rangeland characteristics tend to be homogeneous in rural market areas.

Climatic conditions often relate to terrain and topography. In high elevations and coastal regions, diverse microclimates may be found within a single market area. Frequently, rainfall patterns in coastal areas dictate the boundaries of irrigated cropland and nonirrigated cropland because the amount of rainfall can decrease rapidly as distance from the coast increases.

Typically, no single factor will circumscribe the boundaries of a market area, so appraisers are compelled to consider all of the factors described above as well as any other factors that are appropriate in the area of study. Once the boundaries are established, the market area can be analyzed in terms of demand, supply, and market conditions.

Proper market analysis is critical to the valuation of rural or agricultural properties because most agricultural properties are used to produce a commodity or service that is marketed outside the local area but within a larger region (e.g., nationally or even internationally). Market area boundaries typically encompass one or more counties and, in some instances, may encompass a significant portion of a state or province. The submarket area for a rural property typically includes only a portion of a county or counties.

Submarkets such as microclimate areas or small communities within the larger rural market area may be used to analyze the degree of influence that the local economy has on the subject property. In some cases, such as the fox hunting areas of southern Tennessee, the smaller submarket may have an increased demand and increased property values because the surrounding farms have similar uses, and landowners have agreements to use each other's property during their hunts.

Governmental considerations affecting rural land include factors such as the type and organization of state, county, and local governments, quasigovernmental agencies, and other governing bodies as well as laws, regulations, tax policies, police powers (such as zoning), environmental regulations, USDA commodity programs, and conservation program requirements. These considerations can be quite significant, and appraisers should become familiar with governmental factors operating in the relevant market area. The individual factors typically given the most attention in this portion of the analysis include

- USDA regulations relating to commodity subsidy programs administered by the Farm Service Agency
- USDA regulations regarding conservation compliance programs
- Wetlands regulations and requirements imposed by governmental entities
- Regulations regarding grazing rights on federal or state lands and current fee structures
- Environmental regulations regarding soil and water contamination, agricultural chemicals, air quality, and mineral production activities
- The quality of police and fire protection
- Regulations regarding interest rates and the availability of money for mortgages and operations
- Floodplain designations and attendant restrictions on use and development
- Current laws and regulations governing the domestic, livestock, and irrigation uses of water
- Federal, state, and county game and fish laws that may affect recreational uses
- Eminent domain laws unique to the jurisdiction of the market study

Government regulations have an effect on the value of property, and there appears to be a trend toward increasing governmental regulation in the future. The areas of environmental, wetland, and

Figure 10.1 Market Area and Submarket Boundaries

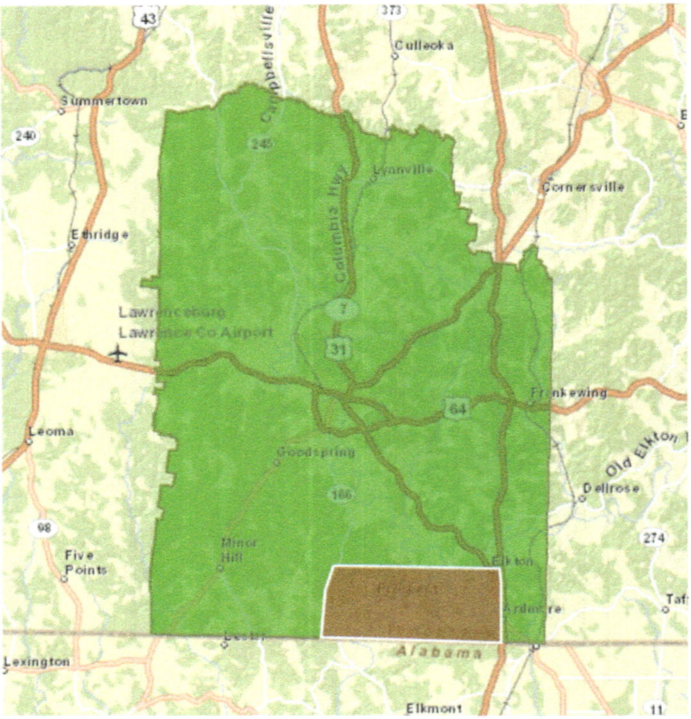

Note: This exhibit was developed using the STDB.com website. It includes a submarket area that could be compared to the larger market area to examine the differences in market conditions.

chemical regulation are growing and changing quickly. Zoning and development ordinances and restrictions are also increasing.

The sources of data on governmental regulation are many and varied. Some of the required data is available at the local level from the appropriate regulatory agencies or from local deed records, while other data must be obtained from the state regulatory agencies charged with administering the regulations. Still other data must be obtained at the federal level. A great deal of data on regulatory requirements is now available online at little or no cost. Appraisers must be aware of any current government regulations that could influence the value of the subject property.

Step 3: Demand Analysis

In its broadest sense, demand serves as a measure of the needs, material desires, and purchasing power of consumers. Demand for rural properties can be influenced by the demographics in the market area. In many rural markets, families who desire privacy and want to grow their own food and raise their own livestock may compete directly with local farmers who make their living producing crops or livestock as a business. When the local market includes local farmers, prices are influenced by the purchasing power of these families. In these cases, population growth and household income can be analyzed and used in the sales comparison approach to compare two properties from different market areas or submarkets.

The personal and social needs of the people who make up a rural community are essentially the same as those evident in other types of communities, and the degree to which these needs are met affects the community's property values. The characteristics of the population often dictate the principal concerns of that population, which may include access to schools for areas with a large number of young families or the availability of off-farm employment in areas where additional jobs are needed. Information about the amenities that make a property desirable for particular buyers is not always quantifiable, but it should be considered in the valuation process, e.g., as an adjustment or in the qualitative analysis of the market condition trends in the application of the sales comparison approach.

Most agricultural uses cannot be directly tied to a demographic growth indicator. Land prices for the production of crops, livestock, and timber are influenced more by anticipated changes in the commodity markets. Small downturns or increases may not significantly affect property prices in the short term, but long-term changes in the market can affect prices.

National and international economic trends affect rural properties in many ways. The state of the US economy influences not only the amount of food products purchased but also the types. High demand for one product over another will filter back to the value of the land that produces the preferred product. Despite government programs that attempt to stabilize prices for agricultural commodities, any sustained downtrend in the economy can have decided effects on rural property value. However, the effects of economic forces are cumulative, and many years of sustained activity may elapse before any specific results occur. A change in government policy that creates favorable export conditions for agricultural commodities will not immediately cause changes in land values. However, if these conditions are sustained, the gradual increase in agricultural productivity and agricultural income will eventually result in higher land values.

Although it is not always possible to quantify the effect of national and international economic trends on rural property value, knowledge of, and sensitivity to, these broad influences will help appraisers understand and respond to changes in the attitudes and behavior of buyers who are responding to the same trends. The anticipation of profit constitutes the essential motivation for purchasing property, so any economic trend that affects a buyer's anticipation of profit will ultimately affect property value.

Changes in supply are fairly obvious because new construction and demolition activity is usually easily observed. Changes in the demand for real property, however, are subtle but can often be anticipated by observant appraisers.

As an economic factor, a change in the demand for agricultural commodities is one of the primary drivers of rural property prices. Changes in consumer preferences in other geographical areas can have significant effects on agriculture too. For example, as consumers have changed their preferences from print media and storage to digital media, the switch has significantly affected the timber pulpwood industry. The decrease in demand has resulted in the closure of sawmills and a drop in property prices for timberland that served the paper industry.

Step 4: Supply Analysis

In the survey and forecast of competition, appraisers analyze the existing and anticipated supply of the property type under investigation. Most appraisers of single-family residential real estate define the existing supply as the properties that are currently available for sale or rent. These are the properties that will most likely affect the decisions of potential buyers considering the subject property. A rural market functions more like the single-family residential market than the urban commercial market. Rural properties are typically held for long periods of time. Many are passed down through generations of family ownership. Competitive properties include all properties–vacant land or improved property–that are actively competing for buyers and renters.

Data on supply in a market may be gathered in various ways:

- field inspection
- review of building permits (issued and acted upon), plat maps, and surveys of competitive sites
- interviews with developers and planners
- interviews with real estate agents and auctioneers

Auctions can be a valuable source of information. In many rural markets, these sales represent market value. Land is typically offered in multiple parcels, and the buyers choose how they are assembled by their bidding. This information can assist appraisers in determining the highest and best use. Auctions can be a valuable source of information that is not readily available through most multiple listing services.

The following characteristics of competitive properties should also be considered:

- availability and price of vacant land
- potential uses of the land available
- construction quality and condition of the improvements found on the available properties
- special features found on the available properties
- availability of construction loans and financing
- costs of construction and development
- owner occupancy versus tenant occupancy
- causes and number of vacancies

- conversions to alternative uses
- effect of building codes, zoning ordinances, and other regulations on construction volume and cost

This information is not only useful in determining the overall health of the real estate market but is also vital to determining the marketability of the subject property.

For urban properties, the existing supply can be correlated with demographics statistics to develop a marginal demand study. Unfortunately, most rural properties cannot be analyzed in the same way. Inferred analysis, which was discussed earlier in this chapter, is often the best tool to indicate changes in supply for rural appraisers.

Step 5: Market Condition Analysis

Rural markets are typically characterized by stable market conditions. Supply and demand remain stable for long periods of time. Major fluctuations are a result of major market changes such as the discovery of new natural resources, manufacturers moving in or out of an area, or new uses for local agricultural products that increase demand. For example, the United States saw an increase in corn prices during the early 2000s due to increases in ethanol production, and land prices in corn-producing areas rose as a result.

In typical rural market areas, appraisers do not have the data sources that may be available in urban market areas. Sales and listings of similar tracts may be few and far between. Therefore, it can be much more difficult to determine how the local economy is performing.

Large commercial agricultural properties are influenced by national and world commodity prices and the stock market. Smaller, family-owned farms and commercial properties in rural areas tend to follow local economic conditions more closely. The values of both large and small properties are determined by the market participants. Local users and purchasers rely on local market conditions. Institutional investors look at commodity prices to determine the long-term potential for their investments.

In the analysis of rural property, economic considerations relate primarily to the financial and business characteristics of an area. The factors of greatest significance are those that relate to the present or future market mechanisms that will affect agricultural production and the delivery of services to the properties in the area. It is critical that appraisers consider the typical operating methodology relative to the fee simple or leased fee ownership of properties. Trends in the prices of property sales and leases and any evidence of changes in production patterns or property use should also be considered.

Appraisers should examine economic trends over a three- to five-year period. Published data on agricultural production by state and county is available online from the Farm Services Agency through the National Agricultural Statistics Service. Demographic information from the US Census Bureau and economic data from numerous governmental and private sources are available in print and online. With this data, appraisers can decide which economic variables contribute significantly to value differences within the region and how these differences influence the value of the subject property.

Land use changes are a major indicator that market conditions may also be changing. If a new producer of a specialty crop enters a market area, that change in supply most likely will not affect the property values in the overall market. However, if multiple producers start to compete for the same properties in the market, then the market value will most likely increase significantly.

Changes in technology and the methodology for growing crops can significantly affect land use in certain areas. Improved methods of irrigation have changed the use of land that once could not produce crops. Laser leveling and land forming have also become common in certain markets.

The construction of new agricultural improvements in a market area can provide rural appraisers with data to analyze the functional obsolescence of older improvements. Vacant and abandoned buildings can also be used to spot changes in user preferences.

Rural appraisers should be also aware of local announcements that may change the existing market conditions. For example, suppose that an appraiser reads in the local newspaper that a large manufacturing plant is closing, and 400 direct jobs as well as 500 jobs at suppliers will be lost. This plant is located in a town with a population of 25,000 people. Table 10.1 lists some possible repercussions of the closure that the appraiser might consider.

The overall health of the property trends in the local market area may follow the single-family market trends. Figure 10.2 provides data on the median sale prices of single-family residential real estate and the number of sales per month. An appraiser could use this data to make inferences about demand for all types of real estate in the market. In this example, the trend seems to be a steady improvement. Based on an impression for the market rather than precisely quantifiable evidence, the appraiser might make judgments about the duration of this upward cycle and the effect on the market conditions for the subject property. (This is an example of Level A inferred analysis, as discussed earlier in the chapter.)

Secondary Data Sources

A common source of data for market analysis is regularly published, area-wide market surveys prepared by private firms or public agencies. The data in these surveys is updated at regular intervals,

Table 10.1 Anticipated Influences of a Manufacturing Plant Closure on Supply and Demand

Loss in demand for housing	Assuming that 900 jobs are lost, 700 of those workers are too young to retire, and 75% of them live in the subject property's market area, there could be about 500 extra homes for sale because 500 families will have to seek employment in other areas. This assumes that only one person per household works in this industry, which may not be true. Of course, some former factory workers will try to find local jobs in other industries, and a few will be successful in their search. Some will commute to other areas for jobs and remain in the same housing market. Some will draw unemployment for a year or two, so this oversupply may not be too bad in the beginning. Many, especially those who work in the skilled trades and management, will have to move to find jobs.
Loss in demand for retail space	A loss of as many as 500 families with 2.5 persons per household would equal a loss in population of 1,250 people. This would not decrease supply but would decrease demand because the stores in the retail space would have fewer customers, and some stores would not survive this downturn.
Loss in demand for office space	A loss in population means a loss in demand for attorneys, accountants, doctors, dentists, insurance and real estate agents, and, of course, real estate appraisers.
Loss in demand for industrial buildings	One large facility and many small ones will no longer be needed. Absent any new demand and no reduction in supply, the closing of the plant should result in an oversupplied market. Owners of companies building industrial buildings would probably need to shift their focus to other trade areas.
Loss in demand for rural properties	Many of the plant's employees may work on family farms or own rural homesteads. They may not be able to keep up the existing mortgage payments because of the loss of a job. As a result, the demand will not only decrease, but the supply may also increase.

Figure 10.2 Median House Prices and Number of Sales per Year

Source: Realtracs.com, chart by Tankersley Appraisal

either quarterly or semiannually. However, even periodic updates may lag behind changes in a dynamic market, and data may become obsolete all too soon. These surveys typically cover whole states or counties. Analysts usually make an inference that the subject property's market area or submarket will perform in a manner similar to the historical performance of the properties in the survey's large geographical area.

The use of broad-based market survey data can increase the reliability of the analysis, but sole reliance on this type of data may have its limitations. Surveys may cover many properties that are not competitive with the subject property. Moreover, the historical pattern of the broader market may not reflect the future prospects of the subject property's submarket or the subject property itself. Analysts must evaluate the data used carefully to determine how well it reflects the actual subject property's type and specific submarket.

One of the most useful publications for market analysis is the USDA Agricultural Census, which is published every five years and includes, among other data points, the number of farms by size and type, inventory and values for crops and livestock, and operator characteristics for the country as a whole and for each state.

The National Agricultural Statistics Service conducts the June Area Survey to collect data on crop acreage, land use, grain stocks, farm rents, and the value of land and sales. The Cash Rents Survey provides the basis for county estimates of the cash rent paid for irrigated cropland, nonirrigated cropland, and pasture. The county-level Cash Rents Survey will be conducted biannually in all states, except Alaska, beginning with a release in September 2016. US and state estimates are released in August every year using the June Area Survey. Previously, an annual survey was conducted from 2008 until 2014.

The USDA, state agencies, and state universities also publish other studies and research that can assist rural appraisers. These studies may include projections for crop yields, cost of operation, or best practices. Useful studies are often available online from these sources.

Sample Analysis Using Secondary Sources

As an example of market condition analysis, consider a farm being appraised in Giles County, Tennessee. According to the 2012 Agricultural Census, the total land in farms in the county was 270,089 acres, a 3% increase since 2007. The average farm size is 160 acres, up from 146 acres in 2007. The average value of agricultural products sold per farm in 2012 was $17,075, a 51% increase from $17,917 in 2007. Crop sales account for 37% of the total sales, and livestock sales make up the remaining 63% of the total sales. The county ranks thirtieth in Tennessee for all crop production. It in the top twenty for livestock.

Figure 10.3 Market Data for Giles County, Tennessee

Top Crop Items

1. Forage land used for all hay and haylage, grass silage, and greenchop
2. Corn for grain
3. Soybeans for beans
4. Wheat for grain, all
5. Winter wheat for grain

Top Livestock Inventory Items

1. Layers
2. Cattle and calves
3. Roosters

Figure 10.4 Farms by Size (2012)

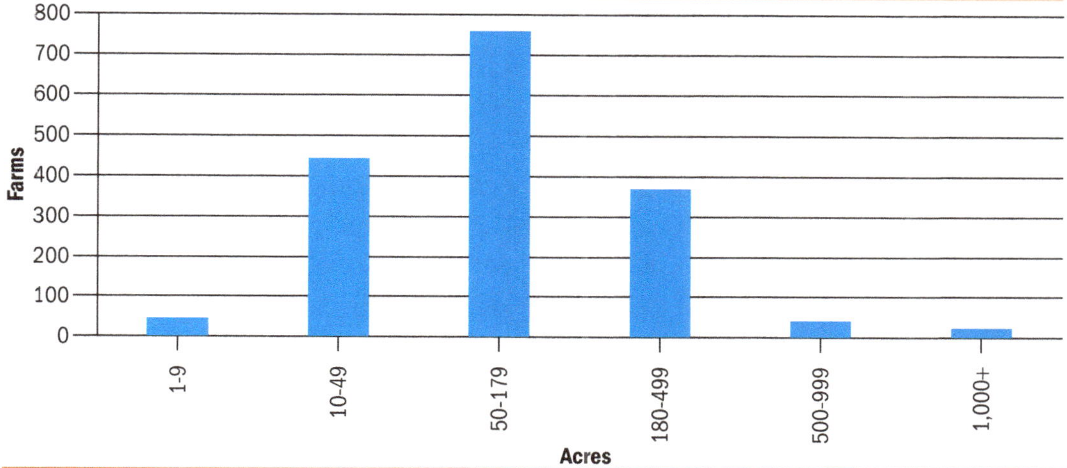

Figure 10.5 Land Use in Farms (2012)

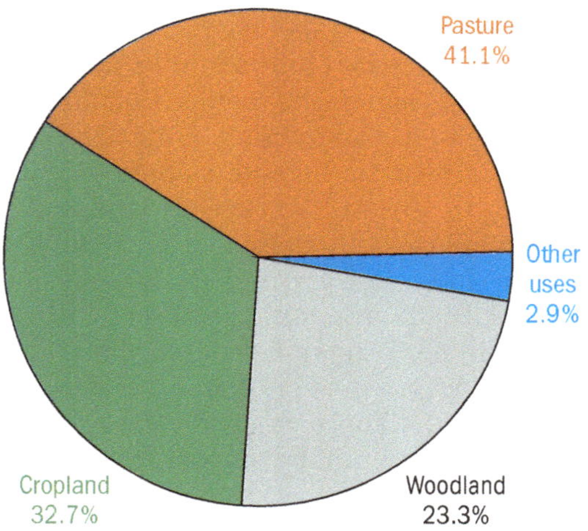

Table 10.2 Tennessee Land Values 2016 Summary (August 2016)

Land Type	Average Price/Acre	Percentage Change (2015-2016)
Crop	$3,600	3.7%
Pasture	$3,540	1.1%

Source: USDA, NASS

Table 10.3 Giles County Cash Rents 2016 Summary (August 2016)

Land Type	Cash Rent/Acre	Percentage Change (2015-2016)
Crop	$77.00	-15%
Pasture	$14.50	1%

Source: USDA, NASS

Market Conditions Conclusions

Inferred analysis of market conditions starts with the investigation of the demand trends that illustrate the current market conditions and direction for the future. Those trends include the following:

- The population is slowly declining.
- Unemployment is decreasing but has historically been higher than the state average.
- Income for the area is well below the state average.

The following local economic indicators were analyzed:

- Single-family home values and the number of sales are increasing.
- Crop land rents are decreasing in the county.
- Pasture land rents are increasing slightly.
- Crop land prices are increasing for the state.
- Pasture land prices are increasing for the state.

The following national economic factor also affects the property value and rents of similar tracts:

- Commodity prices are trending downward.

Most local economic indicators suggest that the market conditions are improving. However, commodity prices have been declining, and no recovery is anticipated in the short run. Local land rents are declining, which may be a leading indicator that land prices may decline in the future.

Step 6: Subject Marketability

In a market study, appraisers do not focus on a specific property, so for most valuation assignments a more detailed marketability study is necessary. In fact, professional appraisal standards require that all appraisal assignments that involve developing an opinion of the market value of a specific property must include a marketability study.

A subject property is affected by the overall market demand and dynamics, but any one property can perform better or worse than the overall market. Marketability analysis explains why this happens.

A marketability conclusion is developed so that appraisers can explain why the subject property will perform at, below, or above market expectations. These conclusions are used to make decisions about data selection and how to apply the approaches to value.

Professional appraisal standards require that exposure time be determined for most market value assignments. Some appraisal clients have additional requirements that require an estimate of marketing time. Marketing time and exposure time are a direct result of property marketability and should be addressed in this section.

If the subject property is one of the best in the market area, it will likely have a shorter-than-average exposure time and marketing time. (This conclusion is typical of an inferred analysis.) If changes to the local economy are anticipated, the marketing time could be longer or shorter than the exposure time.

exposure time. The estimated length of time that the property interest being appraised would have been offered on the market prior to the hypothetical consummation of a sale at market value on the effective date of the appraisal. Comment: Exposure time is a retrospective opinon based on an analysis of past events assuming a competitive and open market.

marketing time. An opinion of the amount of time it might take to sell a real or personal property interest at the concluded market value or at a benchmark price during the period immediately after the effective date of an appraisal. Marketing time is a forecast that is made looking forward from the effective date. Marketing time differs from exposure time, which is always presumed to precede the effective date of appraisal.

Source: Uniform Standards of Professional Appraisal Practice, 2016-2017 ed., Definitions and Advisory Opinion 7: Marketing Time Opinions

All of the characteristics of a rural market area may not be quantifiable, but they should be considered by appraisers in the estimation of property value, particularly in the application of the sales comparison approach. Reference to this type of data often forms part of the narrative of an appraisal report. In reconciling value indications, an appraiser may indicate a value at the higher end of a range because of the property's location in an area where the roads are primarily hard-surfaced as opposed to gravel-surfaced, or where the property is significantly closer to a large shopping center than a comparable property.

Figures 10.6 and 10.7 illustrate the conclusion of inferred analysis from the analysis shown earlier in the chapter. The data shown in Figure 10.7 has been used to analyze the exposure time for active listings and closed sales. Many properties are marketed for more than one year in this market area. Based on this data, the exposure time for the subject property is twelve months. Marketing time is estimated to be the same because no changes in the supply or demand are anticipated in the near future.

Figure 10.6 Subject Marketability Conclusions Using Inferred Methods

Demand Conclusions:	Declining growth trends.
Supply Conclusions:	Very few similar properties available for purchase or rent.
Current Market Conditions:	Overall, local market economic conditions are slowly improving. Farmland rents and prices are increasing in the state and nationally. Commodity prices are decreasing.
Subject Marketability:	The subject property has good soils, and 80% is irrigated with a pivot system and water rights from the river on the north side. Like many farms in the area, it does have a section that is prone to flooding. The subject is better than any of the farms available for sale or rent. It should exceed most user expectations and receive higher rents and ultimately sell for a price per acre that falls on the upper end of the market.

Figure 10.7 Exposure and Marketing Time Analysis and Conclusions

Data Source:	Middle Tennessee Realtor Multiple Listing System	
Search Area:	Giles County	
Time Period:	Past three years	
Property Type:	200+ acres vacant and improved	
Active Listings:	Average DOM	184
	Median DOM	163
Closed Sales:	Average DOM	290
	Median DOM	180

Market Analysis Conclusions and Highest and Best Use Analysis

In summary, the market determines the use and the use affects the value. An existing or proposed rural land use may be put to the test of maximum productivity in highest and best use analysis only after it has been demonstrated that an appropriate level of market support exists for that use.

Chapter 11

Highest and Best Use Analysis

The factors that are considered and evaluated by market participants can generally be described in relation to the use of the property in question. The analysis of use draws on all the concepts, principles, forces, and factors that have been introduced and discussed in previous chapters. Highest and best use analysis is performed to develop a complete understanding of market behavior as it relates to a specific property, a group of properties, or an area.

Analyzing the highest and best use of a rural property may be as simple as understanding what is occurring in a given market and then reporting those observations, or it may require a very specific and detailed study of the market area and all of the forces influencing it. The primary determinants of the scope and depth of the analysis required are the characteristics of the market area and of the property being valued. A feasibility study of an agricultural tract in a transitional or suburban market may require a significant amount of time and analysis, while estimating the use of a tract of pasture in an area dominated by similar uses may be much simpler.

In considering the property as if it were vacant and as currently improved, a valuer must understand and clearly state in the report what the term *improved* means. In a typical appraisal of a rural property, the improvements to be considered may include dwelling units, machinery and equipment storage, grain bins, feed storage units, barns, poultry facilities, or livestock housing. Fencing, corrals, irrigation systems that are permanently affixed to the property, drainage tile, and drainage systems are considered to be improvements in some markets. It is incumbent on valuers to explain clearly in the highest and best use analysis what is and is not considered to be an

Consistent Use

The principle of consistent use holds that land cannot be valued based on one use while improvements are valued based on another. For example, a property might be improved with dairy facilities, but the site could not be valued as cropland instead of the pasture that would accompany the dairy operation. The land would be more valuable as cropland than as pasture, but the cropland use is not consistent with the improved use of the property as a dairy farm.

An improved site is always valued as though vacant and available for its highest and best use. Existing improvements that do not conform with the ideal improvement may reflect an interim use (i.e., not the highest and best use) that contributes some value, contributes no value, or even reduces value if the costs to remove the improvements are substantial.

improvement to the property (i.e., to differentiate between real property and personal property).

Competitive economic and social forces in the market where the property is located must be considered to determine highest and best use. Data collection was discussed in Chapter 5. Both macro-level and micro-level data are used in highest and best use analysis. A valuer must fully analyze the data collected and understand how the data relate to commodity markets, social and economic trends, and property uses prevalent in the market area. Trends in factors that influence use range from supply and demand levels, product prices, and expense levels to changing demographics, local land use regulations, and ongoing changes in environmental legislation. The analysis of these factors will help the valuer determine the ultimate highest and best use of the property being valued. If market factors, principles, and forces form the foundation of an appraisal, the highest and best use analysis can be considered the cornerstone.

In considering a property's highest and best use, the motivations of the buyers who are participating in the market must be addressed. Purchasers of income-producing rural land, including farms and ranches, are much like investors in commercial, industrial, and retail properties–they are usually motivated by the anticipation of future monetary benefits. They are primarily interested in the feasibility of the farming or ranching enterprise that is being purchased.

Highest and Best Use Criteria

Four questions are asked in the analysis of highest and best use, and they establish the criteria or tests that must be met. A highest and best use must be legally permissible, physically possible, financially feasible, and maximally productive. The six-step market analysis process described in the previous chapter provides the data required for the four tests. The criteria are usually considered sequentially, but they are interrelated and may be considered in concert. Logic dictates that the considerations of legal permissibility and physical possibility precede consideration of financial feasibility and maximal productivity. The fact that a property use is financially feasible and maximally productive is irrelevant if the use is either illegal or not physically possible.

If the property zoning prohibits a potential use, it is pointless to consider that use unless there is a reasonable probability that the zoning can be changed. Physical considerations can be overcome by the addition of capital, but the economic impact of adding the capital must be evaluated and the incremental return from the added investment must be carefully measured.

Legally Permissible

A valuer does not need to address all of the legally permitted uses of a property, but it is inappropriate to proceed with a highest and best use analysis if the implied or proposed use is illegal. Greater land use regulation at the local, state, regional, and federal levels in recent years increases the need to address legal permissibility. Local zoning and regional comprehensive land use planning regulations are being implemented in many areas.

Federal, state, and sometimes local regulations control the use and protection of wetlands, highly erodible land, floodplains, and other special areas. Strict enforcement of these regulations can have a significant impact on the potential uses of affected land. Federal and state legislation that is not specifically labeled as land use legislation may also affect property use. Controls that affect the environment such as the Clean Air Act, water quality legislation, and the Endangered Species Act can influence land use in areas surrounding those areas on which the laws primarily focus. For example, the value of timberland in areas surrounding the Northern California habitat of the spotted owl has been impacted by more restrictive timber-cutting regulations. A valuer must clearly document the influence of legal factors in the highest and best use analysis. If valuers feel that they lack the competency to understand or interpret these factors, professional assistance should be obtained.

A title search for deed restrictions and covenants should be made to ensure that there are no legal restrictions on property use in the chain of title. The valuer need not consider whether a use is physically possible, financially feasible, or maximally productive if it is prohibited by a deed restriction or covenant. There may be conservation easements placed on the property or restrictions on development; sometimes the development rights may have been severed from the bundle of rights by a previous owner. The title search will also indicate whether subsurface (mineral) rights have been transferred.

Some common restrictions and covenants applicable to rural property include the following:

- Transfer of development rights to a local, state, or quasi-public agency. This type of conveyance may be found when analyzing rural property in an urban region.
- Life estates. These interests, which typically relate to improved properties such as a dwelling or a farm, are held for a time period that is indefinite on the date the restriction is granted.

- Conservation easements. Easements may be placed on property by a landowner to protect a watershed, a wildlife habitat, or a portion of a property. They can be conveyed to municipalities or other governmental agencies, to land trusts, or to wildlife agencies such as the Audubon Society. Other easements may involve pipelines or transmission towers, among other uses.
- Restrictive covenants. Covenants can be used to prohibit subdivision for a specified period of time, regulate property use, or limit the size, shape, or configuration of a dwelling.

Physically Possible

The utility of real property is affected by its physical characteristics, and this is especially true for rural and agricultural property. Access, drainage, soils, size, shape, terrain, frontage, wetlands acreage, topography, field size, and field configuration are characteristics that must be considered in the determination of highest and best use. The property's characteristics will control the selection and consideration of potential uses.

In the valuation of a specific specialized land use, the property with the most desirable characteristics or combination of desirable characteristics, such as appropriate soils, water, and climate, will be the most physically desirable. Areas of central California and southern Arizona have desirable characteristics, including the appropriate soils, water, and climate for raising a number of crops ranging from alfalfa to citrus crops to vegetables. Some of these crops can be grown nearly year-round. Other areas of the United States have the characteristics necessary for raising other crops or for conducting other agricultural enterprises that are not suitable in the Southwest. For example, the production of cranberries and other fruits, nuts, timber, and specialty crops like sugar cane or tobacco is conducted in other various regions in the United States.

The improvements on a property can be considered part of the tract's physical characteristics. The size, design, utility, and condition of the improvements must be evaluated. If a property is improved with dairy barns that are 30 years old and have stanchion housing and no pipeline milking system, their contribution to the property is likely to be less than that of a modern, free-stall dairy barn with a milking parlor. At times, however, the social forces operating in the area must also be considered. In areas where the farming community is dominated by sectarian groups (such as the Amish in Lancaster County, Pennsylvania), the preferred improvement may be an older stanchion barn rather than the newer complex.

Some physical restrictions can be cured. Land that tends to be wet can be drained, and hedgerows can be removed with heavy equipment. Laser leveling can improve the use of water and make farming on a particular tract easier. The cost to cure a deficiency needs to be recognized if correction is to be undertaken. A valuer cannot assume that fields can be leveled, drained, or otherwise improved to attain a higher use without considering the other three criteria of highest and best use–i.e., the cure must be legally permissible, financially feasible, and maximally productive. An economic analysis may be required to evaluate how a physical restriction can be cured.

Financially Feasible

Considering the criteria of legal permissibility and physical possibility generally results in the elimination of some potential uses. According to appraisal theory, the uses that are legally permissible and physically possible are then tested by applying the remaining two criteria. In conducting actual highest and best use analyses, valuers typically spend more time evaluating the financial feasibility of property uses than the other three criteria.

Maximally Productive

The potential use that satisfies the previous three criteria and results in the highest land value is determined to be the property's highest and best use. The potential highest and best uses considered are generally long-term uses that are expected to be appropriate throughout the economic life of any improvements. The nature of the improvements will tend to dictate the appropriate economic life. For example, specialized livestock housing such as cage-laying poultry buildings typically have a relatively short economic life, which is dictated by the need to periodically replace major components such as cages, feeders, waterers, and egg collection and grading equipment. Technological changes and environmental influences both tend to limit the economic lives of poultry facilities. In contrast, most machinery storage buildings, shop buildings, general utility buildings, and dwellings have relatively long economic lives. As discussed in Chapter 15, economic life can be measured in a number of ways.

Highest and Best Use of Land as though Vacant

Land is valued as if it were vacant. Implicit in the analysis of the highest and best use of the land as though vacant is the assumption that the land is vacant or can be made vacant through the demolition of building improvements that have reached the end of their economic lives. Vacant land is a fundamental concept in valuation theory and serves as a basis and preliminary consideration in the cost approach.

It is not unusual, especially in transitional and suburban markets, to determine that changes are occurring in the market and the uses being considered are temporary in nature or may not be realized until some time in the future. For example, in markets that are adjacent to or surround metropolitan areas, land may have a highest and best use for farming purposes in the near term and an ultimate highest and best use for development as a residential subdivision. In this instance, the property has both an interim and a terminal, or ultimate, highest and best use. An *interim use* is an activity that takes place on a site during a transitional period until the terminal highest and best use is realized. The *terminal* or *ultimate use* is the likely "end use" of the property, which is expected to be realized at the end of the transition period.

Example 1: Highest and Best Use of a Property as though Vacant—Home Site or Agriculture

Consider a 250-acre property suited to dryland crop production that could also be used as a rural home with excess acreage. Use for crop production would require the construction of a machinery storage building at an estimated cost of $125,000 and would result in an estimated market value of $215,000. If the property is to be used as the site of a rural residence, a dwelling would have to be constructed at an estimated cost of $210,000 and the property would have an estimated market value of $285,000.

The estimated costs of constructing the improvements on the property for the different scenarios and the market values of the two uses can be compared in the analysis of highest and best use. The calculations used in the analysis are shown in Table 11.1. Use for agricultural production is suggested by the analysis (if alternative uses have been eliminated based on other criteria). This use results in the highest residual land value. The residual values are $300 per acre for the home site and $360 per acre for the agricultural use.

Table 11.1 Different Types of Land Use

	Home Site	Agricultural Production
Market value	$285,000	$215,000
Cost to construct improvements	– 210,000	– 125,000
Residual land value	$75,000	$90,000

Example 2: Highest and Best Use of a Property as though Vacant—Various Crops

In the highest and best use analysis of an income-producing property, the competing uses are generally considered in terms of their ability to generate positive net operating income, the potential rates of return, and the perceived stability of the income. For example, consider an agricultural property consisting of approximately 100 irrigated acres that can be used to raise more than one crop. Analysis has shown that the property is currently used to produce corn and soybeans in rotation, and it generates a stabilized income of $65 per acre per year. It could also be used to raise vegetables, which would produce a stabilized income of $75 per acre per year. Finally, the land is suited to the production of specialized Chinese vegetables and small fruit such as strawberries in rotation; this use would result in a stabilized net income of $100 per acre per year. The overall rates of return for the three alternatives reflect their relative risk:

- 7.5% for corn and soybeans
- 8.0% for vegetables
- 10.5% for Chinese vegetables and strawberries

The potential net operating incomes, overall rates of return, and indicated land values for the three options are developed in Table 11.2. Of the three possible uses for the subject property, raising Chinese vegetables in rotation with strawberries results in the highest residual land value for the property. Thus, it is determined to be the highest and best use.

Table 11.2 Different Types of Crops

	Potential *NOI*	Overall Rate	Capitalized Value
Corn and soybeans	$6,500	7.5%	$86,667
Vegetables	$7,500	8.0%	$93,750
Chinese vegetables	$10,000	10.5%	$95,238

The Ideal Improvement

If an appraiser concludes that a building improvement is appropriate for the highest and best use of a parcel of vacant land, the appraiser then identifies and describes the type and characteristics of the ideal improvement to be constructed. The ideal improvement should meet the following criteria:

- It takes maximum advantage of the potential market demand for the site's highest and best use.
- It conforms to current market standards and the character of the market area.
- It contains the most suitably priced components.

If a new improvement is considered capable of supporting the highest and best use of the land as though vacant, it presumably will have no physical deterioration or obsolescence—meaning that it would be neither an underimprovement nor an overimprovement. Any difference in value between the existing improvement and the ideal improvement would be attributable to depreciation. Thus, the ideal improvement identified in highest and best use analysis helps the appraiser estimate depreciation in the application of the cost approach.

An appraiser's conclusion as to the ideal improvement should be as specific as the market suggests, such as to the level of estimating the number of bushels or the throughput capacity. The market might recognize the use of the ideal improvement of a particular site as a grain elevator or a cotton gin, for example. Specificity of the ideal improvement affects the choice of comparable properties to be analyzed in the application of the approaches to value.

Highest and Best Use of a Property as Improved

In general, the highest and best use of a property as improved is tested for three types of alternative uses: (1) demolition of the existing improvements and redevelopment of the site, (2) alterations and other modifications to the existing improvements, or (3) continued use of the existing improvements for their current purpose. It is necessary to identify the property use that can be expected to produce the greatest overall return for the investment made. Most rural property is purchased with the expectation of generating income for the owner, so it is reasonable to assume that the potential buyer is economically motivated and informed. If a property is currently being used as a dairy, for example, will this produce the maximum potential return? Would the return increase or decline if the improvements were remodeled and converted to horse stalls to conform to the changing demands of the surrounding suburban market? The market value of the property would likely differ under these two scenarios, and the use that results in the higher present value will be determined to be the highest and best use as improved.

An important reason to determine the highest and best use of a property as improved is to assist in the identification and selection of comparable sales to be used in the application of the sales comparison approach, and comparable rental properties to be used in the application of the income capitalization approach. The highest and best use of the vacant land sales and improved property sales used for comparison should be the same or similar to the highest and best use of the property being valued.

All four tests of highest and best use are relevant to the analysis of the property as improved. It is self-evident that the current use of a property as improved is physically possible, and the legal permissibility of the current use is often nearly as obvious and easy to confirm. But a valuer needs to test whether the existing improvements contribute value, rather than simply assume that the current use is the highest and best use because the improvements are already in place.

Example 3: Highest and Best Use of a Property as Improved—Dairy Farm

An analysis of the highest and best use of a property as improved may suggest that little or no capital improvements are required. Alternatively, significant expenditures may be required to convert the property to a different use or to remodel or rehabilitate it for its existing use.

Consider a free-stall-style dairy barn and its associated land base, which have been used for milking 150 cows a day. The property has been sold for $240,000, subject to mortgage approval. The dairy, including its supporting buildings, has a capacity of 200 milking cows, but a wastewater treatment facility would be required to operate the property at capacity levels. The estimated cost of constructing this type of facility is about $150,000. Without the wastewater treatment capability, the property cannot be operated at more than 75% of its total capacity–i.e., at a 150-cow level. Analysis of the market reveals that dairy farms similar to the subject sell for about $1,500 per cow without wastewater treatment facilities. Farms with wastewater treatment facilities sell for about $1,800 per cow. In considering the highest and best use of the dairy farm, the valuer must assess the financial feasibility of constructing the necessary improvements to expand the property's capacity. The calculations are shown in Table 11.3. The analysis suggests that the expansion of the dairy to 200-head capacity from the existing 150-head capacity is legally permissible and physically possible, but it is neither financially feasible nor maximally productive.

Table 11.3 Estimated Value of Property without Wastewater Treatment Facility

Estimated value of property with treatment facility	
200 cows @ $1,800 per cow	$360,000
Less cost to install treatment facility	– 150,000
Value of site with water treatment facility	$210,000
Value of site as is	
150 cows @ $1,500 per cow	$225,000

Highest and Best Use Conclusions

Many rural properties are capable of supporting more than one highest and best use. There often are different uses on different tracts or parcels or complementary uses on the same tract. Depending on the market and market forces, multiple uses can interact as (1) augmenting uses, (2) complementary uses, or (3) conflicting uses.

Augmenting uses generally have a synergistic effect on value. The uses seem to work well together and enhance the value of the property. The property value is greater with the combination of uses than the value supported by each individual use. Some examples of augmenting uses are ranches that have scenic amenities and provide recreational opportunities as well as commercial enterprises and crop farms that serve as hunting plantations subsequent to harvest and prior to planting season.

Complementary uses are uses that add to value. The uses work together and result in a total property value that equals, or sometimes exceeds, the value supported by each of the uses. Examples of complementary uses are parcels that include cropland, pasture, and timberlands, plantation properties that have various types of cropland, and timberland that can support timber production as well as hunting.

Conflicting uses can have a negative effect on value. These uses are inherently in conflict with each other and result in a value for the total property, when all uses are considered, that is less than the sum of the parts. An example of conflicting uses would be a residential property in a rural area near a landfill.

A valuer may conclude that the highest and best use of a property is one that will last for a period of time and then change to another use at some point in the future. The use identified for the short term, generally three to seven years, is an interim use. As social, economic, and demographic forces and factors change over time, land use can also change. The areas of change are known as *transitional areas*, and it is in the transitional market that the valuer will likely encounter interim uses.

In some transitional markets, buyers invest in real estate in anticipation of profiting from increasing property prices. In those markets, purchasers and landowners retain their real estate assets in anticipation that real estate price levels will increase significantly. Typically, there is little or no income generated by the property over the holding period. However, if the real estate has agricultural potential and a rental market exists, there may be sufficient income generated to offset the fixed costs of ownership–i.e., real estate taxes and some portion of the interest on borrowed capital. This type of ownership often carries increased risk, especially when the market is comprised of property that has high holding costs (for repairs, maintenance, and similar expenses).

Property that is appropriate for a limited number of uses presents practical problems for valuers. The highest and best use of a special-purpose property as improved is probably the continuation of the current use, if that use is economically viable. For example, the highest and best use of a grain elevator is probably continued use as a grain elevator, unless it is not economically feasible to continue that use.

It may be appropriate to analyze and value a special-purpose property on the basis of two highest and best uses–i.e., continuation of the existing, special-purpose use and conversion to an alternative use. This is an analysis of the property's value in use and the potential property value in exchange. For example, a poultry facility may first be analyzed on the basis of that highest and best use. If the demand for similar poultry units is low, an analysis of a highest and best use as warehouse or storage space may also be completed.

Chapter 12

The Sales Comparison Approach

For rural and urban properties alike, the sales comparison approach is applied in five basic steps:

1. Research the market to obtain information about sales transactions and offerings of properties similar to the subject property.
2. Verify the information obtained in Step 1. Appraisers verify that the data obtained is factually accurate and that the transactions were arm's-length market transactions. Verification is accomplished by obtaining information from credible sources, such as a party to the transaction.
3. Identify relevant units of comparison (such as acres, animal unit, or crop productivity unit) and develop a comparative analysis for each relevant unit.
4. Adjust the sale price of each comparable property, as appropriate, or eliminate the property as a comparable. Compare the subject property and comparable properties on the basis of the relevant elements of comparison, including property rights conveyed, financing terms, conditions of sale, time, location, and physical characteristics.
5. Reconcile multiple value indications derived from the comparables into a single value indication, which can be a point value, a value range, or a relationship to a benchmark value.

In the first step–the research of the market for sales of comparable properties–the major consideration in the selection of properties in rural areas is the degree of their comparability with the subject property and similarity of highest and best use. Appraisers seek sales of comparable properties that are representative of the probable market for the subject property.

In the second step, the process of comparison and adjustment must be based on sales transactions that represent typical interactions within the market where the subject property would be offered and sold. To develop valid market value conclusions, each transaction should be verified with a party involved in the transaction. In the interview process, an appraiser's goal is to obtain the information that the parties used in arriving at the final sale price. Appraisers seek to have the best possible understanding of the thought processes that were involved in the transactions. Appraisers also ask questions that relate to the items of consideration that influenced the sale price. Crops, equipment, livestock, growing costs, and other items of consideration are often included in the purchase price, along with the motivation of the buyer or seller.

Another item to be considered is the production capability of the property, such as the soil classification, yield per acre, or carrying capacity. For special-purpose properties, information relating to the special use–including the operating history and yield data–should be obtained.

Appraisers of rural property typically must compare properties with physical characteristics as well as land and building components that vary significantly. Consequently, it is common practice in rural appraising to reduce the total property to a single unit that is representative of the whole. In the third step of the sales comparison approach, various units of comparison are used in analyzing rural properties, but the unit of comparison most commonly applied to rural properties, particularly farms, is price per acre (such as gross acres, acres net of roads, or acres in production). Applying this unit of comparison allows the total property to be represented by a single acre. When analyzing comparable sales, the total price is converted into the price per acre by dividing the total sale price by the number of acres that make up the property.

A similar procedure is used to derive a unit price for livestock operations, but the value is usually expressed in terms of carrying capacity or feed equivalents. The units of comparison used in these cases are the AU (animal unit) or the AUM (animal unit month). For dairy farms, the unit of comparison may be number of milk cows, number of lock ups, or linear footage of feed lane. For timber properties, a common unit of comparison is thousand board feet (abbreviated as MBF) per acre of merchantable timber. The unit of comparison applied to properties that store or process agricultural commodities could be based on the holding capacity, the amount of the commodity that can be stored or processed, or the number of livestock that can be fed within the facility.

The sale price of some agribusiness properties may include business assets, in which case price per throughput (the price divided by the annual flow of product through the property) may be appropriate.

For instance, sales of grain terminals may be compared using price per bushel throughput as the unit of comparison. Similarly, price per animal sold yearly might be the appropriate unit of comparison for livestock properties, and price per tonnage sold annually may be appropriate for the sales of bulk fertilizer terminals. If buyers and sellers in those rural property markets include business assets, then valuers must pay careful attention to the realty and non-realty components of the property trading hands.

After determining the appropriate unit of comparison, appraisers review the tabulated data to identify which characteristics of the properties and sales transactions to use in applying the sales comparison approach. No matter which unit of comparison is selected, the total property is represented by a single unit. The unit of comparison is used in two ways: to derive a price per unit for all comparable sales and to convert the indicated unit value of the subject property into an estimated total property value.

Unit sales prices that have not been adjusted are generally not applied directly to the subject property. Once a unit of comparison has been applied to all the comparable sales, an appraiser has a range of unit prices that provides a basis for calculating an appropriate unit value for the subject property. To obtain an appropriate unit of value, however, the data for each comparable sale property must be adjusted to reflect differences between it and the subject property.

When properties are located within the same rural region, location does not usually have a major influence on value. There are, however, location-related factors that may differ among properties and make the consideration of location more complex.

Distances from community services, markets, and conveniences are locational features that can affect the marketability and, therefore, the value of a rural property. For example, the distance that a livestock ranch is located from transportation terminals and auction facilities or the distance from a timber property to a lumber mill may affect the property's value. The availability and types of roads that service the area, zoning, and the proximity of utilities may also be considered by potential market buyers. Each of these factors alone might not necessitate a location adjustment but, as in the case of other adjustments, the need for an adjustment is indicated by the market data collected and analyzed in the valuation.

The other most significant element of comparison for rural properties–physical characteristics–requires the most extensive comparison and produces the greatest number of variations to consider. There are two kinds of physical characteristics:

1. Land features and improvements
2. Building improvements

In the appraisal of a farm, land features and improvements may be more significant than building improvements, but all differences between the improvements on the comparable sales and the subject property should be considered.

Depending on the significance and degree of variation in the physical characteristics of the properties and their influence on value, appraisers may choose to deal with each factor separately or in aggregate classifications. For example, soil, water, topography, yields, drainage, elevation, field size, field shape, and hazards could be considered individually or together. When determining how to proceed, appraisers should ascertain the significance of each individual factor and the degree to which the data supports the separate consideration of each.

Identification and Measurement of Adjustments

After the significant differences between the subject property and all comparable sales have been analyzed, appraisers measure these differences in terms of their influence on property value. The validity of the sales comparison approach depends on an adequate data sample representative of the market and comparable to the subject property as well as the ability of an appraiser to measure the differences between the comparable properties and the subject property. Valuers use various analytical techniques to identify and measure adjustments and to compare sales of similar properties. The techniques can be quantitative or qualitative, as illustrated in Figure 12.1.

Comparative analysis is the process in which one technique or a combination of techniques is used to convert each comparable sale price into a separate indication of value for the subject property. The first step in comparative analysis is to identify which elements of comparison affect values in the subject property's market. Each relevant element of comparison is then considered in relation to each comparable sale property to decide if an adjustment is necessary.

Figure 12.1 Comparative Analysis

Quantitative Techniques

- paired data analysis
- statistical analysis
- graphic analysis
- cost approach techniques

Qualitative Techniques

- trend analysis
- relative comparison analysis
- ranking analysis

Quantitative Techniques

Paired data analysis is a process in which two properties are compared to measure the amount of difference between them with respect to a particular element of comparison. For this technique to be accurate, the properties must be similar in every respect except for the element of comparison being measured. When this type of property is not available, dissimilar properties may be used if sufficient market evidence is available to adjust the sales prices for all dissimilar elements beforehand except the one for which an adjustment is sought. However, the more adjustments that have to be made prior to the paired data analysis, the more suspect the results of the analysis become.

Qualitative Techniques

Good comparable data often lends itself to effective qualitative analysis. When the data supports a narrow range of value (even if some data is more reliable and credible than other data), market participants understand well-presented qualitative conclusions as well as or better than quantitative analysis, in which the depth of support for specific adjustment amounts may not be present.

Trend analysis involves the use of large amounts of data. Trend analysis is especially useful for analyzing trends in value levels over time. Total sales activity for regions or states can be entered into a computer database and cataloged according to various market characteristics, including highest and best use.

Relative comparison analysis is a way to study the relationships indicated by the market data without making mathematical adjustments. This technique (sometimes called *bracketing*) is sometimes used in complex real estate markets. To apply this technique, elements of comparison are first identified. The characteristics of comparable properties are then compared with the characteristics of the subject property and ranked as superior, inferior, or equal. This analysis is similar to paired data analysis, except that it does not result in specific dollar or percentage amounts. Bracketing is especially useful when productivity factors (such as price per animal unit, milking cow, sow, or productivity index point) are used as units of comparison.

A simple example of relative comparison analysis taken from a hog facility appraisal is shown as Table 12.1. The unit of comparison is price per sow. All properties are compared to the subject property for the following elements of comparison:

- time (date)
- size (number of sows)
- effective age of facilities
- location

Once the relative comparability of the comparable properties is estimated, the transactions are arrayed with the superior properties above the subject property and the inferior properties below. In this example, the subject property is nicely bracketed between Property C ($1,210 per sow) and Property D ($990 per sow).

Table 12.1 Relative Comparison Analysis with Bracketing

	Subject Property	Property A	Property B	Property C	Property D
Date		Current	Current	Current	Current
Price		$1,584,000	$792,000	$1,452,000	$1,386,000
Number of sows	1,200	1,200	900	1,200	1,400
Price per sow		$1,320	$880	$1,210	$990
Effective age	10 years	8 years	12 years	9 years	11 years
Location	Good	Good	Average	Good	Average
Comparison Grid					
Time		Equal	Equal	Equal	Equal
Size		Equal	Inferior	Equal	Equal
Effective age		Superior	Inferior	Superior	Inferior
Location		Equal	Inferior	Equal	Inferior
Comparability		*Superior*	*Inferior*	*Superior*	*Inferior*
Summary of Analysis	**Price/Sow**	**Comparability**			
Property A	$1,320	Superior			
Property C	$1,210	Superior			
Subject property					
Property D	$990	Inferior			
Property B	$880	Inferior			
Conclude estimated value of subject property: $1,100 per sow					

Bracketing is used when the market data is not sufficient to derive dollar or percentage adjustments. To use this method, appraisers must fully explain the logic employed in selecting the elements of comparison and in determining the relative rankings of superior, inferior, and equal.

Ranking analysis, a variant of relative comparison analysis, focuses on individual property characteristics when adjustment amounts cannot be estimated. To judge the comparability when some factors are inferior and others are superior, a valuer places subjective weights on the factors that are most important in the specific market. For example, suppose that size is considered the most significant element of comparison among market participants in the market for hog facilities illustrated above. The size of Property B is inferior to the subject property, whereas the size of Property D is equal to the subject property. This suggests that Property B is more inferior to the subject property than Property D, which refines the conclusion reached in the relative comparison analysis.

Steps in Comparative Analysis

In comparative analysis, appraisers employ either paired data analysis or bracketing (or a combination of the two) to derive a value indication. When quantitative adjustments are applied, they are expressed either as dollar or percentage adjustments. In the adjustment grid, dollar adjustments are preferred, although the analysis to estimate the adjustment may be based on percentages. When qualitative analysis (bracketing) is applied, terms such as *inferior*, *superior*, *greatly inferior*, *greatly superior*, and *equal* are used.

Land Class Ratio Analysis and Land Adjustment Process

When analyzing sales of comparable properties to derive a land adjustment, appraisers must remember that this "land adjustment" may encompass all characteristics of the land, including productivity, topography, open land versus wooded land, and other characteristics. This technique is especially appropriate in the appraisal of a mixed-use rural property that is made up of various productive ranges of cropland, pastureland, woods, and waste. Conversely, if an appraiser is working in an area where there is typically only one class of land, a productivity or land quality adjustment may adequately reflect the degree of comparability between the land component of a comparable sale property and the subject property. An accepted technique for estimating the contributory value of multiple land characteristics is land class ratio analysis.

The first step in developing a land class ratio analysis is to place all unimproved comparable sales in a summary format that reflects their total sale price, date of sale, overall size, sale price per acre, and acreage distribution among the predominant land classes recognized in that market (see Table 12.2). The land classes specified may be (a) Class 1, 2, 3, or 4 cropland, (b) open or wooded land, (c) cultivated land or pasture, (d) good or poor rangeland, or (e) any other observable land classification recognized in the market where the subject property is located.

The eight capability unit classifications established by the USDA Soil Conservation Service may or may not be relevant measures of

Table 12.2 Summary of Unimproved Land Sales for Land Class Ratio Analysis

						Contributory Value			
Sale No.	Date	Sale Price	Acres	Location	$/Acre	Crop Acres	Crop $/Acre	Pasture Acres	Pasture $/Acre
1	6 months	$675,000	450	Avg.	$1,500	0		450	
2	6 months	$1,500,000	500	Avg.	$3,000	500		0	
3	6 months	$1,085,000	525	Avg.	$2,067	200		325	
4	Current	$960,000	375	Avg.	$2,560	300		75	
5	1 year	$2,100,000	700	Avg.	$3,000	600		100	
6	Current	$1,550,000	690	Good	$2,246	300		390	
11	6 months	$1,200,000	526	Good	$2,281	200		326	
12	1 year	$2,000,000	594	Good	$3,367	509		85	

land class in particular areas. For example, a Class IV capability rating on pasture could be more or less valuable than a Class VI capability rating on similar land due to other factors. In recreational land markets, appraisers may encounter situations in which scenic Class V floodplain or river bottom lands are much more valuable than good-quality Class II cultivatable cropland.

In areas where the production capabilities of the soil are well-recognized by market participants, productivity indexes may serve to identify the various classes of land in the subject property and comparable sale properties. Detailed and generally accurate soil data is readily available and used by nearly all market participants to a much greater degree than in the past. However, caution should still be exercised when relying on productivity indexes. Sometimes the production capabilities of a given soil type or land class do not enhance its value. Productivity indexes should only be used when buyers and sellers in the market appear to readily recognize the production capabilities associated with the farms purchased. If productivity estimates are common in the market area, discussion of these factors with buyers and sellers should be a part of the verification process.

Table 12.2 is an unimproved land sales summary chart, which might be used in an appraisal report to illustrate the development of a market-derived land class ratio. Appraisers can use this type of chart to derive the per-acre values of cropland and pasture. The completed grid is shown in Table 12.3. Note that this table shows the analysis of unimproved comparable sales properties only (Sales 1, 2, 3, 4, 5, 6, 11, and 12). The sales of improved properties (Sales 7, 8, 9, 10, and 13) are analyzed later in the chapter. Note also that many farms include multiple classes of land, and the analysis of land with mixed uses and productive capabilities requires analysis like the following.

Sales 1 and 2 in Table 12.2 have only one class of land. Sale 1 suggests that pasture in an average location had a value of $1,500 per acre six months ago. Sale 2 reflects a value of $3,000 per acre for cropland in an average location six months ago. The land values by land class for Sale 3 are determined by simple extraction. Assume that, based on Sale 2, the cropland in Sale 3 has a value of $3,000 per acre. Note that Sales 2 and 3 are similar in terms of size, location, and date of sale.

Analysis of Sale 3	
Total sale price of Sale 3	$1,085,000
Less cropland: 200 acres @ $3,000	– $600,000
Residual attributable to pasture value	$485,000
$485,000/325 acres of pasture = $1,492 per acre	
Conclude that Sale 3's cropland value is	$3,000 per acre
Conclude that Sale 3's pasture value is	$1,492 per acre

The first three sales shown in the table give a strong indication that a 100% to 50% relationship exists between cropland and pasture in this market–that is, cropland is worth twice as much as pasture. In this first example, the two sales of land of a single class, Sales 1 and 2, were used to indicate the value of cropland and pasture. Then, based on this analysis, it was assumed that the market was indicating a value of $3,000 per acre for the cropland in Sale 3 and a residual value for the pasture of $1,500 per acre (rounded). It should be emphasized that the first three sales all occurred in the same time frame, are approximately the same size, and are similar in terms of location. Sales 4, 5, 6, and 12 are either current or one-year-old transactions. Sale 11 is similar to Sales 1, 2, and 3 in terms of time (six months ago) but differs in terms of location.

Analysis of Sales 4 and 5 indicates that the $3,000 per acre for cropland would not be relevant to either of these sales. If a $3,000-per-acre value is assumed for the cropland in Sale 4, then a value of approximately $275 per acre would be indicated for the pasture in Sale 4. Conversely, if the $3,000-per-acre cropland value is applied to Sale 5, then a residual value of $3,000 per acre is suggested for the pasture in Sale 5. Both of these value indications for pasture are inconsistent with the information revealed in the first three sales. This suggests that time may be an important factor to consider in the analysis of properties in this market.

To develop a value opinion for the two classes of land in Sales 4 and 5, the 100% to 50% land class ratio derived from Sales 1 through 3 is applied. The table below illustrates how an appraiser can apply the land class ratio analysis reflected by the first three sales to estimate the value contribution of each class of land in Sales 4 and 5. The same procedure is used to derive land class values for Sales 6, 11, and 12.

Analysis of Sale 4

Cropland	300 acres × 100%	=	300.0
Pasture	75 acres × 50%	=	37.5
Total	375 acres		337.5 net crop equivalent acres
Total sale price of Unimproved Sale 4			$960,000
Then: $960,000/337.5 net crop equivalent acres		=	$2,844.44
$2,844 per acre × 50%		=	$1,422.22
Conclude			
Sale 4's cropland value			$2,844 per acre (rounded)
Sale 4's pasture value			$1,422 per acre (rounded)
Proof			
300 acres	Cropland @ $2,844 per acre	=	$853,200
75 acres	Pasture @ $1,422 per acre	=	$106,650
375 acres		=	$959,850, rounded to $960,000
Total	$960,000/375 acres	=	$2,560 per acre

In this second example, the sale price of $960,000 is simply allocated over what is known as *net crop equivalent acres.* In simple terms, if Sale 4 had contained only cropland (and no pasture at all), then the property would have had no more than 337.5 acres. The $960,000 sale price spread over 337.5 crop equivalent acres yields a value indication of $2,844 per acre for the cropland in Sale 4. Since pasture in this market sells for 50% of the value of cropland, then it is reasonable to conclude that the pasture in Sale 4 would be valued at 50% of the $2,844-per-acre value indication for cropland. This results in a land value of $1,422 per acre for pasture.

The important part of this example lies in the proof shown. This proof indicates that if the $2,844 per acre is applied to the 300 acres of cropland and the $1,422 per acre is applied to the 75 acres of pasture, then the total sale price should equal $960,000, after accounting for rounding. This confirms that the analysis of Sale 4 is correct.

This same procedure can be applied to Sale 5 to reflect a value indication for the cropland and pasture in that sale, as shown below. Remember that the proof is the most important part of the analysis, at least until the appraiser becomes totally comfortable with land class ratio analysis.

Analysis of Sale 5

Cropland	600 acres × 100%	=	600.0
Pasture	100 acres × 50%	=	50.0
Total	700 acres		650.0 net crop equivalent acres
Total sale price of Unimproved Sale 5			$2,100,000
Then: $2,100,000/650.0 net crop equivalent acres		=	$3,230.77
$3,230.77 per acre × 50%		=	$1,615.38
Conclude			
Sale 5's cropland value			$3,231 per acre (rounded)
Sale 5's pasture value			$1,615 per acre (rounded)
Proof			
600 acres	Cropland @ $3,231 per acre	=	$1,938,600
100 acres	Pasture @ $1,615 per acre	=	$161,500
700 acres		=	$2,100,100
Total	$2,100,000/700 acres	=	$3,000 per acre

After completing similar land class analyses for the remaining unimproved land sales (Sales 6, 11, and 12), estimated values per acre for each respective class of land can be incorporated into the grid (see Table 12.3).

Appraisers use this process in the cost approach to accomplish the following two tasks:

1. Develop an opinion of value for each land class on the subject property based on the process illustrated in Table 12.3 (i.e., the product of the number of acres in each land class on the subject property and the value per acre of that land class extracted from analysis of the comparable sale properties).
2. Add up the subtotals into a total value indication for the land.

However, if appraisers use a unit of comparison such as price per acre or a unit of comparison tied to some sort of improvement (e.g., price per square foot of improvement area), the process of adjusting for land mix and capability involves calculating a weighted average land class composite for each comparable sale, which can be supported with linear regression analysis to arrive at an adjustment amount.

In the example above, cropland had a value averaging $3,000 per acre, and pasture had a value averaging $1,500 per acre. Therefore, cropland would be assigned a 100% weight ($3,000 ÷ $3,000) and pasture a 50% weight ($1,500 ÷ $3,000). For each comparable property, the acreage in each class is multiplied by the respective weight of that class. The resulting subtotals are then summed and divided by the acreage to arrive at a land class composite. For a property with all cropland, the composite would be 100% or 1.00, while a farm having all pasture would have a land class composite of 50% or 0.50.

Figure 12.2 illustrates the relationship between the land class composite and the per-acre sale price. As the table and graph show, there is a high correlation (i.e., a correlation coefficient of 0.93 and coefficient of determination, R^2, of 0.87) between the two variables, strongly indicating that an adjustment of $33.08 times each percentage point of difference of land class could be applied for the difference in land class composite.

This technique is useful in that it can be used to supplement and complement the land valuation derived in the application of the cost approach as well as serve as an adjustment for land quality in the application of the sales comparison approach.

Table 12.3 Calculation of Each Comparable Sale's Land Class Composite

Sale No.	(Crop Acres	×	Weight	=	Subtotal)	+	(Pasture Acres	×	Weight	= Subtotal)	=	Total Weighted Average	Divided by Total Acres	Land Class Composite
1	0	×	100%	=	0	+	450	×	50%	225	=	225	450	50.0%
2	500	×	100%	=	500	+	0	×	50%	0	=	500	500	100.0%
3	200	×	100%	=	200	+	325	×	50%	162.5	=	362.5	525	69.0%
4	300	×	100%	=	300	+	75	×	50%	37.5	=	337.5	375	90.0%
5	600	×	100%	=	600	+	100	×	50%	50	=	650	700	92.9%
6	300	×	100%	=	300	+	390	×	50%	195	=	495	690	71.7%
11	200	×	100%	=	200	+	326	×	50%	163	=	363	526	69.0%
12	509	×	100%	=	509	+	85	×	50%	42.5	=	551.5	594	92.8%

Figure 12.2 Linear Regression Results and Data Plot

Sale No.	Land Class Composite	Price per Acre
1	50.00%	$1,500
3	69.00%	$2,067
11	69.00%	$2,281
6	71.70%	$2,246
4	90.00%	$2,560
12	92.80%	$3,367
5	92.90%	$3,000
2	100.00%	$3,000
	Slope	3308.31692
	Correlation	0.934435851
	R^2	0.87317036

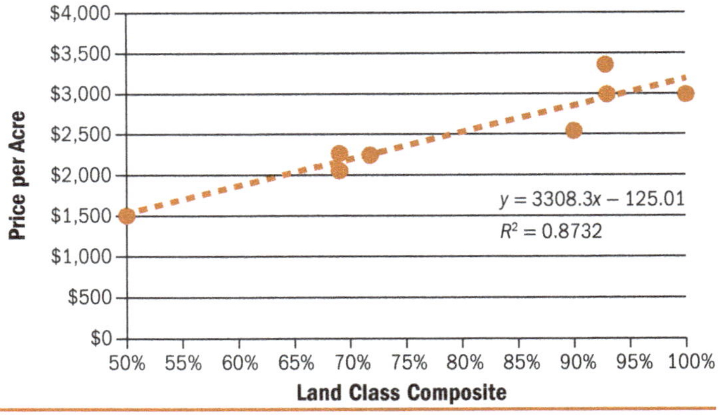

Analysis of Improved Property Sales: Estimating the Building Contribution

Once an appraiser has calculated the land class values for the sales of improved properties in the subject property's market, the next step is to estimate the building contribution reflected in the sales of improved properties within the data set. The land values developed for the sales of improved properties are based on the land class values reflected in the unimproved land sales. In analyzing land-intensive properties (i.e., properties in which land value represents the major part of total property value), appraisers should estimate the land value first. Then the remaining value is attributable to the building improvements.

It has been common practice to simply "estimate" the building's contribution to value using an abbreviated cost approach technique and then deduct this building value from the sale price to arrive at the land value. **This is an unacceptable procedure** because buildings on agricultural properties often have little or no contributory value. While conducting research on comparable properties, appraisers should ask the parties being interviewed to what extent, if any, the presence of buildings on the land influenced the price.

The following example is presented to illustrate an appropriate methodology. Note that Sale 7 (shown in Table 12.5) is a sale of an improved property that is similar in terms of location, size, and date of sale to unimproved Sale 3 (shown in Table 12.4).

In Table 12.5, all the sales of improved properties have been analyzed in the same manner shown in the previous example. Note that improved Sale 8's per-acre land class values are the same as those shown for unimproved Sale 4 in Table 12.4. Sale 4 and Sale 8 are very similar in terms of date of sale, location, and size. Also, improved

Table 12.4 Unimproved Land Sales Summary for Land Class Ratio Analysis

Sale No.	Date	Sale Price	Acres	Location	$/Acre	Contributory Value: Crop Acres	Crop $/Acre	Pasture Acres	Pasture $/Acre
1	6 months	$675,000	450	Avg.	$1,500	0	N/A	450	$1,500
2	6 months	$1,500,000	500	Avg.	$3,000	500	$3,000	0	N/A
3	6 months	$1,085,000	525	Avg.	$2,067	200	$3,000	325	$1,500
4	Current	$960,000	375	Avg.	$2,560	300	$2,844	75	$1,422
5	1 year	$2,100,000	700	Avg.	$3,000	600	$3,231	100	$1,615
6	Current	$1,550,000	690	Good	$2,246	300	$3,131	390	$1,566
11	6 months	$1,200,000	526	Good	$2,281	200	$3,206	326	$1,653
12	1 year	$2,000,000	594	Good	$3,367	509	$3,626	85	$1,813

Table 12.5 Summary of Improved Property Sales

Sale No.	Date	Sale Price	Acres	Location	$/Acre	Land Class Acres: Crop Acres	Crop $/Acre*	Pasture Acres	Pasture $/Acre	Building Acres†	Bldg. $/Acre
7	6 mos.	$900,000	300	Avg.	$3,000	200	$3,000	100	$1,500	H	$500
8	Current	$1,175,000	450	Avg.	$2,611	350	$2,844	100	$1,422	S	$83
9	1 year	$1,100,000	330	Good	$3,333	255	$3,626	75	$1,813	S/B	$119
10	6 mos.	$1,372,500	450	Good	$3,050	300	$3,206	150	$1,653	H/S/B	$362
13	1 year	$1,614,000	500	Avg.	$3,228	392	$3,231	108	$1,615	H/S/B	$346

* Values for each land class are derived from those sales in Table 12.4 that are most similar in terms of location and date of sale to each improved sale in this chart. The allocated values have been rounded to the nearest five dollars.

† H = house; S = shed; B = barn

Sale 9's land class values (shown in Table 12.5) are similar to those of unimproved Sale 12 (shown in Table 12.4). Improved Sale 10's land class values are taken from the values indicated by unimproved Sale 11. The land class values for improved Sale 13, which is reasonably comparable to unimproved Sale 5 in terms of most elements of comparison, are based on Sale 5's land class values.

The key to this land valuation method lies in the appraiser's ability to recognize the similarities in size, location, and date of sale between the improved properties and the sales of unimproved lands for which values have been previously defined. Once a total land value has been developed for each improved property, this land value is deducted from the total sale price to reflect the residual value attributable to the buildings or improvements. These residual improvement values are then reduced to a per-acre value contribution and inserted in the last column of Table 12.5.

Table 12.6 is an example of how the total building value of each improved property sale shown in the last column of Table 12.5 might

Table 12.6 Allocation of Buildings on Improved Property Sales

Sale No.	**7**		**8**		**9**		**10**		**13**	
Sale date	6 months		Current		12 months		6 months		12 months	
Total improvement value*	$150,000		$37,400		$39,245		$162,750		$173,028	
Improvements	**Cond./ Utility† Size**	**$/Unit**	**Cond./ Utility Size**	**$/Unit**	**Cond./ Utility Size**	**$/Unit**	**Cond./ Utility Size**	**$/Unit**	**Cond./ Utility Size**	**$/Unit**
Dwelling (sq. ft.)	G/G 2,500	$150,000 $60		None		None	A/A 2,500	$125,000 $50	G/G 2,425	$145,500 $60
Equip. shed (sq. ft.)		None	G/G 4,000	$37,400 $9.35	G/G 3,200	$29,920 $9.35	F/F 4,000	$28,750 $7.19	A/A 3,200	$27,528 $8.60
Grain bins (bu. cap.)		None		None	G/G 5,000	$9,325 $1.87	G/A 6,000	$9,000 $1.50	P/P 5,000	$0 $0
Total improvement contributions		$150,000		$37,400		$39,245		$162,750		$173,028

* These total values were derived by applying the market extraction technique to each improved sale.

† Condition/utility ratings: G = good; A = average; F = fair; P = poor

be allocated to each of the respective improvements or buildings on the individual properties.

The analysis of Sale 7 below indicated that the buildings (in this case, the dwelling) contributed $150,000 to the sale price. Table 12.6 shows the dollars-per-square-foot contributory value of the 2,500-sq.-ft. house on Sale 7, which is in good condition and has good utility.

Analysis of Sale 7	
Total sale price of 300 acres with buildings	$900,000
Less crop value of 200 acres @ $3,000 per acre (Sale 2)	– $600,000
Less pasture value of 100 acres @ $1,500 per acre (Sale 1)	– $150,000
Residual attributable to Sale 7's buildings	$150,000
$150,000/300 acres =	$500 per acre building value

Sales 8, 9, 10, and 13 are shown to have total building contributions of $37,400, $39,245, $162,750, and $173,028, respectively. To prepare the table, a detailed analysis of these sales was conducted in the same manner shown for Sale 7. All of these residual building or improvement values are "as of the date" of each sale.

As shown in Table 12.6, Sale 8 is improved with only one building–a 4,000-sq.-ft. equipment shed in good condition with good utility. This shed represents all of the $37,400 building contributory value, indicating a value of $9.35 per square foot for a good/good equipment shed.

Analysis of Sale 8	
Unit value of equipment shed	$37,400/4,000 = $9.35 per square foot

Sales 9, 10, and 13 each have more than one building. The value information revealed by Sales 7 and 8 was used to allocate each sale's total improvement value to the individual components or structures on each property.

Sale 13 took place within six months of Sale 7. Both Sale 7 and Sale 13 have good/good dwellings that are relatively similar in size. Using the $60 per square foot value suggested in Sale 7's analysis, a dwelling value of $145,500 can be estimated for Sale 13. Because the total improvement value of Sale 13 is $173,028, a residual of $27,528 is left for Sale 13's average/average equipment shed and its poor/poor grain bins. It is not unreasonable to assume that Sale 13's poor grain bins have no contributory value. This assumption suggests that the $27,528 residual is attributable to Sale 13's equipment shed. The 3,200-sq.-ft. equipment shed thus has a unit value of $8.60 per square foot. An $8.60-per-sq.-ft. value for the average/average equipment shed in Sale 13 not only lends support for the good/good shed value of $9.35 in Sale 8, but also supports the $60-per-sq.-ft. value ascribed to Sale 13's good/good dwelling, which was derived from the analysis of Sale 7.

Analysis of Sale 13

Value of dwelling	2,425 square feet × $60 per square foot = $145,500
Residual value of equipment shed	$173,028 – $145,500 = $27,528
Unit value of shed	$27,528/3,200 = $8.60 per square foot

Because Sales 8 and 13 both give good indications of value for equipment sheds, Sale 9 can be analyzed to suggest a residual value for its good condition/good utility grain bins. Sale 9 is one year old, as is Sale 13. However, Sale 13's equipment shed is judged to be only average in utility and condition, while Sale 9's shed has good utility and is in good condition. Sale 13's analysis reveals that the equipment shed on Sale 9 must be worth more than $8.60 per square foot. The Sale 8 equipment shed is very similar to the Sale 9 equipment shed in terms of utility and condition, and Sale 8 indicates that a good/good shed has a value of $9.35 per square foot.

In the analysis of these sales, no determination has been made about the influence of time or market conditions on land or building values. Although Sale 13, which is one year old, sets an $8.60 parameter for average/average shed value, Sale 8 suggests a value of $9.35 per square foot for good/good sheds. Focusing only on condition and utility, it is reasonable to assume that the good/good shed on Sale 9, which is one year old, has a value near $9.35 per square foot.

The total building value for Sale 9, after deducting the land contribution from its actual sale price, was estimated to be $39,245. The $9.35 per square foot equipment shed value estimate reflects a $29,920 contribution for that building ($9.35 × 3,200 square feet). Deducting the shed contribution of $29,920 from the total building contribution of $39,245 leaves a residual of $9,325 for the grain bins on Sale 9. These grain bins have good condition/good utility and a capacity to store 5,000 bushels, thereby reflecting a value of $1.87 per bushel of grain bin capacity.

Analysis of Sale 9	
Value of shed	$9.35 per square foot × 3,200 square feet = $29,920
Residual for grain bins	$39,245 – $29,920 = $9,325
Value of grain bin capacity	$9,325/5,000 bushels = $1.87 per bushel

It should be emphasized that common sense and a working knowledge of new costs for various farm improvements are important in the allocation of building value. For example, if the analysis of Sale 9 had suggested that the equipment shed value should have been $7.50 per square foot rather than $9.35 per square foot, a total equipment shed contribution of only $24,000 ($7.50 × 3,200 square feet) would have been indicated. With the market-derived residual value of $39,245 for all of Sale 9's buildings, a value of $15,245 would have been indicated for the grain bins on Sale 9. This $15,245 value would have resulted in a per-bushel capacity value of $3.05 ($15,245/5,000 bushel capacity). If the cost new of similar grain bins one year ago was $2.50 to $2.70 per bushel of capacity, it would be evident that the equipment shed had been undervalued at $7.50 per square foot. The grain bins cannot contribute more than their cost new at the time the sale occurred.

The analysis of Sale 10 requires significantly more judgment than the other four improved comparable sale allocations. Since most of Sale 10's total building value of $162,750 is likely to be represented by the dwelling, it is reasonable to use the dwelling analyses of Sales 7 and 13 to aid in the valuation of Sale 10's average condition/average utility dwelling. The good/good dwellings on Sales 7 and 13 confirm that an average/average dwelling similar in size should have a value somewhat less than $60 per square foot. Selecting a value of $50 per square foot for Sale 10's dwelling is very reasonable. At $50 per square foot, Sale 10's dwelling has a contributory value of $125,000 ($50 × 2,500 square feet).

It requires slightly less judgment to compare good/good grain bins to good/average grain bins than it does to compare an average/average or good/good equipment shed to a fair/fair equipment shed. Therefore, it is sensible to estimate Sale 10's grain bins based on the results of the Sale 9 analysis. If the good condition/good utility bins on Sale 9 have a value of $1.87 per bushel, it is reasonable to assume that the good/average bins on Sale 10 have a value of about $1.50 per bushel of capacity. This suggests a value contribution of $9,000 for the 6,000-bushel storage facility on Sale 10 ($1.50 × 6,000 bushel capacity).

Once the values of the dwelling and grain bins on Sale 10 have been estimated, the remainder, or residual, of the $162,750 total improvement value can be attributed to Sale 10's fair/fair equipment shed. The table below summarizes the method of extraction used to estimate Sale 10's shed value.

Analysis of Sale 10	
Total building/improvement value	$162,750
Less value of dwelling ($50 × 2,500)	– $125,000
Less value of grain bins ($1.50 × 6,000)	– $9,000
Residual to equipment shed	$28,750

The $28,750 residual for Sale 10's equipment shed equates to a unit value of $7.19 per square foot. This finding is well within reason for a fair condition/fair utility shed, since average/average sheds and good/good sheds have been shown to contribute value at a rate of $8.60 and $9.35 per square foot, respectively.

The building allocation process can become very complex when an appraiser is not fortunate enough to have at least one or two good sales of only one building. However, good judgment is invaluable in allocating the total building value to the individual buildings and improvements. As long as the individual building values are *reasonable* and supported by analyses of other comparable sales, no fault will be found with an appraiser's building allocation methodology.

Compare Competitive Properties to the Subject Property's Land and Buildings

Note that up to this point little has been said about the hypothetical subject property. The subject property's characteristics are important in the basic research insofar as the sales of comparable properties to be used in the analysis must have the same highest and best use as the subject property, and other characteristics of these sales should be "reasonably" comparable to the subject property. However, when analyzing sales to derive land class ratios and building value contributions, the subject property must almost be left out of consideration. Once the land class ratios and the building allocations have been developed for the sales of improved properties, they should be compared directly to the subject property's land and buildings to arrive at an adjustment to reflect each comparable sale's land mix and building contribution.

When an appraiser has arrived at land and building values for each improved sale, several paired data analyses can be conducted. The improved sales are adjusted for elements of comparison such as financing terms, market conditions, size (land area), location, and other characteristics affecting property values in the subject property's market.

For the purposes of this presentation, assume that the subject property has the following basic characteristics:

Date of appraisal	Current
Location	Good
Acres	510
Cropland	400
Pasture	110
Buildings	
Dwelling: average condition, good utility, 2,250 square feet	
Equipment shed: average condition, average utility, 4,200 square feet	
Grain bins: fair condition, fair utility, 10,000 bushel capacity	

When comparing the improved sales to the subject property in terms of land and buildings, all other elements of comparison are ignored or assumed equal.

The land adjustment process is predicated on the following assumption: If all other elements of comparison for the subject property and a given sale were equal, then the subject property's cropland should be similar on a per-acre basis to the cropland value of the property being compared to the subject property.

The building adjustment is accomplished by comparing the individual buildings on each comparable sale property directly to the subject property's buildings on a per-building, per-square-foot, per-bushel, or overall basis. Judgment may be required to compare a good condition/good utility building on a given sale property to a similar subject building with only fair condition and utility. Detailed market research and proper building allocation of the comparable sales will set excellent parameters and inform an appraiser's judgment in this direct comparison process.

The tables that follow illustrate how to compare the improved sales to the subject property for land quality, land mix, and buildings. The sales analyzed are those shown in Table 12.5. Note that the details of each comparable sale are shown on the left side of each table, while the subject property's characteristics are shown on the right. The per-acre land class values for each improved sale are also taken from Table 12.5, while the building/improvement values of the properties are taken from Table 12.6.

Although Sales 7, 8, and 9 have fewer buildings than the subject property, they have been included to properly "spread" the building values on properties that have more than one building. In Table 12.7, the indicated land adjustment suggested by this comparison is $176 per acre to Sale 7's $2,500-per-acre land value ($2,676 – $2,500 =

Table 12.7 Land Adjustment Analysis

Improved Sale 7				Subject Property			
Cropland	200 acres @	$3,000	$600,000	Cropland	400 acres @	$3,000	$1,200,000
Pasture	100 acres @	$1,500	$150,000	Pasture	110 acres @	$1,500	$165,000
Total	300 acres @	$2,500	$750,000	Total	510 acres @	$2,676	$1,365,000
Indicated land adjustment to Sale 7 = +$176 per acre							

$176). The building adjustment required to make Sale 7 more like the subject property in terms of buildings is the difference between Sale 7's $500-per-acre building value and the subject property's $334-per-acre building value ($500 – $334 = $166).

In Table 12.8, the good/good dwelling on Sale 7 is judged to be somewhat superior to the average/good house on the subject property. A value estimate of $54 per square foot for the subject property's dwelling is within reason, especially if one reviews the overall findings for dwellings shown in Table 12.6. Since Sale 7 has no equipment shed or grain bins, one must again rely on the allocation process completed in Table 12.6 for comparable sales of properties with grain bins and sheds. The $8.60 per square foot for the subject property's average/average equipment shed is in line with the values for equipment sheds in Table 12.6, and the $1.25 per bushel for the subject property's fair/fair grain bins is well supported by the Table 12.6 analyses.

Land and building adjustment analyses for Sale 8 are shown in Tables 12.9 and 12.10. The subject property's land value of $2,537 per acre less Sale 8's land value of $2,528 per acre indicates that a + $9-per-acre adjustment is required to make Sale 8 like the subject property in terms of land quality. The $83-per-acre building value shown for Sale 8 is inferior to the subject property's suggested building value of $334 per acre, requiring a positive adjustment of $251 per acre.

In Table 12.9, the subject property is shown to be modestly superior in terms of land mix or quality when compared to Sale 8. Sale 8 has no dwelling or grain bins, so value estimates for these two improvements

Table 12.8 Building Adjustment Analysis

Buildings on Sale 7	Cond./ Utility	Size	$/Unit	$	Buildings on Subject Property	Cond./ Utility	Size	$/Unit	$
Dwelling	G/G	2,500 @	$60	$150,000	Dwelling	A/G	2,250 @	$54	$121,500
Horse shed	None				Horse shed	A/A	4,200 @	$8.6	$36,120
Grain bin	None				Grain bin	F/F	10,000 @	$1.25	$12,500
Total				$150,000	Total				$170,120
Building value per acre				$500	Building value per acre				$334
Indicated building adjustment to Sale 7 = –$166 per acre									

Table 12.9 Land Adjustment Analysis

Improved Sale 8				Subject Property			
Cropland	350 acres @	$2,844	$995,400	Cropland	400 acres @	$2,844	$1,137,600
Pasture	100 acres @	$1,422	$142,200	Pasture	110 acres @	$1,422	$156,420
Total	450 acres @	$2,528	$1,137,600	Total	510 acres @	$2,537	$1,294,020

Indicated land adjustment to Sale 8 = +$9 per acre

Table 12.10 Building Adjustment Analysis

Buildings on Sale 8					Buildings on Subject Property				
	Cond./ Utility	Size	$/Unit	$		Cond./ Utility	Size	$/Unit	$
Dwelling	None				Dwelling	A/G	2,250 @	$54	$121,500
Horse shed	G/G	4,000 @	$9.35	$37,400	Horse shed	A/A	4,200 @	$8.75	$36,750
Grain bin	None				Grain bin	F/F	10,000 @	$1.25	$12,500
Total				$37,400	Total				$170,750
Building value per acre				$83	Building value per acre				$334

Indicated building adjustment to Sale 8 = +$251 per acre

can be interpolated from the analyses in Table 12.6. If it is then assumed that there is no other sale to compare to the subject property for equipment shed values, then the good/good shed on Sale 8 might suggest a value of about $8.75 per square foot for the average/average shed on the subject property. This Sale 8 building adjustment analysis suggests that Sale 8 requires a positive $251-per-acre adjustment to make it more like the subject property in terms of building characteristics.

Analysis of Sale 9 is presented in Tables 12.11 and 12.12. It is assumed in Table 12.12 that no other sales were available for comparison to the subject property. The good/good equipment shed value of $9.35 per square foot for Sale 9 might suggest a value of $8.75 per square foot for the subject property's average/average shed. Furthermore, Sale 9's good/good grain bins, at $1.87 per bushel of capacity, strongly support a value of $1.25 per bushel for the subject property's fair/fair grain bins. The subject property's $54-per-sq.-ft. dwelling value is taken from an overview of the dwelling values in Table 12.6.

Sale 10 is analyzed in Tables 12.13 and 12.14. Table 12.13 illustrates that Sale 10 has an inferior land mix to the subject property. A quick review of the percentage of the higher-valued cropland in each property lends support to the need for a land adjustment. Because approximately 78.50% of the subject property's 510 acres are cropland and Sale 10 is only 66.70% cropland, a moderate land adjustment is appropriate.

The building adjustment analysis for Sale 10 in Table 12.14 is predicated on the assumption that this is the only comparable sale

Table 12.11 Land Adjustment Analysis

Improved Sale 9			
Cropland	255 ac. @	$3,626	$924,630
Pasture	75 ac. @	$1,813	$135,975
Total	330 ac. @	$3,214	$1,060,605

Subject Property			
Cropland	400 ac. @	$3,626	$1,450,400
Pasture	110 ac. @	$1,813	$199,430
Total	510 ac. @	$3,235	$1,649,830

Indicated land adjustment to Sale 9 = +$21 per acre

Table 12.12 Building Adjustment Analysis

Buildings on Sale 9	Cond./Utility	Size	$/Unit	$
Dwelling	None			
Horse shed	G/G	3,200 @	$9.35	$29,920
Grain bin	G/G	5,000 @	$1.87	$9,350
Total				$39,270
Building value per acre				$119

Buildings on Subject Property	Cond./Utility	Size	$/Unit	$
Dwelling	A/G	2,250 @	$54	$121,500
Horse shed	A/A	4,200 @	$8.75	$36,750
Grain bin	F/F	10,000 @	$1.25	$12,500
Total				$170,750
Building value per acre				$335

Indicated building adjustment to Sale 9 = +$216 per acre

Table 12.13 Land Adjustment Analysis

Improved Sale 10			
Cropland	300 ac. @	$3,206	$961,800
Pasture	150 ac. @	$1,653	$247,950
Total	450 ac. @	$2,688	$1,209,750

Subject Property			
Cropland	400 ac. @	$3,206	$1,282,400
Pasture	110 ac. @	$1,653	$181,830
Total	510 ac. @	$2,871	$1,464,230

Indicated land adjustment to Sale 10 = +$183 per acre

Table 12.14 Building Adjustment Analysis

Buildings on Sale 10	Cond./Utility	Size	$/Unit	$
Dwelling	A/A	2,400 @	$50	$120,000
Horse shed	F/F	4,000 @	$7.19	$28,760
Grain bin	G/A	6,000 @	$1.50	$9,000
Total				$157,760
Building value per acre				$351

Buildings on Subject Property	Cond./Utility	Size	$/Unit	$
Dwelling	A/G	2,250 @	$55	$123,750
Horse shed	A/A	4,200 @	$8.60	$36,120
Grain bin	F/F	10,000 @	$1.25	$12,500
Total				$172,370
Building value per acre				$338

Indicated building adjustment to Sale 10 = –$13 per acre

available for direct comparison to the subject property's buildings or improvements. Sale 10's average/average dwelling, valued at $50 per square foot, would suggest a value of about $55 per square foot for the subject property's average/good dwelling. The fair/fair equipment shed on Sale 10 at $7.19 per square foot lends support to a value estimate of $8.60 per square foot for the subject property's average/average shed. And finally, the good/average grain bins on Sale 10, which reflect a unit value of $1.50 per bushel of storage capacity, support a $1.25 per bushel value for the fair/fair grain bins on the subject property.

Tables 12.15 and 12.16 present the analysis of Sale 13. The result of Sale 13's land adjustment analysis shown in Table 12.15 reinforces the fact that land class ratio analysis really works. Sale 13 requires virtually no land adjustment because the subject property has 78.43% of its 510 total acres in the cropland class, while Sale 13 has a nearly equal 78.40% of its total acreage in the cropland class. This relationship suggests that the properties are nearly equal in land mix, and therefore no land adjustment is warranted.

Sale 13 is also very comparable to the subject property in terms of improvements. The good/good dwelling valued at $60 per square foot in Sale 13 supports a value estimate of about $55 per square foot for the subject property's slightly inferior average/good dwelling. The equipment sheds on the subject property and Sale 13 are of average condition and utility, so the $9.35 per square foot value for Sale 13's shed is appropriate for the shed on the subject property. The $0 value for Sale 13's poor/poor grain bins and the grain bin results shown in

Table 12.15 Land Adjustment Analysis

Improved Sale 13				**Subject Property**			
Cropland	392 ac. @	$3,231	$1,266,552	Cropland	400 ac. @	$3,231	$1,292,400
Pasture	108 ac. @	$1,615	$174,420	Pasture	110 ac. @	$1,615	$177,650
Total	500 ac. @	$2,882	$1,440,972	Total	510 ac. @	$2,882	$1,470,050
Indicated land adjustment to Sale 13 = $0 per acre							

Table 12.16 Building Adjustment Analysis

Buildings on Sale 13					**Buildings on Subject Property**				
	Cond./ Utility	**Size**	**$/Unit**	**$**		**Cond./ Utility**	**Size**	**$/Unit**	**$**
Dwelling	G/G	2,625 @	$60	$157,500	Dwelling	A/G	2,250 @	$55	$123,750
Horse shed	A/A	3,200 @	$9.35	$29,920	Horse shed	A/A	4,200 @	$9.35	$39,270
Grain bin	P/P	5,000 @	$0.00	$0	Grain bin	F/F	10,000 @	$1.25	$12,500
Total				$187,420	Total				$175,520
Building value per acre				$375	Building value per acre				$344
Indicated building adjustment to Sale 13 = –$31 per acre									

the analyses of the other improved sales in Table 12.6 lend support to a value of approximately $1.25 per bushel of capacity for the subject property's fair/fair grain bins.

Once the land and building adjustment analyses shown in Tables 12.7 through 12.16 are completed, all sales of comparable improved properties can be compared directly to one another and the subject property for any other elements of comparison determined to be appropriate, which may include time, location, size, and other characteristics. The elements of comparison found to be significant can be analyzed, and individual pairings can be developed and applied in the sales comparison approach.

Application of Quantitative Comparative Analysis

When an appraiser has arrived at the adjusted land and building values for each comparable improved property sale, several paired data analyses can be conducted to isolate elements of comparison such as terms, time, size (land area), location, and other characteristics affecting property value in the subject property's market. The five fully analyzed sales of improved properties shown in Tables 12.17 and 12.18 represent the data to be processed in this example of quantitative comparative analysis.

Step 1: Adjust All Comparable Properties to the Subject Property's Land and Building Mix

Earlier in the chapter, Sales 7, 8, 9, 10, and 13 were adjusted to the subject property's land mix and buildings. When the sales of improved properties were compared to the subject property for land and buildings, all other elements of comparison were ignored or assumed to be equal between the comparable properties and subject property. In Step 2, adjustments are made for these other elements of comparison.

Table 12.17 Summary of Improved Properties

						Land Class Acres					
Sale No.	Sale Date	Sale Price	Acres	Location	$/Acre	Crop Acres	Crop $/Acre	Pasture Acres	Pasture $/Acre	Building Acres	Bldg. $/Acre
7	6 mos.	$900,000	300	Avg.	$3,000	200	$3,000	100	$1,500	H	$500
8	Current	$1,175,000	450	Avg.	$2,611	350	$2,844	100	$1,422	S	$83.11
9	12 mos.	$1,085,000	330	Good	$3,288	255	$3,626	75	$1,813	S/B	$74
10	6 mos.	$1,372,500	450	Good	$3,050	300	$3,206	150	$1,653	H/S/B	$362
13	12 mos.	$1,614,000	500	Avg.	$3,228	392	$3,231	108	$1,615	H/S/B	$346

Table 12.18 Allocation of Buildings on Improved Properties

Sale No.	7		8		9		10		13	
Sale date	6 months		Current		12 months		6 months		12 months	
Total improvement value	$150,000		$37,400		$39,245		$162,750		$173,028	
Improvements	**Cond./ Utility Size**	**$/Unit**	**Cond./ Utility Size**	**$/Unit**	**Cond./ Utility Size**	**$/Unit**	**Cond./ Utility Size**	**$/Unit**	**Cond./ Utility Size**	**$/Unit**
Dwelling (sq. ft.)	G/G	$150,000		None		None	A/A	$125,000	G/G	$145,500
	2,500	$60					2,400	$50	2,625	$60
Equip. shed (sq. ft.)		None	G/G	$37,400	G/G	$29,920	F/F	$28,750	A/A	$27,528
			4,000	$9.35	3,200	$9.35	4,000	$7.19	3,200	$8.60
Grain bins (in bus. cap.)		None		None	G/G	$9,325	G/A	$9,000	P/P	$0
					5,000	$1.87	6,000	$1.50	5,000	$0
Total imp. contrib.		$150,000		$37,400		$39,245		$162,750		$173,028

Step 2: Compare the Sales of Improved Properties Adjusted for Land and Building Mix to Each Other and to the Subject Property (Paired Data Analysis)

Now the sales of improved properties can be compared directly to one another and to the subject property for other elements of comparison such as time, location, and size. Table 12.19 shows an adjustment grid. Using this grid, appraisers gradually make each improved comparable property more like the subject property, adjusting for each successive element of comparison down each column.

Market Conditions Adjustment

The first adjustments shown in Table 12.19 are the land and buildings adjustments derived earlier in this chapter. There is considerable debate about *when* it is technically correct to make a market conditions adjustment. Traditionally, property rights conveyed and financing terms adjustments are made first, and then market conditions adjustments are made. All other adjustments generally follow the market conditions adjustment.

However, in some markets land values may be changing at a different rate and possibly in a different direction than building values. If this is occurring, then deriving a market conditions adjustment from sales of unimproved land first will definitely not produce an appropriate rate to apply to improved land. Similarly, if all the sales of improved land are not almost equal to the subject property in terms of land and building ratios, then deriving a market conditions adjustment directly from the pairing of two or three sales of improved land

Table 12.19 Sales Comparison Adjustment Grid

Sale No.	7	8	9	10	13
Date	6 months	Current	1 year	6 months	1 year
Size	300 acres	450 acres	330 acres	450 acres	500 acres
Financing	Cash equivalent	Cash equivalent	Cash equivalent	Cash equivalent	Cash equivalent
$/Acre	$3,000	$2,611	$3,288	$3,050	$3,228
Adjustments					
Land	+ 176	+ 9	+ 21	+ 183	+ 0
Buildings	− 166	+ 252	+ 216	− 13	− 31
Adjusted $/acre	$3,010	$2,872	$3,525	$3,220	$3,197
Location	Average	Average	Good	Good	Average
Location adjustment	+ 400	+ 400	+ 0	+ 0	+ 400
Adjusted $/acre	$3,410	$3,272	$3,525	$3,220	$3,597
Size (in acres)	300 acres	450 acres	330 acres	450 acres	500 acres
Size adjustment	− 225	+ 0	− 225	+ 0	+ 0
Final indicated value	$3,185	$3,272	$3,300	$3,220	$3,597

may not give an accurate reading of the influence of changing market conditions over time.

Thus far, all land allocations and adjustments and all building allocations and adjustments have been made "as of the date" of each respective sale. By taking the land mix and buildings adjustments derived and applying them in Table 12.19, all sales of improved properties are made equal to the subject property in terms of land and buildings "as of the date" each sale occurred. Once all comparable sales have been made equal in terms of land mix ratios and buildings, a more accurate assessment of the time influence can be made.

Financing Adjustment

If a financing terms adjustment should be made, it should be applied to the gross sale price of a particular property prior to any land or building allocations, thereby reflecting the cash-equivalent value of the entire property. The properties shown in this example all reflect cash-equivalent sales prices, so no terms or financing adjustments are required.

The detailed discussion that follows the adjustment analysis in Table 12.19 describes how the market conditions, size, and location adjustments were derived. This direct comparison process serves as a presentation of paired data analysis. Note that the adjusted values become more tightly clustered further down the column for each property. This reflects the fact that more of the variations or differences between the comparable sales and the subject property are being accounted for with each successive adjustment as the appraiser moves toward the final indicated value for each sale.

Location Adjustment

Sales 7 and 9 are within 30 acres (10%) of being the same size. Once all sales have been adjusted toward the subject property for land, buildings, and market conditions, these two properties are equal in terms of all other elements of comparison except location. Sale 7 has an average location, while Sale 9 and the subject property have good locations. The $515 difference in the sales prices of Sales 7 and 9 after adjustment for land and building differences suggests that properties with average locations require a $515-per-acre adjustment to make them more like properties with good locations.

Sales 10 and 13 are equal to one another in terms of land and buildings once all sales of comparable properties have been adjusted toward the subject property for these elements of comparison. However, Sale 10 has a good location, and Sale 13 has an average location. The $339 difference in the market conditions–adjusted values of Sales 10 and 13 suggests that properties with average locations require a $339-per-acre adjustment to make them more like properties with good locations.

Based on these two pairings for location, all properties with average locations require a positive $400-per-acre adjustment to their market conditions–adjusted values when compared to the subject property, which has a good location.

Size Adjustment

A review of the location-adjusted sales prices in Table 12.19 suggests that size has a modest influence on per-acre sales prices of properties in the 300-acre (Sale 7) to 500-acre (Sale 13) range. Sales 8, 10, and 13 are the largest of the five comparable properties, with an average size of 466.66 acres. The average location-adjusted value of these three properties is $3,242 per acre. Sales 7 and 9 are the smallest properties with an average size of 315 acres and an average location-adjusted value of $3,467 per acre. This analysis shows a moderate difference in value of about $225 per acre in favor of the smaller properties.

Step 3: Derive a Final Value Conclusion

The five indicated values suggested by the sales of improved properties shown in Table 12.19 can now be summarized. The two properties requiring no location adjustment suggest a value of $3,220 to $3,300 per acre. The property requiring no market conditions adjustment suggests a value of $3,272 per acre. The three properties requiring the lowest gross dollar adjustment suggest a value of $3,185 to $3,300 per acre. The midpoint of the two extreme value indications is $3,228 per acre. The average of the five indicated values is $3,315 per acre. The average exclusive of the two extremes is $3,264 per acre. The median indicated value is $3,272 per acre.

Indicated value via the sales comparison approach
510 acres @ $3,250 per acre = $1,657,500
Rounded to $1,660,000

This complete sales comparison approach analysis has been presented to give appraisers another tool to use in presenting a methodical analysis of the available comparable sales and arriving at a well-supported final opinion of value for the subject property. An appraiser who chooses this or a similar method of analysis can feel comfortable even with erratic data. The key to an accurate sales comparison approach lies in allowing the market to do the talking. While judgment is always necessary, "rules of thumb" and preconceived notions should be eliminated from the appraisal process. Proper data analysis places clear parameters on the appraiser's judgment.

Quantitative Comparative Analysis Using Productivity-Based Units of Comparison

The unit of comparison applied in the previous example was based on physical size. The same technique can be applied using productivity-based units of comparison (e.g., per animal unit, per sow, per dairy cow, per bushel or barrel). The land mix and buildings of all the properties in the data set are first adjusted to be similar to the subject property. Table 12.20 shows the land mix adjustment on an animal unit (AU) basis for the comparison of two ranch properties. The carrying capacity of irrigated pasture is two acres per animal unit. Rangeland has a carrying capacity of 25 acres per animal unit. The adjusted value of the subject property's land is derived by applying per-acre values, but the actual land mix adjustment is based on dollars per animal unit.

The final building totals in the building adjustment grid would also be divided by the animal unit capacity of the comparable and subject properties, and these adjustments would be made on an animal unit basis. Working down a sales comparison grid for this ranch property (similar to Table 12.19), the paired data and adjustments for

Table 12.20 Animal Unit Land Adjustment Analysis (Productivity-Based Unit of Comparison)

Sale X (650 Animal Units)				Subject Property (700 Animal Units)			
Irr. Pasture	500 acres @	$1,800	= $900,000	Irr. Pasture	1,000 acres @	$1,800	$1,800,000
Rangeland	10,000 acres @	$300	= $3,000,000	Rangeland	5,000 acres @	$300	$1,500,000
Total	650 AUs @	$6,000*	= $3,900,000	Total	700 AUs@	$4,714†	$3,300,000
Indicated land adjustment to Sale X = –$1,286 per animal unit (AU)							

* $3,900,000 total divided by 650 AUs

† $3,300,000 total divided by 700 AUs

each successive element of comparison would reflect the animal unit (AU) unit of comparison. This same procedure can be used to apply any productivity-based unit of comparison.

Chapter 13

The Income Capitalization Approach

The income capitalization approach is predicated on the principle of anticipation, which holds that present value is indicated by the expectation of future benefits. In this chapter, two methodologies are explained:

- direct capitalization
- yield capitalization

The distinction between the types of income analyzed in this approach is important. Rural properties are either (1) owner-operated or (2) rented for cash or a share of the crop. Owner-operators look to the farming enterprise as the source of income. These properties are not developed to collect income in the form of rent. Consequently, the income and expense analysis of owner-operated properties includes other components that contribute to income, such as equipment and entrepreneurial ventures. Analysis of the income and expenses of an owner-operated property is an excellent tool for testing the feasibility of the enterprise, but it is a difficult method to employ in the valuation of real property due to the complexities and subjectivity of the analysis. Landowners who choose not to farm their land often use cash rental or crop share rental arrangements to lease the land to someone who will operate a farming enterprise. An important step in preparing an appraisal is ascertaining the applicability of the income capitalization approach to the type of property being appraised. The approach is certainly appropriate in appraising rural properties that typically are leased. However, an owner-operator analysis may be more appropriate in the valuation of properties that are predominantly owner-operated, such as orchards and vineyards. An analysis of owner-operator income and expenses customarily employs capitalization rates and yield rates abstracted from owner-operator income and expense statements.

Income Estimates

The primary purpose of estimating the gross income of a property is to project the probable net income benefits from its ownership. To apply direct or yield capitalization, a reliable estimate of income expectancy must be developed. Rental income may be in the form of cash or a share of the crop, depending on local customs and the crops grown. Cash rent is generally a specified amount per acre. In a crop share arrangement, the owner receives a percentage of the commodity produced on the property. Before a rural appraiser can estimate the potential gross income that will be used to develop a reasonable net income projection, the appraiser must consider two unique aspects of farm income:

1. The income stream of an agricultural property may be cyclical due to the variation in cropping patterns, including rotation.
2. Lease periods are generally shorter than they are for urban properties. If the property being appraised is leased, the appraiser must be aware of all the lease terms. Leases will vary in type, terms, and conditions from community to community. Our society is increasingly complex, and appraisers should not make general assumptions as to the terms of a lease.

Income and expense forecasts begin with lease analysis. *Gross rental leases* are leases in which the lessor pays all operating expenses of the real estate. In *net rental leases*, the tenant pays all operating expenses.

The analysis of the gross income of a farm property involves identification of the typical income arrangement for the area–cash rent, crop share rent, or owner-operator income. The investigation of crop share rent also includes

- the crops most likely to be produced in the area
- yields based on property-specific data or local area data

Table 13.1 Classifying Leases

Type of Lease	Characteristics
Flat rental lease	A specific level of rent continues for the duration of the lease.
Graduated rental lease	The rent undergoes specified changes at one or more points during the lease term.
Revaluation lease	Rent is adjusted at periodic intervals based on a revaluation of the real estate.
Index lease	Periodic rent adjustments are tied to a specific index, such as commodity indexes or the general inflation rate.
Percentage or crop share lease	Rent is set at a percentage of the crops grown by the tenant, which may vary depending on local custom and the land owner's contribution of seeds, fertilizer, harvesting costs, irrigation, water supply, etc.

Figure 13.1 Typical Lease Data

- The lease date
- A legal description of the leased property
- The names of the lessor and lessee
- The lease term
- The dates of occupancy
- The commencement of payments
- The amount and source of the rent
- Escalation clauses
- Rent concessions
- The right of assignment or right to sublet
- Renewal options
- Land and tenant covenants regarding the division of expenses
- Purchase options
- Escape clauses
- Stipulated advance rent or bonding
- Provisions for casualty or condemnation loss
- Continued occupancy clauses
- Specifications concerning tenant improvements
- Revaluation clauses

Table 13.2 Classifying Types of Rent

Type of Rent	Characteristics
Contract rent	The actual rental income specified in a lease
Market rent or economic rent	The rental income a property would most likely command in the open market
Excess rent	The amount by which contract rent exceeds market rent at the time of the appraisal
Percentage rent	Rental income received in accordance with the terms of a percentage clause in a lease
Overage rent	Rent paid over and above the guaranteed minimum rent

- commodity prices
- the crop share for percentage rents

The analysis of cash rent also includes property-specific terms and market rent terms.

Cash Rent

Cash rent is generally fixed for a specified period of time. Many cash rents are established on a year-to-year basis for agricultural properties, but rents can be specified for extended periods. Cash rents can be based on total acreage, crop acreage, base acreage, or any other basis typical for the area. They can be paid in advance or in arrears, or payment can be split between the beginning and end of the crop year. Cash rentals for specialized agricultural buildings often have the same characteristics as rentals of typical commercial buildings. Appraisers must take care to analyze the local market to ascertain the cash lease arrangement common in the area. Long-term leases must be analyzed because there will often be a substantial difference between contract rent and market rent.

Buildings on farms are often rented to tenants other than those who rent the land. Rental arrangements for these buildings can take the form of simple residential rentals or more complicated arrangements in which the home, barns, and grain bins are rented separately. Another source of farm income that is becoming common is rent from off-season use of the agricultural property for recreation such as hunting, fishing, or resort-related activities. These arrangements are typically specified in cash leases. As in appraisals of other income-producing property, obtaining and carefully reviewing all lease terms is a prerequisite to developing a value indication by this method. In many agricultural markets, there are long-term relationships–which may or may not be acknowledged–that affect lease pricing and terms.

Table 13.3 shows an income analysis for a hypothetical property. In this case, differing cash rentals are charged for the different types of land on the farm, and the buildings are rented separately to an individual who is not involved in the farming operation. Other land not associated with the buildings is leased to a recreational tenant, who uses the property for hunting. Hunting is typically confined to the fall and winter and does not interfere with the agricultural enterprise.

Table 13.3 Cash Lease Income Analysis

Land Type		Rent per Unit	Total Cash Rent
Cropland A	120 acres	$90 per acre	$10,800
Cropland B	300 acres	$60 per acre	$18,000
Pasture	130 acres	$30 per acre	$3,900
Buildings		$1,500 per month	$18,000
Recreational lease	550	$30 per acre	$16,500
Total			$67,200

Crop Share Rental

In many areas, crop share rentals are common. Consequently, annual income per acre fluctuates. It is not a stable, guaranteed annual sum as is typical in urban leases or in the cash rent arrangements previously described. When commodity prices are on an upswing, many cash rent leases include adjustments for the prices obtained by the grower at harvest.[1]

To analyze crop share rentals, a rural appraiser must have a working knowledge of production practices and prepare an estimate of gross income on an average annual basis over a period of time. A five-year period is commonly used. By using an extended time period and making necessary adjustments for economic realities, appraisers can level out the cyclical effect of crop rotations and commodity prices and estimate an average income stream.

1. For example, cropland might be leased for a base rent of $50 per acre with a contractual adjustment for any increase above an agreed-upon price per bushel, say, $2 per acre for every increase of $0.10 over the base crop price.

There are accepted procedures for developing the income and expense projections needed to derive a gross income estimate. The gross income estimate is based on the amount of acreage and the type of crop grown on the property in a rotation cycle on an annual basis. An entire agricultural property may be devoted to a single crop (e.g., dryland wheat production) in any single year. However, over the rotation cycle, a variety of crops can be grown on the property. For example, irrigated land may have a typical rotation of potatoes, barley, corn, and alfalfa. To estimate the gross income over the rotation cycle, appraisers ascertain the amount of acreage that will be devoted to a particular crop on an average annual basis and estimate the probable production (yield) of the crop, the probable price that will be received for the crop, and the percentage of the crop that will be paid to the landowner. Appraisers should employ assumptions that are typical for the area and property type and analyze the historical income and expenses of the subject property.

To complete a gross income estimate, appraisers include other actual or potential income from the property. Gross income may be derived from renting improvements separate from the farm operation and recreational leases. The income will typically be specified in cash leases. When there are different tenants on a farm, the lease may actually be a combination of percentage and cash leases. If this is the typical rental method in the area, appraisers should have no problem with the mixture of percentage and cash leases.

Table 13.4 describes a crop share rental of the same hypothetical property presented in Table 13.3. On this farm, 200 acres are typically planted to Crop A, and the typical share is one-fourth of the total production to the owner as rent. Three hundred acres are typically planted to Crop B, and the typical share is one-third of the total production to the owner as rent. The pasture is typically rented for cash, possibly to a different tenant, and the buildings are rented to yet another tenant for a monthly rent. The recreational lease is also assigned to another tenant for cash.

Table 13.4 Crop Share Income Analysis

		Yield per Acre	Total Yield	Price per Unit	Share	Owner Income
Crop A	200 acres	100 bushel	20,000	$12	¼	$60,000
Crop B	300 acres	40 bushel	12,000	$9	⅓	$36,000
Pasture	50 acres	N/A	N/A	$30	All	$1,500
Buildings		N/A	N/A	$1,200 per mo.	All	$14,400
Recreation	550 acres	N/A	N/A	$15	All	$8,250
Total						$120,150

Owner-Operator Income Statement

To analyze the income for an owner-operator property, the entire income attributable to the operation of the farm must be estimated (see Table 13.5). It is assumed that the owner operates the farm and no rental income is received, except possibly income from a recreational lease. The home and buildings are occupied by the owner, and the rent attributed to these buildings is actually the rental savings the owner is receiving because no financial outlay is required for a separate residence or base of operations.

Owner-operator income estimation may be required in appraising a standard crop operation if leases are not available for analysis. Because this form of income estimation is the most difficult and most susceptible to error, it is usually not attempted when lease information is available. For specialized agricultural endeavors such as permanent plantings, poultry operations, grain elevators, and feed lots, lease information is not readily available, so the owner-operator method of income estimation must be used.

In the following example, 200 acres of the farm are typically planted to Crop A, and 300 acres are planted to Crop B. The pasture is operated by the owner for hay production, and the buildings are occupied by the owner. There is a recreational lease to a tenant who pays in cash.

Table 13.5 Owner-Operator Income Analysis

		Yield per Acre	Total Yield	Price per Unit	Owner Income
Crop A	200 acres	100 bushel	20,000	$12	$240,000
Crop B	300 acres	40 bushel	12,000	$9	$108,000
Pasture	50 acres	1 ton	50	$60	$3,000
Buildings		N/A	N/A	$1,200 per mo.	$14,400
Recreation	550	N/A	N/A	$15	$8,250
Total					$373,650

Effective Gross Income

Effective gross income is the portion of income that remains after an allowance has been made for vacancy and collection loss. Although vacancy is not typically considered in the income statement of a farm, appraisers should be aware that vacancy can be a problem for some types of agricultural properties. If vacancy or collection loss is typical for a particular type of property, then a charge should be shown in the expense statement.

Cash leases or a portion of the annual rent is paid in advance, so vacancy and collection loss does not occur. In the case of a crop failure in a crop share arrangement, *vacancy* is not the appropriate term.

Although vacancy is not a concern in the analysis of a typical crop production agricultural unit, vacancy can affect other agricultural endeavors and must be considered. Vacancy may occur for a variety of reasons, such as a disease found in the animals housed in specialized buildings, the inability to rent rural buildings, a lack of quality irrigation water, poor drainage, or an irrigation system in disrepair. Credit loss stems from the inability of a tenant to pay the agreed-upon rent.

An appropriate percentage factor may be applied to estimate vacancy, and this percentage factor should reflect current market conditions. If the property being appraised has a history of rental income, this is excellent property-specific data from which to estimate the gross and actual income received. Appraisers should also check other data sources such as area farm managers, other landlords, tenants, and lenders. Table 13.6 illustrates the procedure for deducting an anticipated 5% vacancy and collection loss from potential gross income to arrive at an estimate of effective gross income.

Table 13.6 Estimating Effective Gross Income

Potential gross income	$300,000
Less vacancy and collection loss @ 5%	– 15,000
Effective gross income	$285,000

Analyzing Expenses

To derive the net income from the operation of a rural property, appraisers begin with the gross income estimate and deduct all the expenses required to secure and produce the income stream and maintain and manage the property. In analyzing the expense statement, appraisers should ensure that the charges included by the landowner are directly related to the operation and management of the property. Deductions for depreciation and mortgage interest expense may be proper on the owner's income tax statement, but they are not included in the net income from operations.

The expenses deducted from income should be consistent with the nature of the income–i.e., cash rent, crop share rent, or owner-operated expenses. Under some rental agreements, most notably cash leases and crop share arrangements, expenses may vary from crop to crop, area to area, or year to year. The expenses deducted should be typical for the area at the time of the appraisal.

A landlord-tenant agreement (income in the form of rent) is the preferred basis for the income and expense estimate. The landlord's income and expense statement isolates that part of the net income that is directly attributable to the real property.

Whether the rental terms specify cash or crop share rent, the landlord expenses in a gross rental lease, also called a *full-service lease*, usually include expenses for basic items such as management, real estate taxes, insurance, and building maintenance. Other expenses for irrigation assessments, electrical power, and irrigation system maintenance may or may not be included depending on the typical lease arrangement in the area. In certain areas and for certain improvements, a reserve for replacement of these items must be included in addition to standard maintenance expenses.

Appraisers should be aware of all the lease terms applicable to the property being appraised as well as the lease terms for the comparable properties. All data must be handled in a consistent manner. Obviously, not all expense items will be included in every situation. However, consistent data on the subject and comparable properties must be collected. Sources of expense data vary, and the records of the property being appraised are a good place to start.

Table 13.7 presents an income and expense statement for the cash-rented farm described previously. (See Table 13.3.)

Table 13.7 Cash Rental Income and Expense Summary

Land Type	Acreage	Rent per Unit	Total Cash Rent
Cropland A	120	$90 per acre	$10,800
Cropland B	300	$60 per acre	$18,000
Pasture	130	$30 per acre	$3,900
Buildings		$1,500 per month	$18,000
Recreational lease	550	$30 per acre	$16,500
Total gross income			$67,200
Expenses			
Taxes			$6,060
Insurance			$5,280
Maintenance			$7,200
Management			$5,232
Total expenses			$23,772
Net operating income			$43,428

Management fees are incurred whether the property is professionally managed or managed by the owner-operator. In any case, a property that is leased generates the need for management services. Management fees can be estimated with information from area farm managers and trust departments.

Real estate taxes are recorded at the county or city assessor's office. Often the property owner has a current tax statement. Obtaining the assessor's parcel numbers helps to expedite the investigation process.

Typically real estate taxes are estimated annually. Some counties tax personal property such as machinery, tools, and livestock at the same time, and appraisers might separate the two types of taxes or other tax-related charges such as municipal fees related to irrigation, drainage, or weed control if the scope of work of the assignment warrants. The real estate tax history of the property can be analyzed to help identify trends because appraisers must project this expense, especially when yield capitalization is applied. Many states allow for real estate tax deferral or abatement on qualified agricultural properties (e.g., open space assessment). A change in use or failure to comply with state law may cause a farm to lose its favorable tax status. In some instances, taxes that have been waived or deferred for a specific number of years can become a potential lien on the property. These accumulated deferred taxes may be an adjustment item in the sales comparison approach, particularly when valuing properties in transition.

Insurance rates for various classes of buildings are available from local insurance agents. Insurance documents show a value estimate for the buildings, but this may reflect replacement cost, which could exceed the contributory value of the improvements. Appraisers should determine if the buildings are or can be insured. It is frequently difficult to insure unused or unoccupied farm buildings. The expenses attributable to property improvements, including insurance and building maintenance, depend on local practice. In some areas, typical rental agreements call for the tenant to provide a specified level of insurance on improvements.

Depending on the extent of improvements and community practice, building and other maintenance expenses can range from almost zero to a significant portion of the landlord's income. *These expense estimates must reflect the actual rental terms.* The tenant may assume all maintenance duties or be responsible for repair only, with the general maintenance burden falling on the landlord. For example, the rental contract may require the tenant to replace a specified number of fences and irrigation system components annually.

A reserve for replacement expense is often applicable to short-lived items or components of long-lived items in urban properties, but it is not a typical expense in agricultural appraisal. A center-pivot irrigation system, for example, may be owned by the tenant or the landowner. Establishing a reserve for replacement of this type of asset may be appropriate, but often in agricultural appraisal the expense item remains in the income statement and is therefore left in the capitalization rate. The key is local custom and consistency in the derivation and application of the expense estimate.

When estimating maintenance expenses, appraisers must consider all property improvements:

- buildings

- irrigation and drainage facilities, including pumps, motors, and sprinkler equipment
- specialty improvements such as trellises or permanent plantings
- other non-building improvements such as fences, corrals, and livestock watering facilities

Appraisers ascertain whether the maintenance required is sufficient and, if not, whether additional costs must be included to meet market expectations. In the preceding example, a percentage of the building value is allocated to annual maintenance.

Irrigation expenses are generally available from the irrigation district, county offices, or individual irrigation company staff. These charges will usually cover annual operation and maintenance as well as any debt repayments. The charge can be expressed as an amount per acre, per acre foot of water delivered, or per farm unit, or a combination of these bases can be used. For some irrigation systems, the power charges for pumping must be added to a district charge or considered in the irrigation expense, as is the case with irrigation wells or pumping stations on lakes or rivers. The division of the irrigation charge between the owner and the tenant may vary with the specific rental terms and should be identified.

In a crop share situation, the landlord may participate in growing costs. Growing costs vary by crop and community, and specific farm records are useful. The suppliers and dealers of inputs such as seed and fertilizers should have current product costs as well as delivery and application charges. Under a crop share rental agreement, the landlord and the tenant may share the cost of fertilizer, herbicide, insecticide, harvesting, and hauling. The landlord may participate in labor costs as well.

Some expense items frequently encountered in an owner's expense statement should not be included in the estimate of expenses related to the income-producing capability of the property. These expenses include depreciation, mortgage interest, amortization, income taxes, additions to capital, and unusual owner expenses not directly related to income production and operations.

Net Operating Income

The next step in the income capitalization approach is to derive net operating income. Table 13.7 (shown previously) itemizes the income and expenses for the subject farm property. The effective gross income minus expenses equals the net operating income. This is a good point in the analysis for appraisers to test the reasonableness of the estimated expenses. The expense ratio for the sample property is 35%–i.e., expenses amount to 35% of effective gross income. Is this information consistent with the income, expense, and rental data gathered from comparable properties? Are the expenses treated consistently? Consistency in the treatment of income and expenses for the subject property and the comparable properties is imperative in the income capitalization approach. It is critical in the derivation of an overall capitalization rate.

Direct Capitalization

Direct capitalization is a method used in the income capitalization approach to convert a single year's income estimate into a value indication. This conversion is accomplished in one step, either by dividing the income estimate by an appropriate income rate or multiplying it by an appropriate income factor. In direct capitalization no precise allocation is made between the return *on* and the return *of* capital because the method does not simulate investor assumptions or expectations concerning the holding period, the pattern of income, or changes in the value of the original investment. However, a satisfactory rate of return for the investor and recapture of the capital invested are implicit in the rates or factors applied in direct capitalization because they are derived from similar investment properties.

Direct capitalization may be based on potential gross income, effective gross income, net operating income, equity income, mortgage income, land income, or building income. The income rates used in direct capitalization include

- the overall (property) capitalization rate (R_O)
- the mortgage capitalization rate (R_M)
- the equity capitalization rate or equity dividend rate (R_E)
- the land capitalization rate (R_L)
- the building capitalization rate (R_B)

Income factors include

- the potential gross income multiplier (*PGIM*)
- the gross rent multiplier (*GRM*)
- the effective gross income multiplier (*EGIM*)

Income rates and factors reflect the relationship between income and value and are derived from market data. It is essential that the properties used as comparable sales reflect risk, income, expense, and physical and location characteristics that are similar to the property being appraised. Consequently, income multipliers and rates must be extracted from properties that reflect similar income-to-expense ratios, land-to-building ratios, risk characteristics, and expected changes in income and value over a typical investment holding period.

Derivation of Overall Capitalization Rates

Deriving capitalization rates from comparable properties is preferred when sufficient data on sales of similar, competitive properties is available. Data on each property's sale price, income, expenses, financing terms, and market conditions at the time of sale is needed. In addition, appraisers must be certain that each comparable property's net operating income is calculated and estimated in the same way that the subject property's income is estimated and that non-market financing terms or different market conditions have not affected the prices of the comparable properties. When these requirements are met, appraisers can estimate an overall rate by dividing each property's net operating income by its sale price. The formula is

$$\text{Overall Rate} = \frac{NOI}{\text{Sale Price}}$$

Data requirements for direct capitalization are stringent and sometimes difficult to meet. However, because rural property uses (e.g., crop farms, dairy farms, cattle ranches, orchards) tend to be consistent in a given area, buyers and sellers have similar expectations about the direction and character of income changes, resale prices (or property appreciation rates), tax benefits, and holding periods. Therefore, direct extraction of rates from sales is a viable method of capitalization for rural properties.

Table 13.8 illustrates the derivation of overall capitalization rates from comparable properties. If all three transactions are equally reliable and comparable, an appraiser might conclude that an overall rate of 4.70% to 5.10% is reasonable. In this example, however, Sale 3 is a larger property. As a result, a narrower range of 4.90% to 5.10% is more appropriate, assuming that the subject property is similar in size to Sales 1 and 2. The final rate selected depends on an appraiser's judgment as to how comparable each sale is to the subject property.

If there are differences between a comparable property and the subject property that could affect the overall capitalization rate, the appraiser must account for those differences in the selection process. When rates derived from comparable properties are used, the overall capitalization rate is applied to the subject property in a manner consistent with the manner in which the rate was derived. In other words, if the subject property is appraised using its cash rental

Table 13.8 Capitalization Rate Derivation

	Sale 1	Sale 2	Sale 3
Sale price	$900,000	$825,000	$1,100,000
Net operating income	$45,900	$40,425	$51,700
Indicated overall rate	0.0510	0.0490	0.0470

income, the capitalization rates must be derived on the same basis. Rates derived from crop share income cannot be used to value a cash rental property, and a combination of rates is not appropriate.

Table 13.9 illustrates the final step in arriving at a value opinion by direct capitalization. The net operating income is divided by the overall capitalization rate. A 5% rate was selected based on the rates extracted from the comparable properties. The net operating income calculation is $47,028/0.05 = $940,560. The estimate is rounded to $941,000.

This simple technique for estimating overall capitalization rates is preferred and will produce a reliable indication of value by the income capitalization approach if three conditions are met:

1. Income and expenses must be estimated on the same basis for the subject property and all comparable properties.
2. Market expectations concerning resale prices, tax benefits, and holding periods must be similar for all the properties.
3. The financing terms and market conditions that affect the comparable sales must be similar to those affecting the subject property, or an adjustment must be made for any dissimilarities.

Yield Capitalization

Yield capitalization can be a useful tool when warranted by the property type being appraised. When not warranted by the property type or local market expectations, application of this method should

Table 13.9 Value Indication by the Income Capitalization Approach

Land Type	Acreage	Rent per Unit	Total Cash Rent
Cropland A	120	$90 per acre	$10,800
Cropland B	300	$60 per acre	$18,000
Pasture	130	$30 per acre	$3,900
Buildings		$1,800 per month	$21,600
Recreational lease	550	$30 per acre	$16,500
Total gross income			$70,800
Expenses			
Taxes	$6,060		
Insurance	$5,280		
Maintenance	$7,200		
Management	$5,232		
Total expenses			$23,772
Net operating income			$47, 028
Overall capitalization rate—5%			
Indicated value by the income capitalization approach		$47,028/0.05 = $940,560	
Rounded to:		$941,000	

be considered with care. When yield capitalization is used, future benefits are converted into present value by applying an appropriate yield rate. To select an appropriate yield rate for a market value appraisal, appraisers analyze market evidence of the yields anticipated by typical investors. When investment value is sought, the yield rate used should reflect the individual investor's requirements, which may differ from the requirements typical in the market.

To perform yield capitalization, appraisers

1. select an appropriate holding period,
2. forecast all future cash flows or cash flow patterns,
3. choose an appropriate yield, or discount, rate, and
4. convert the future benefits of property ownership into present value by discounting each annual future benefit or by developing an overall rate that reflects the property's income pattern, value change, and yield rate.

The application of capitalization rates that reflect an appropriate yield rate, the use of present value factors, and discounted cash flow analysis are all features of yield capitalization. Mortgage-equity formulas and yield rate or value change formulas may be used to derive overall capitalization rates. Like direct capitalization, yield capitalization may employ residual techniques. To apply the discounting procedure, appraisers must be familiar with income patterns, capital return concepts, the mathematics of discounting, and investor requirements or assumptions (e.g., holding period, anticipated market growth, inflation) and select appropriate discount rates.[2]

Sample DCF Analysis

The property being appraised is a permanent planting. Table 13.10 illustrates the extraction of an internal rate of return from a comparable property that was purchased and developed as a permanent planting. The analysis is performed on a before-tax basis and with no consideration of debt service. The comparable property was purchased for $544,000. The costs to establish the permanent planting were $844,800 in Year 1 and $627,200 in Year 2. Year 3 is the first year in which a crop is produced. By Year 5, full production is reached. Expenses are incurred during Years 3 through 5 for operations, maintenance, and harvesting. The property is resold at the end of Year 5 for $3.2 million.

Table 13.11 illustrates the application of the market-derived internal rate of return to the cash flows of the subject property. The subject property is a permanent planting with establishment costs

2. A more detailed discussion of yield capitalization can be found in *The Appraisal of Real Estate*, 14th ed. (Chicago: Appraisal Institute, 2013), 509–540.

Table 13.10 Discount Rate Derivation from Comparable Property

		Year 0	Year 1	Year 2	Year 3	Year 4	Year 5
Gross income			$0	$0	$192,000	$384,000	$576,000
Expenses for development			-$844,800	-$627,200	$0	$0	
Production expenses			$0	$0	-$206,720	-$237,920	-$269,120
Land acquisition		-$544,000					
Net operating income		-$544,000	-$844,800	-$627,200	-$14,720	$146,080	$306,880
Reversionary value							$3,200,000
Cash flow		-$544,000	-$844,800	-$627,200	-$14,720	$146,080	$3,506,880
*IRR**	16.04%						

* Calculated using a financial calculator.

Table 13.11 Discounted Cash Flow—Subject Property

	Year 1	Year 2	Year 3	Year 4	Year 5	Total
Gross income			$240,000	$480,000	$720,000	
Expenses for development	-$680,000	-$1,056,000	$0	$0	$0	
Production expenses	$0	$0	-$258,400	-$298,000	-$336,400	
Net operating income	-$680,000	-$1,056,000	-$18,400	$182,000	$383,600	
Reversionary value					$4,000,000	
Cash flow	-$680,000	-$1,056,000	-$18,400	$182,000	$4,383,600	
PV factors @ 16%	0.862069	0.743163	0.640658	0.552291	0.476113	
Present value	-$586,207	-$784,780	-$11,788	$100,517	$2,087,089	$804,831

of $680,000 in Year 1 and $1,056,000 in Year 2. The first harvest is in Year 3, and mature production is reached in Year 5. Expenses include operating, maintenance, and harvesting costs. The reversionary value at the end of Year 5 is $4 million.

Chapter 14

Land Valuation

In rural property appraisal, land may be valued as a component of the cost approach application, or a parcel of rural land may in itself be the property being appraised. In either case, the opinion of land value must be based on the land value of the subject property as though vacant, or unimproved, and available for its highest and best use. Comparable sales of unimproved land within the subject market area should be used to derive land value.

A search for sales of comparable properties in the subject property's market area must be conducted to provide data. The comparable properties should be as much like the subject property as possible, particularly with regard to highest and best use. For example, consider a 500-acre subject property with a good location. Using the categories established by the Natural Resources Conservation Service, the property consists of 300 acres of Class I land and 200 acres of Class IV land. The yield index on the Class I land for both the comparable properties and the subject property is 45, with normal production at 80 to 90 bushels per acre for small grains. Searching the market area reveals sales of eight unimproved properties that are similar to the subject property. These sales are of various mixtures of Class I and IV land, as summarized in Table 14.1. The summary grid shows the calculations of land value per acre for Sales 1 and 2. However, the remaining sales have two types of land, and the per-acre contributory values for these sales cannot be completed without further analysis.

It is apparent that there are a number of differences between the comparable properties and the subject property. These differences include

- sales dates ranging from current to one year ago
- favorable seller financing
- good to average locations
- various mixtures of land

To provide a reliable indication of the value of the subject land, these differences need to be reconciled.

The first difference to be considered is land mix. In this market, the productivity and earnings capacity of Class I and Class IV land differ. This difference is reflected in sales prices. Sale 1, with all Class IV land, sold for $500 per acre. Sale 2, which is all Class I land, sold at the same time for $1,000 per acre.

When identifying land types, appraisers must take care to reflect the market. Different market areas will indicate different types of land categories. Some markets are very sensitive to the land classifications of the Natural Resources Conservation Service, while others are not. For example, the difference between Class I land and Class III land in a corn- or soybean-producing area would most likely be very great. However, the market may perceive the difference between Class IV and VI pasture to be very slight or nonexistent. In some areas, productivity indexes such as the crop productivity index or corn suitability rating are the recognized measures of land categories. Some areas are so homogeneous as to provide no indications of differing land types. In locations with wetlands, the per-acre sales prices of land may differ greatly based on the amount of tillable land. In other areas, much broader land types are recognized, such as simply cropland and pasture or irrigated land and dry cropland. Appraisers must identify how the local market differentiates between various types of land. An adjustment should be made only if there is a market-derived basis for the adjustment.

Table 14.1 Sales Summary

Sale	Date	Sale Price	Acres	Terms	Location	$/Acre	Class I Acres	Class I $/Acre	Class IV Acres	Class IV $/Acre
1	6 mos.	$125,000	250	Cash	Average	$500	n/a		250	$500
2	6 mos.	$500,000	500	Cash	Average	$1,000	500	$1,000	n/a	
3	6 mos.	$362,500	525	Cash	Average	$690	200		325	
4	Current	$320,625	375	Cash	Average	$855	300		75	
5	1 year	$696,100	700	Cash	Average	$994	600		100	
6	Current	$517,500	690	Cash	Good	$750	300		390	
7	6 mos.	$400,000	526	Cash	Good	$760	200		326	
8	1 year	$650,000	594	Seller finan.	Average	$1,094	509		85	

In this demonstration, Sales 1 and 2 each have only one type of land. These sales form the basis for value allocations by land class for the sales with a mix of land classes. Sale 3 is the mixed land class sale that is most similar to Sales 1 and 2 in other respects, so it is the first to be analyzed. If the $1,000-per-acre value for Class I land indicated by Sale 2 is applied to Sale 3, the remaining $500 per acre can be allocated to the Class IV land in Sale 3, as shown below.

Sale price	$362,500
Less: 200 acres Class I × $1,000 per acre =	– 200,000
Residual to Class IV land	$162,500
$162,500/325 acres Class IV = $500 per acre	

The application of data from Sales 1 and 2 to Sale 3 supports the similarities between the sales of property of a single land class and the sales of properties of mixed land classes. The ratio analysis method is applied to the remaining sales. Sales 1 through 3 are basically only different in terms of land mix.

The comparisons of Sales 1, 2, and 3 indicate that Class I land sells for $1,000 per acre and Class IV land sells for $500 per acre. Thus, Class IV land has a value of 50% of Class I land, calculated as $500/$1,000. This ratio is used to derive the value contribution of Class I and IV land on the rest of the sales. Consider Sale 8. The procedure for allocating the sale price between the Class I and IV land can be set up as shown below.

Class I land: 509 acres × 100% (1) =	509.0
Class IV land: 85 acres × 50% (0.5) =	42.5
	551.5

This calculation converts all the property's land to the equivalent of Class I land. Sale 8 has 551.5 net Class I–equivalent acres. With a ratio of 50%, the logic behind the procedure can be stated another way: If the Class IV land is worth half as much as the Class I land, then it would take two acres of Class IV land to equal one acre of Class I land. The 85 acres of Class IV land divided by two is 42.5 acres. In other words, 85 acres of Class IV land is the same as 42.5 acres of Class I land. To complete the ratio analysis calculations, the net Class I-equivalent acreage is divided into the total sale price to provide the indication of the Class I contributory value per acre. The Class IV land allocation is based on the ratio of Class IV to Class I land multiplied by the Class I value. The analysis of Sale 8 is completed as follows:

$650,000/551.5 =	$1,178.60 per acre (value of the Class I land)
$1,178.60 × 50% (0.5) =	$589.30 per acre (value of Class IV land)

Algebraically, the calculation to find the net Class I-equivalent acreage is shown below:

$509\ (1x) + 85\ (0.5x) =$	\$650,000
$509x + 42.5x =$	\$650,000
$551.5x =$	\$650,000
$x =$	\$1,178.60 per acre for Class I land
$\$1,178.60 \times 50\% =$	\$589.30 per acre for Class IV land

The proof of the ratio calculation is summarized as follows:

509 acres Class I × \$1,178.60 per acre =	\$599,907.40
85 acres Class IV × \$589.30 per acre =	50,090.50
Total	\$649,997.90

The total sale price indicated by the "proof" varies only slightly from the actual sale price of \$650,000. The \$1,178.60 could be rounded to \$1,179 per acre. If the ratio of 50% is applied to the rounded figure, the allocation for Class IV land in Sale 8 is \$590 per acre. Rounding the indications will cause the proof to vary somewhat more from the actual sale price, but will make little difference in the total valuation process.

The contributory values of the land in each class for the other properties are calculated in the same manner, using the 50% ratio with the different acreage of Class IV land. The results are shown in Table 14.2.

Table 14.2 Land Sale Analysis

Sale	Date	Sale Price	Acres	Terms	Location	\$/Acre	Class I Acres	Class I \$/Acre	Class IV Acres	Class IV \$/Acre
1	6 mos.	\$125,000	250	Cash	Average	\$500	n/a		250	\$500
2	6 mos.	\$500,000	500	Cash	Average	\$1,000	500	\$1,000	n/a	
3	6 mos.	\$362,500	525	Cash	Average	\$690	200	\$1,000	325	\$500
4	Current	\$320,625	375	Cash	Average	\$855	300	\$950	75	\$475
5	1 year	\$696,100	700	Cash	Average	\$994	600	\$1,071	100	\$535
6	Current	\$517,500	690	Cash	Good	\$750	300	\$1,045	390	\$523
7	6 mos.	\$400,000	526	Cash	Good	\$760	200	\$1,102	326	\$551
8	1 year	\$650,000	594	Seller finan.	Average	\$1,094	509	\$1,179	85	\$590

Sale	Date	Financing	Class I	Class IV
5	1 year	Cash	\$1,071 per acre	\$535 per acre
8	1 year	Owner	\$1,179 per acre	\$590 per acre

After the land allocations are made, indications of the effects of favorable seller financing and market conditions can be developed. Sale 8 sold with favorable seller financing. Sale 5 occurred at the same time under the same market conditions and is similar except that it is a cash sale.

The differences between allocated values within the respective classes provide indications of the effect of favorable terms. The Class I comparisons indicate a difference of $108 per acre, or -9.16% ($108/$1,179). The Class IV comparisons indicate a difference of $55 per acre, or -9.32% ($55/$590). From this information it is reasonable to conclude that a minus 9% adjustment for favorable financing terms should be applied to Sale 8 when compared to the rural property being appraised. Sales 4 and 6 are current. Sales 3 and 5 are older sales with locations similar to Sale 4. Sale 7 is a six-month-old sale with a location similar to Sale 6. Comparisons between these sales are shown in Table 14.3.

Table 14.3 Market Conditions Adjustments

Sale	Date	Class I	Class IV	Compound Change
4	Current	$950 per acre	$475 per acre	
3	6 months	$1,000 per acre	$500 per acre	-0.85% per month
5	1 year	$1,071 per acre	$535 per acre	-0.99% per month
6	Current	$1,045 per acre	$523 per acre	
7	6 months	$1,102 per acre	$551 per acre	-0.88% per month

In this text, market conditions adjustments are measured and applied on a compound basis. Using a financial calculator, the keystrokes for the comparison of Sales 3 and 4 using Class I land are as follows:

1000 [CHS] [PV]
950 [FV]
6 [n]
solve for [i]

The change can be measured using the Class I or Class IV land. Rounded to the nearest 1/100th of 1%, the indications are the same. With the range from -0.85% to -0.99% per month, a time adjustment of -0.9% per month is appropriate.

The indicated adjustments for favorable financing and time are applied to the selected comparable property sales. The adjustments can be applied by one of two methods–the total sale price can be adjusted for favorable financing and market conditions, after which each sale is reallocated, or the adjustments can be applied directly to the allocated values of each land class on the eight sales. The indications produced using these methods are essentially the same, although there may be variances of a few dollars per acre due to rounding.

Sale 7 is a six-month-old sale of a comparable property with a total sale price of $400,000, cash. If this sale is adjusted to the current time period using a time adjustment of -0.9% per month, the time-adjusted price is $378,880. Sale 7 has 363 net Class I-equivalent acres. Dividing 363 into the adjusted price of $378,880 provides an allocation of $1,044 per acre for the Class I land. The Class IV land, at 50% of the value of Class I land, is allocated at $522 per acre. Alternatively, the time adjustment of -0.9% per month can be applied to the Class I allocated price of $1,102 per acre, yielding a time-adjusted price of $1,044 per acre. When the time adjustment is applied to the Class IV allocation, the indicated time-adjusted price is $522 per acre.

Sale 8 requires adjustments for favorable financing and time. The adjustment for favorable financing is applied first. The time adjustment is then applied to the sale price that has already been adjusted for financing, as shown below.

Sale price	$650,000
Less 9% for favorable financing	– 58,500
Cash-equivalent sale price	$591,500

The HP-12C calculator keystrokes are

$591,500	[CHS] [PV]
12	[n]
0.9	[CHS] [i]
solve for [FV] = $530,687	
$530,687/551.5 net acres = $962 per acre	

Alternatively, the following calculations can be used.

Sale price Class I	$1,179/acre
Less 9% for favorable financing	– 106/acre
Cash-equivalent price Class I land	$1,073/acre

The HP-12C calculator keystrokes are

$1,073	[CHS] [PV]
12	[n]
0.9	[CHS] [i]
solve for [FV] = $963/acre	

Table 14.4 shows that the value indications for Class I land vary from $947 to $1,045 per acre. Class IV land ranges from $474 to $522 per acre. Sales 6 and 7 have good locations like the subject property, while the other comparable properties have average locations. The Class I indications from Sales 6 and 7 are $1,045 and $1,044 per acre. Class IV

Table 14.4 Sales Adjusted for Favorable Financing and Time

Sale	Date	Adjusted Sale Price	Acres	Terms	Location	$/Acre	Class I Acres	Class I $/Acre	Class IV Acres	Class IV $/Acre
1	6 mos.	$118,000	250	Cash	Average	474	n/a		250	$474
2	6 mos.	$473,600	500	Cash	Average	947	500	$947	n/a	
3	6 mos.	$343,360	525	Cash	Average	654	200	$947	325	$474
4	Curr.	$320,625	375	Cash	Average	855	300	$950	75	$475
5	1 year	$624,533	700	Cash	Average	892	600	$961	100	$480
6	Curr.	$517,500	690	Cash	Good	750	300	$1,045	390	$523
7	6 mos.	$378,880	526	Cash	Good	720	200	$1,044	326	$522
8	1 year	$530,687	594	Seller finan.	Average	893	509	$963	85	$482

indications from these comparable properties are $523 and $522 per acre. An opinion of land value for the subject property is developed from these similar sales of comparable properties as shown below.

300 acres Class I land × $1,045 =	$313,500
200 acres Class IV land × $523 =	104,600
Land value	$418,100

In some market areas, sales will exhibit many differences beyond land mix, terms, and time. In the example, it is apparent that location is also a factor, but focusing on those properties most similar to the subject property eliminated the need to adjust for location. When there are additional characteristics that affect value, such as size, shape, access, and crop allotments, it may be necessary to complete a more detailed analysis of land value that includes adjustments for these important market characteristics. A direct comparison technique like the one shown in Chapter 12 can be used to develop an opinion of land value for the cost approach. Using the techniques explained in Chapter 12, a direct comparison grid for the hypothetical subject property can be created (see Table 14.5).

Table 14.5 Direct Comparison Grid

	Subject	Sale 3	Sale 4	Sale 5	Sale 6	Sale 7	Sale 8
Date	Current	6 months	Current	1 year	Current	6 months	1 year
Acres	500	525	375	700	690	526	594
Financing	Cash	Cash	Cash	Cash	Cash	Cash	Terms
Location	Good	Average	Average	Average	Good	Good	Average
$/Acre		$690	$855	$994	$750	$760	$1,094
Land adj.		+ 110	− 95	− 137	+ 86	+ 121	− 152
Adjusted $/acre		800	760	857	836	881	942

For the financing terms adjustment, compare Sale 5 to Sale 8, indicating $85, or 9% (rounded):

Adjustment	0	0	0	0	0	– 85
Subtotal	800	760	857	836	881	857

For the market conditions adjustment, compare Sale 3 to Sale 4, indicating $40, or -0.851% per month; compare Sale 4 to Sale 5, indicating $97, or -0.996% per month; and compare Sale 6 to Sale 7, indicating $45, or -0.87% per month; for a final adjustment of -0.9% per month:

Adjustment	– 42	0	– 88	0	– 47	– 88
Subtotal	758	760	769	836	834	769

For the location adjustment, compare Sale 5 to Sale 6, indicating $67; compare Sale 3 to Sale 7, indicating $76; and compare Sale 7 to Sale 8, indicating $65; for a final adjustment of $70 per acre:

Adjustment	+ 70	+ 70	+ 70	0	0	+ 70
Adj. $/Acre	828	830	839	836	834	839

Sales 6 and 7 are most similar to the subject and provide the most reliable indications of value. An appropriate land value for the subject property is: 500 acres × $835 = $417,500

Alternative Land Valuation Techniques

The most reliable way to develop a land value opinion is through the comparison of sales of similar unimproved properties. However, when few sales are available or additional support is desired, other techniques can be used. Because the alternative techniques are not widely used in rural appraisals, mainly due to the difficulty appraisers face in finding the significant amount of data needed for them, they are only described briefly in Table 14.6.

Table 14.6 Alternative Land Valuation Techniques

Technique	Characteristics	Applicability to Rural Property
Land residual method	The residual income to the land (total *NOI* less income attributable to the improvements) is capitalized.	Generally most applicable to highly improved properties such as extensive feeding facilities for hogs or poultry, processing plants, or storage facilities in which a substantial portion of the property value lies in the improvements.
Allocation	Sales of comparable properties are analyzed, with sale prices allocated between the land and building components, with the ratio applied to the estimated value of the subject property.	Generally not used as the primary method of land valuation, rather as a secondary test of reasonableness.
Extraction	The land value is developed by subtracting the cost (plus depreciation) of the improvements from the sale price.	Most applicable when improvement values are low or easily identified, as is the case in appraising a large ranch or farm with few improvements.
Subdivision development analysis	The land value is estimated at points along the development timeline using discounted cash flow analysis.	Most often used when the property is in a transitional stage from agricultural to urban use.
Ground rent capitalization	Land value is developed through direct capitalization of ground rent.	May be used to value land for airports, garbage transfer sites, and power or telephone substations.

Chapter 15

The Cost Approach

Land is usually the major component of value in a rural property, and the value of buildings and other improvements can be relatively minor in relation to the land value. As a result, the cost approach, which focuses on the current cost of reproducing or replacing improvements, often has less applicability than the sales comparison and income capitalization approaches in the valuation of rural properties.

The cost approach is applicable, however, in the valuation of rural properties on which the structural improvements account for a significant allocation of the market value of the property as a whole. For example, concentrated animal feeding operations (CAFOs), dairy facilities, grain processing facilities, and certain other agricultural properties involve specialized improvements that have much in common with industrial facilities, for which the cost approach can be particularly useful.

Also, cost approach techniques are applicable for allocating value between land and improvements, when the scope of work of a valuation assignment involves that sort of allocation.

Cost-Estimating Techniques

The techniques used to estimate reproduction cost or replacement cost include

1. The quantity survey method (or segregated cost method)
2. The unit-in-place method
3. The comparative-unit method (or calculator method)

The quantity survey method is the most detailed, complex, and costly method of estimating costs. It requires expert analysis and is used in rural appraisal when exact figures are needed. The quantity survey method involves adding together all types of labor and materials, subcontractor fees, and equipment required to reproduce or replace an improvement. Each item of cost is expressed in terms of current local prices and wages and calculated by the hour, pound, or other unit of measurement. Overhead, entrepreneurial profit, and other indirect costs may be added as a lump sum or as a percentage. This method requires a detailed knowledge of all the components of the construction process and precise measurement and calculation of the quantities used. The quantity survey method is used to estimate the cost of highly customized properties that have unique features and in situations when other information is not available. A quantity survey is shown in Table 15.1.

When applying the unit-in-place method, appraisers estimate the unit costs of materials or components of the structure as installed. The measurement includes both the cost of materials and the labor necessary to put each unit in place. The structure is generally broken down into large units: exterior walls, roof, foundation, floors, and other elements. A unit cost expressed in dollars per square foot or linear foot for installation is added to larger unit costs. Equipment and fixtures are usually added as a lump sum, which includes expenses for installation. Other indirect costs and overhead are then

Table 15.1 Reproduction Cost—Quantity Survey Method

Description	Quantity	Cost
42,500-bushel galvanized steel grain bins with ladders	2 @ $27,625	$55,250
1,500-bushel-per-hour, 50-ft. elevator leg with 9-in. x 16-in. grain buckets	1	8,500
100-ft., 6-in. steel discharge pipe	1	2,250
75-ft. discharge augers and steel pipe from bins to elevator leg	2 @ $3,187.50	6,375
10-ft. discharge auger and steel pipe from dump pit to elevator leg	1	850
Steel dump grate spouts	1	450
Concrete work - cone-bottom floors and dump pit		22,950
Subtotal - materials		$96,625
Labor for installation		71,000
Electrical subcontract (motors, controls, installation)		2,425
Fees and permits		1,575
Ground preparation and excavation		1,850
Subtotal - materials and installation		$173,475
Sales tax 5%		8,674
Total new cost		$182,149
Entrepreneurial incentive 5%		9,107
Reproduction cost		$191,256

added either as lump sums or percentages. Accurate measurements and descriptions are essential for the successful use of this method. A unit-in-place cost estimate is shown in Table 15.2.

Table 15.2 Reproduction Cost—Unit-in-Place Method

Description	Quantity	Cost per Quantity	Cost
Two cylindrical steel grain bins with 34-ft. vertical height	68 feet	$1,375.00 per foot	$93,500
Concrete cone-bottom floors	2,512 square feet	$13.50	33,912
Discharge augers	75 feet	$127.00	9,525
Grain elevator leg with cages, ladders, and distributor	50 feet	$285.00	14,250
Dump pit construction			8,500
6-in. downspouts	100 feet	$25.00	2,500
Electrical motors and controls			3,500
Site excavation and preparation	20 hours	$75.00	1,500
Subtotal			$167,187
Current cost multiplier			× 1.03
Subtotal			$172,203
Local cost multiplier			× 0.95
Total direct costs			$163,592
Indirect costs of architecture, engineering, appraisal, interest during construction, insurance, and permits			× 1.10
Total cost new			$179,952
Entrepreneurial incentive 5%			8,998
Reproduction cost			$188,950

In the comparative-unit method, which is often referred to as the *square foot method*, all components of the structure are combined and their costs are estimated based on the cost per unit (e.g., square foot) of building area or cost per volume (e.g., bushels) of building capacity. The costs used are usually completed construction costs, including all installation expenses, builder's overhead, and profit. Most cost estimating services provide information on a square-foot or cubic-foot basis. This method may be the least accurate of the three, but it is the simplest to apply, the least time-consuming, and the least expensive. If an appraiser applies the comparative-unit method carefully, it provides a cost estimate that is sufficiently detailed and accurate for estimating the cost of new rural improvements. Some in-office quality control should be exercised to ensure that the cost manual provides a cost indication similar to the costs quoted by local contractors. The comparative-unit method is illustrated in Table 15.3.

Table 15.3 Reproduction Cost by the Comparative-Unit Method

Description	Cost per Bushel
Base cost for two, 42,500-bushel, galvanized steel grain bins with concrete cone bottoms	$1.50
Unloading augers: 12-ft., U-trough, covered	0.13
50-ft., 1,500-bushels per hour elevator leg with boot pit	0.20
Concrete, steel-lined dump pit	0.12
Spouting with 4-hole distributor	+ 0.05
Subtotal	$2.00
Current cost multiplier	× 1.03
Subtotal	$2.06
Local cost multiplier	× 0.95
Total direct costs	$1.96
Architect's fee @ 8%	+ 0.16
Total cost new per square foot	$2.12
Total cost for 85,000-bushel facility	$180,200
Entrepreneurial incentive 5%	9,010
Reproduction cost	$189,210

Depreciation

Depreciation can be caused by physical factors such as breakage, deterioration, and simple aging, such as in the case of a barn that has lost value through wear and tear over a decade of use. A rural property may have additional depreciation due to functional obsolescence, which is a market resistance caused by an internal characteristic such as outmoded or inadequate dairy equipment, inefficient access to a livestock facility, or old and undersized grain storage bins. External obsolescence is caused by conditions outside the property, such as an inharmonious user nearby, lack of demand, changing uses in the area, and unfavorable economic conditions.

A depreciation rate can be developed through analysis of several improved sales within the subject property's market. Market support for depreciation using the extraction technique is preferred but other, less precise, procedures such as the age-life method can also be used.

To derive a depreciation rate from improved sales, one must first estimate the reproduction or replacement cost for the buildings on each comparable sale. The difference between the costs of the buildings and their actual contributory value is attributable to the overall depreciation they have suffered from all causes (physical deterioration and functional and external obsolescence).

The contributory value of the buildings on each sale property is derived by first estimating land value for the sales. These land value estimates should come from unimproved sales in the same market. Remember that each improved sale's building contribution must be estimated *as of the date of sale.* This will eliminate any confusion that might arise in markets where land is increasing or decreasing in value at a different rate than buildings. Because it is necessary to estimate depreciation as of the date of sale, it is imperative to also estimate costs as of the date of sale.

Once the building contribution "as of the date of sale" is estimated for each improved sale, that contributory value can be subtracted from each sale's respective cost estimate to reflect the total overall depreciation. These overall depreciation rates may be refined to reflect a separate depreciation rate for each building based on its effective age and contributory value versus its cost new. Remember, this overall depreciation rate encompasses all forms of depreciation that may exist in the buildings:

- physical deterioration
- functional obsolescence
- external obsolescence

Physical Deterioration

Physical deterioration is the adverse effect on value caused by deterioration or impairment of the improvement's condition as a result of disintegration or wear and tear. Physical deterioration is divided into two major categories:

- curable
- incurable

Some causes of physical deterioration include the elements, wear and tear, use, damage, breakage, pest or insect infestation, and deferred maintenance.

Curable physical deterioration is also referred to as *deferred maintenance.* In estimating curable physical deterioration, appraisers should not include any excess cost to repair or prepare the surface for the item that is included in the reproduction or replacement cost estimate. For example, if panels on a grain bin need replacing, the cost to remove the old panels and prepare the remaining surface to accept the new panels will exceed the reproduction cost of the new panels if they were installed during original construction. Therefore, the curable physical deterioration will include the cost of the panels as if installed during original construction plus an excess cost for preparation (see Table 15.4). For future depreciation estimates, only the cost actually included in reproduction cost should be deducted from reproduction cost to obtain the base for estimating incurable physical deterioration in long-lived items.

Table 15.4 Deferred Maintenance	Construction Cost	Cost to Cure
Installation of new panels	$3,500	$3,500
Removal of old panels	N/A	+ $1,000
Total	$3,500	$4,500
Curable physical deterioration: $4,500		

Incurable physical deterioration refers to items whose physical deterioration either cannot be corrected or can be corrected only at a cost in excess of the amount the item contributes to the value of the property. Incurable physical deterioration is divided into short-lived and long-lived components.

A short-lived item is a component with an economic life that is shorter than the structure's economic life. Short-lived items such as heating and cooling systems, floor and roof coverings, and appliances will typically be replaced several times during the life of the improvement. These items are not economically feasible to replace as of the date of appraisal, but will most likely move into the physical curable category at some future date. This form of depreciation is typically estimated with an age-life analysis. Any items that are considered short-lived and physically curable would not be depreciated at this step in this analysis, as shown in Table 15.5. Note that the figure for cost new should include all costs (direct, indirect, and entrepreneurial incentive).

Table 15.5 Estimate of Short-Lived Items of Incurable Physical Deterioration

Item	Cost New	Actual Age	Physical Life	Ratio	Depreciation
Floor covering	$3,000	10	20	50%	$1,500
Appliances	$2,000	6	10	60%	$1,200
Roof covering	$4,000	6	20	30%	$1,200
Total incurable physical deterioration in short-lived items					$3,900

Improvement components that are not and will not be economically feasible to replace in the future are categorized as long-lived items of incurable physical deterioration. These items, which are expected to last throughout the entire economic life of the improvement, include the structure itself, foundation walls, roof structure, and other components. This form of depreciation is typically measured (a) with an age-life analysis, (b) with a modified age-life analysis, or (c) through application of a market-derived depreciation rate. The reproduction or replacement cost of the items depreciated in the curable and incurable short-lived categories must be deducted from reproduction cost to avoid double depreciation. The procedure is illustrated in Table 15.6.

Table 15.6 Estimate of Long-Lived Items of Incurable Physical Depreciation

Reproduction cost	$125,000
Less cost new of curable items included in reproduction cost	– 1,000
Less cost new of incurable short-lived items	– 11,000
Remaining cost new to be depreciated	$113,000
Long-lived percentage (by method estimated)	40%
Less dollar amount of depreciation ($113,000 × 40%)	$45,200
Depreciated reproduction cost	$67,800

Functional Obsolescence

The major test used to determine whether an item of functional obsolescence is curable or incurable is the relationship between the cost to cure and the increased income or benefits to be derived by curing the obsolescence. Additionally, some superadequacies such as an excess foundation or an excessive grain handling system may be classified as incurable because it is not physically practical to cure them. Most farm improvements are not rented separately from the farm operation, and it is difficult to actually compare economic benefits with the cost to cure. In valuing specialized agricultural buildings in areas where rental markets do exist, the relationship between cost and increased rental income can be measured. For the rural appraiser, the question is often: Would a prudent operator spend the money necessary to cure the obsolescence to receive the additional benefit from the improvement? Curable functional obsolescence may be divided into three subcategories:

1. deficiency requiring addition
2. deficiency requiring substitution or modernization
3. superadequacy

A *deficiency requiring addition* is measured by how much the cost of the addition exceeds the cost of the item if it were installed new during construction. The curable functional obsolescence calculation shown in Table 15.7 concerns a barn that was built with a dirt floor, while current building design requires a concrete floor.

A *deficiency requiring substitution or modernization* is measured as the cost of the existing component included in the cost estimate minus the physical deterioration already charged against the existing component and salvage value, plus the cost to remove the existing component and the added cost of installation. *Salvage value* is defined as the value attributed to an item that is to be removed from the property, usually for use elsewhere.

The curable functional obsolescence calculation shown in Table 15.8 concerns equipment in a commercial broiler facility that is out-

Table 15.7 Estimate of Curable Functional Obsolescence—Deficiency Requiring Addition

Step 1.	Cost of dirt floor	$0
Step 2.	Depreciation previously charged	– $0
Step 3.	Cost to install a concrete floor in existing barn	$5,500
Step 4.	Cost to install a concrete floor in existing barn if the structure were being built on the date of the appraisal	– 5,000
Step 5.	Depreciation	$500

Table 15.8 Estimate of Curable Functional Obsolescence—Deficiency Requiring Substitution or Modernization

Step 1. Cost new of existing equipment already included in cost new estimate	$24,000
Step 2. Less physical deterioration previously charged	– 21,200
Step 3. Cost to cure ($1,500 removal cost – $1,000 salvage value + $26,500 cost of installing new equipment)	27,000
Step 4. Plus cost to install new equipment	– 26,500
Step 5. Depreciation	$3,300

dated due to changes required by the integrator. There are no problems with the building structure. The question as to whether or not the depreciation is curable is easily answered because the integrator will not place chickens in the house until this modernization is completed.

A *superadequacy* is measured as the current reproduction cost of the item minus any physical deterioration already charged, plus the cost of removal, minus any salvage value. A superadequacy is only curable if it can be removed and value is added (or costs reduced) to the property–including any salvage value–by its removal.

Table 15.9 relates to a farm that is deficient in hay storage and includes an old barn with numerous mangers and stalls that could be converted into open space storage if these structural components were removed. The test of feasibility is obviously met because a prudent operator would most likely convert the old barn rather than build a new barn.

Table 15.9 Estimate of Curable Functional Obsolescence—Superadequacy

Step 1.	Current reproduction cost of the manger and stalls	$16,000
Step 2.	Less physical deterioration previously charged	– 12,650
Step 3.	Cost to cure (i.e., removal of stalls and mangers)	2,200
Step 4.	Cost of proper item if installed originally (i.e., open storage)	– 0
Step 5.	Loss in value	$5,550

Incurable functional obsolescence may be caused by a deficiency or a superadequacy. An item that should have been included in the cost new but was not causes obsolescence due to a deficiency; excess items included in construction that are included in reproduction cost cause obsolescence due to a superadequacy. The test of curability or incurability can be whether it is physically possible or financially feasible to add or remove the item that causes the obsolescence.

Incurable functional obsolescence due to a deficiency occurs when a typical building of the same type includes a feature or component that is not present in the subject building. The addition of this feature to the subject would enhance the market value, but not to a level that equals the cost to cure the deficiency. If a defined rental market exists for the type of building being appraised, the determination of feasibility is easier than it would be if there were no market rental structure, which is the case with most farm buildings.

An example of incurable functional obsolescence caused by a deficiency would be a hay barn constructed for use with square hay bales (see Table 15.10). In many areas, square bales have been replaced with round bales, and the type of barn required has also changed. The barn in question has a 14-ft. interior height and three round bales stacked 17 feet high. The barn can be used for round bale storage, but it is not as efficient as a barn with a 17-ft. interior height that allows the stacking of three bales. The deficiency in question here is height, and it is measured by comparing barns with sufficient height to barns with insufficient height. The obsolescence is incurable, at least from a feasibility standpoint, although it may not be physically impossible to correct. The market in this area indicates similar physical deterioration for both types of barns, but a barn with deficient height sells for 25% less than the physically depreciated value of the higher barn.

Incurable functional obsolescence due to a superadequacy relates to features within the building that are in excess of current standards. They may be physically impossible to remove, as is the case with a superadequate foundation, or they may be removable only at a cost that exceeds the added value. Incurable functional obsolescence due

Table 15.10 Estimate of Incurable Functional Obsolescence—Deficiency

Step 1.	Current reproduction cost of the barn	$50,000
Step 2.	Less physical deterioration previously charged	– 25,000
Step 3.	Value of the loss (25%)	6,250
Step 4.	Cost of proper item if installed originally	– 20,000
Step 5.	Depreciation	$11,250

to a superadequacy is measured differently depending on whether a reproduction or replacement cost estimate is used. If reproduction cost is the base, appraisers must consider the cost to construct a building that contains the same superadequacy. The superadequacy is measured as the reproduction cost of the superadequate component, minus any physical deterioration already charged, plus the present value of any added cost of ownership due to the presence of the superadequacy, less any value added such as increased rent. The charge for increased operating expense is based on items such as additional taxes, insurance, maintenance, and utility charges. Replacement cost does not include the cost to construct the superadequacy, so no loss in the value of a superadequacy due to physical deterioration is considered when replacement cost is used. However, a charge is required if the superadequacy causes a monetary burden of ownership reflected in additional taxes, insurance, maintenance, and utility charges. The capitalized value of these additional expenses, over and above any increase in market rent resulting from the superadequacy, represents the loss in value caused by incurable functional obsolescence due to the superadequacy. This type of functional obsolescence can also be measured by market extraction, as illustrated in the case study later in this chapter.

Assume that a farm has a superadequate grain handling system (see Table 15.11). A grain handling system is necessary, but not one as elaborate as the system included in the subject property. Comparing typical systems to the elaborate system indicates that the functional obsolescence can be estimated at 40% of the physically depreciated value.

Table 15.11 Estimate of Curable Functional Obsolescence—Superadequacy

Step 1.	Current reproduction cost of the facility including the superadequacy	$150,000
Step 2.	Less physical deterioration previously charged	– 42,500
Step 3.	Value of the loss (40%)	43,000
Step 4.	Cost of proper item if installed originally	– 75,000
Step 5.	Depreciation	$75,500

External Obsolescence

External obsolescence is the adverse effect on value from influences outside the property itself. External obsolescence affects both land and improvements. The result of the adverse influence on the subject's land is considered in the opinion of land value; only the loss imputable to the improvements is considered depreciation. External obsolescence results in some degree of market rejection of a property, and appraisers must measure the market rejection to estimate this type of depreciation. If rental data is available, capitalization of the rental loss is one way to measure external obsolescence. However, because there is usually no defined rental market for agricultural buildings, another technique must be used. One reliable technique for estimating external obsolescence is to compare rural property sales that have external obsolescence to sales that do not. External obsolescence is almost always incurable.

Causes of external obsolescence include the following:

- intrusion of inharmonious uses or users
- heavy traffic
- unpleasant odors
- physical hazards
- a declining market or economy, purchasing power, or demand–all of which result in declining values

As an example, assume that, due to an inharmonious use, buildings on farms in the subject neighborhood contribute less value than similar buildings in areas farther away from the use. The decline attributable to the building has been extracted from market data, which indicates the value decrease to be 30% of the physically and functionally depreciated value (see Table 15.12).

Table 15.12 Estimate of External Obsolescence

Reproduction cost		$150,000
Less physical deterioration	–	42,500
Physically depreciated reproduction cost		$107,500
Less functional obsolescence @ 40%	–	43,000
Physically/functionally depreciated reproduction cost		$64,500
Less external obsolescence @ 30%	–	19,350
Depreciated reproduction cost		$45,150

Methods of Estimating Depreciation

Depreciation is a loss in value from all causes. It is measured from market data and categorized as physical, functional, and external. The separate components of all categories of depreciation are deducted from the new cost of the improvements to estimate their contributory value as of the date of the appraisal. As stated previously, either reproduction or replacement cost can be used, but the handling of certain categories of depreciation may differ depending on the base used. Some forms of functional obsolescence are automatically eliminated when replacement cost is used. Removal costs and added operating costs must be accounted for in the estimation of curable and incurable functional obsolescence, regardless of whether reproduction or replacement cost is used.

In rural appraisal, depreciation is generally estimated using one of four methods:

1. The economic age-life method
2. The modified economic age-life method
3. The breakdown method
4. Market extraction of depreciation rates

The techniques applied within each method also provide the basis for extracting depreciation from market data.

Economic Age-Life Method

In the economic age-life method, a ratio of effective age to economic life is applied to the new cost of the improvements. This method of estimating depreciation lumps all forms of depreciation into a single amount, as summarized in the formula:

$$\text{Depreciation} = \frac{\text{Effective Age}}{\text{Economic Life}} \times \text{Reproduction or Replacement Cost}$$

A set of relationships between economic life, effective age, and remaining economic life can be established based on the definitions of these terms:

$$\text{Economic Life} = \text{Effective Age} + \text{Remaining Economic Life}$$
$$\text{Effective Age} = \text{Economic Life} - \text{Remaining Economic Life}$$
$$\text{Remaining Economic Life} = \text{Economic Life} - \text{Effective Age}$$

A major weakness of the economic age-life method is that it obscures many types of depreciation by grouping them together. The curable items are not treated separately, and no provision is made for the short-lived items, which typically depreciate at a rate much greater than the structure as a whole. This method estimates overall depreciation, but it does not consider functional and external obsolescence and physical deterioration separately, nor does it differentiate between the causes. The economic age-life method is illustrated in Table 15.13.

Modified Economic Age-Life Method

The effect of curable items of depreciation is recognized in the modified economic age-life method. To apply this method, appraisers first estimate the cost to cure all curable items of physical deterioration and functional obsolescence. This sum is deducted from the reproduction or replacement cost as of the date of appraisal. Then the ratio of effective age to economic life is estimated and applied to the remaining

Table 15.13	Estimating Depreciation by the Economic Age-Life Method
Reproduction cost	$100,000
Effective age	10 years
Economic life	50 years
Remaining economic life	40 years
Ratio applied to cost 10/50	20%
Total depreciation	− 20,000
Depreciated value of improvements	$80,000
Land value	250,000
Indicated value by the cost approach	$330,000

portion of the building cost to be depreciated. When curable items are recognized separately, the effective age may decrease and the remaining economic life may be extended. As with the age-life method, appraisers apply the ratio to the entire remaining portion of the structure. If there are many items that depreciate more rapidly than the basic structure (i.e., short-lived items), this method may not be appropriate. Table 15.14 illustrates the modified economic age-life method.

Table 15.14	Estimating Depreciation by the Modified Economic Age-Life Method
Reproduction cost	$100,000
Less physical and functional curable items	− 3,000
Effective age	9 years
Economic life	50 years
Remaining economic life	41 years
Ratio applied to cost less physical and functional curable items: 9/50	18%
Reproduction cost less curable items	$97,000
Less other depreciation (all causes) 18%	− 17,460
Depreciated value of improvements	$79,540
Land value	+ 250,000
Indicated value by the cost approach	$329,540
Rounded to	$329,500

Breakdown Method

To apply the breakdown method, appraisers analyze each cause of depreciation separately, measure the amount of each, and then total the estimates to derive a lump sum that is deducted from the replacement or reproduction cost.

Valuations of many types of farm buildings and most single-family residences simply do not require the detail afforded by the breakdown method. An estimate derived using an age-life method or an overall market-extracted rate is sufficient for many of these properties. However, as the complexity of the property to be appraised increases, the breakdown method becomes more applicable. If a property has many depreciable short-lived items, substantial functional or external obsolescence, or both, the breakdown method may be mandated to furnish the client with the amount of detail required by the assignment. Appraisers should be familiar with this exacting and detailed method of estimating depreciation (see Table 15.15).

Market Extraction of Depreciation

The market extraction method of depreciation estimation provides appraisers with the ability to "prove" the depreciation estimate. To extract depreciation from the market, comparable sales similar to the

Table 15.15 Cost Approach—Breakdown Method

Reproduction cost						$100,000
Depreciation						
Physical deterioration						
Curable physical					$2,000	
Incurable physical short-lived items						
Item	**Cost**	**Age**	**Life**	**Depr.**		
Roof cover	$3,000	10	20	$1,500		
Appliances	$2,000	2	10	$400		
Floor cover	$3,000	2	10	$600		
Total	$8,000			$2,500	$2,500	
Incurable physical long-lived items (10-year effective age/50-year economic life)						
($100,000 – $2,000 – $8,000 = $90,000 × 20%)					$18,000	
Functional obsolescence						
Curable excess ($2,000 – 20% physical charged)					$1,600	
Curable deficiency					$3,000	
Incurable excess ($10,000 – 20% physical charged)					$8,000	
Incurable deficiency					$2,000	
External obsolescence						
$62,900 × 30%						
(Physically and functionally depreciated reproduction cost × appropriate percentage)					$18,870	
Total depreciation (all causes)						– 55,970
Depreciated reproduction cost						$44,030
Land value						400,000
Value indicated by the cost approach						$444,030
Rounded to						$444,000

subject must be available, and appraisers must have sufficient data to develop an opinion of the land value under the comparable sales. Land value is subtracted from the sale price, and the residual is the contributory value of the comparable improvements as of the date of sale. Once the reproduction or replacement cost is estimated, the contributory value is deducted from the reproduction cost to derive a dollar amount of depreciation. This depreciation amount is then divided by the reproduction cost to yield an overall rate of depreciation (see Table 15.16).

The very simple example in Table 15.6 indicates that the improvements exhibit depreciation of 30% from all causes. This method is sufficient to extract an overall rate of depreciation for many types of farm improvements, especially when improvements in the area are homogeneous. The improvements on the subject property must be very similar to the improvements on the sale properties in terms of age, utility, and market acceptance.

Now assume that the improvements described in Table 15.16 are similar to the subject improvements, except that the sale improvements have an effective age of 10 years and the subject improvements have an effective age of 8 years. The overall rate of depreciation can be further reduced to an annual percentage by dividing 30% by the 10 years of effective age to yield a rate of 3% per year. This rate is applied to the subject property's effective age of 8 years to indicate that the subject improvements should be depreciated 24%. This additional step eliminates the need for the improvements to be similar to the subject in terms of age, but does not eliminate the need for the improvements to be similar in terms of short-lived items, utility, and market acceptance. If the farms in the area exhibit differences in these characteristics, application of the breakdown method is more appropriate.

Table 15.16 Market Extraction of Overall Depreciation Rate

Sale price	$720,000
Less land value	– 650,000
Contributory value of improvements	$70,000
Reproduction cost of improvements	$100,000
Less contributory value	– 70,000
Depreciation	$30,000
Overall depreciation percentage ($30,000/$100,000)	30%

Most farms have multiple improvements, and these improvements often exhibit varying rates of depreciation. The depreciation rate for individual improvements can be measured by first analyzing a farm that has only one improvement, such as a residence. After extracting an annual rate of depreciation for the single building, a farm with two buildings can be measured by using the rate for the single building and the residual for the second building. Groups of buildings such as barns can often be lumped together, and there are always some buildings that can be eliminated from consideration due to condition, age, or other factors. This process is illustrated in Tables 15.16, 15.17, and 15.18.

Table 15.17 Extracting Depreciation from a Single-Improvement Farm

Sale price	$350,000
Less land value	– 300,000
Contributory value of improvements	$50,000
Reproduction cost of improvements	$75,000
Less contributory value	– 50,000
Depreciation	$25,000
Overall depreciation percentage ($25,000/$75,000)	33%
Effective age of 11 years	
Annual depreciation rate	3%

Assuming that the sale in Table 15.17 contained only a residence and that other sales will support this indication, it can be concluded that residences in this market exhibit depreciation at the rate of 3% per year. This depreciation rate should only be applied to physical depreciation unless all sales in the market have similar functional problems or external obsolescence is present in all improvements in the market. If functional or external obsolescence is present in some improvements but not all, the process requires additional steps to break out the obsolescence.

The process illustrated in Table 15.18 uses an annual depreciation rate from one sale to analyze another sale and provides an indication that barns depreciate at an annual rate of 5%. This process can be continued for other improvements and will substantially improve with additional data. If appraisers keep this type of information on file, substantial support for market-derived depreciation rates will be available. Table 15.19 summarizes numerous depreciation rates from sales derived over a period of time.

As appraisers develop additional data from additional sales, these indications can be added to the summary and provide additional support for depreciation rates for different types of buildings. If the rates change substantially, appraisers will be alerted to changes in the market acceptance of certain types of buildings.

Table 15.18 Extracting Depreciation from a Multiple-Improvement Farm

Sale price	$360,000
Less land value	− 300,000
Contributory value of improvements	$60,000
Reproduction cost (residence)	$75,000
Less market depreciation for residence (3%/year for 15-year effective age × $75,000)	− 33,750
Contributory value of residence	$41,250
Contributory value of improvements	$60,000
Less contributory value of residence	− 41,250
Contributory value (barn)	$18,750
Reproduction cost (barn)	$25,000
Less contributory value (barn)	− 18,750
Depreciation (barn)	$6,250
Overall depreciation percentage (barn) ($6,250 depreciation/$25,000 reproduction cost)	25%
Effective age of 5 years	
Annual depreciation rate	5%

Table 15.19 Depreciation Rate Summary (Annual Depreciation Rates)

Property	House	Barn	Grain Bins	Farm Shop
1	3.0%	N/A	N/A	N/A
2	3.0%	5.0%	N/A	N/A
3	2.8%	N/A	N/A	N/A
4	3.2%	N/A	N/A	N/A
5	N/A	N/A	5.7%	N/A
6	N/A	5.2%	5.7%	N/A
7	2.8%	N/A	N/A	4.4%
8	N/A	N/A	5.5%	N/A
9	3.0%	N/A	N/A	4.0%
10	N/A	5.0%	5.6%	4.0%

Depreciation Case Study

In the appraisal of a rural property with substantial improvements that exhibit depreciation from a variety of sources, the market extraction process can be used to estimate depreciation with the breakdown method. Appraisers can allocate depreciation derived from the comparable sale properties into the categories appropriate to the individual buildings on the subject property, and the residual process can be applied to estimate the remaining types of depreciation.

The following case study illustrates the methods employed to extract depreciation from sales via the breakdown method.

Market Area

The subject property and the comparable sales are located in an area where agriculture is the prime use and motivating factor in all purchases. There is little market pressure for other uses in the area, so agricultural production and improvements are the most significant factors in all sales. Land in the area is typically planted to small grain, and typical yields range from 80 to 90 bushels per acre. In the 1990s, there was a change in government programs that affected the desirability of on-farm grain storage. Prior to the change, it was common for farms to have the capacity to store a two-year supply of grain production. Currently, one year's storage is sufficient and, due to the design of the storage bins, it is impossible to lease out the excess storage capacity. Grain storage in the area is of two different designs. The most prevalent design is an upright circular steel bin that has a flat-bottom concrete floor with a metal grate for the unloading auger and a portable sweep auger for cleanout. These bins are loaded using a portable auger powered by either an electric or gasoline motor. Each bin has one or two air fans depending on size. The more elaborate systems in the area are cone-bottom, upright steel bins with a 1,500-bushel-per-hour elevator leg with ladder and cage assembly, dump pit, a conveyor to the leg, and electric motors and controls. The cost of these systems is substantially higher, and current buyers are unwilling to compensate sellers for the full, physically depreciated value of these systems.

Sales

The following sales are located in the subject area and are similar to the property being appraised.

- Property 1 sold for $523,000 one year ago. The land contribution at the time of sale was $472,350. The only improvement on the property is a 38,000-bushel grain storage bin, which has an effective age of three years and is of flat-bottom, upright design. There is no physical curable deterioration, and the reproduction cost of the short-lived items (motors and fans) is $6,600. The short-lived items are three years old and have a projected life of 10 years. The total reproduction cost as of the date of sale is $58,900. The storage is adequate for one year's production on the 450-acre farm.
- Property 2 sold for $700,000 one year ago. The land contribution at the time of sale was $639,150. The only improvement is a 52,000-bushel grain storage bin, which has an effective age of five years and is of flat-bottom, upright design. The physical curable

deterioration is estimated at $2,000, of which $1,750 is included in reproduction cost and $250 is excess cost. The reproduction cost of the short-lived items is $6,600. These items are five years old and have a projected life of 10 years. The total reproduction cost as of the date of sale is $80,600. The storage is adequate for one year's production on the 610-acre farm.

- Property 3 sold for $547,775 one year ago. The land contribution at the time of sale was $500,000. The only improvement is a 42,000-bushel grain storage bin. This storage bin has an effective age of five years and is of cone-bottom, upright design with a 55-ft. elevator leg with a dump pit, conveyor, and electric motors and controls. The physical curable deterioration is $1,000, all of which is included in the reproduction cost. The reproduction cost of the short-lived items is $7,800. These items are five years old and have a projected life of 10 years. The total reproduction cost as of the date of sale is $89,250. The storage is adequate for one year's production on the 575-acre farm.
- Property 4 sold for $716,400 three years ago. The land contribution at the time of sale was $635,000. The only improvement is a 110,500-bushel grain storage bin with an effective age of 10 years, which is of cone-bottom, upright design with a 55-ft. elevator leg with a dump pit, conveyor, and electric motors and controls. The physical curable deterioration is $20,000, of which $17,500 is included in the reproduction cost and $2,500 is excess cost. The curable deterioration includes some damage to the bin and short-lived items. These short-lived items are fully depreciated and in need of replacing, so they are included as physical curable deterioration. The total reproduction cost as of the date of sale is $215,475. The storage is adequate for two years' production on the 650-acre farm.
- Property 5 sold for $683,000, and the sale date is current. This is a resale of Property 4. Land values in the area have not changed, so the land contribution at the time of sale is $635,000. The current effective age of the grain storage bin is 13 years. There are no curable items, and the reproduction cost of the short-lived items is $12,750. These items are three years old and have a projected life of 10 years. The total reproduction cost as of the date of sale is $232,050. The storage is adequate for two years' production on the 650-acre farm.
- Property 6 sold for $532,900, and the sale date is current. The land contribution at the time of sale is $465,000. The only improvement is an 80,000-bushel grain storage bin, which has an effective age of two years and is of flat-bottom, upright design. There is no physical curable deterioration. The reproduction cost

of the short-lived items is $7,400. The short-lived items are two years old and have a projected life of 10 years. The total reproduction cost as of the date of sale is $124,000. The storage is adequate for two year's production on the 650-acre farm.

Analysis of the Comparable Properties

The properties are analyzed to estimate the market depreciation of grain bins with various characteristics. The curable items are considered first, followed by the incurable short-lived items. If there is no functional or external obsolescence, the remainder is incurable physical deterioration, long-lived. After an annual percentage for this type of depreciation is developed, the remainder can be classified as either functional or external obsolescence depending on the situation.

Property 1

The analysis of Property 1 illustrated in Table 15.20 includes an age-life analysis of the short-lived components and extraction of an annual rate of depreciation for the incurable physical deterioration in long-lived items. The grain bin is functional and does not represent an overimprovement. It is of sufficient size to hold one year's production but not two, so no external obsolescence is present. There are no items of curable physical deterioration, therefore only incurable short-lived and long-lived depreciation is considered.

Table 15.20 Analysis of Property 1

	Sale price	$523,000
Less	Land contribution	– 472,350
Equals	Contributory value of the improvements	$50,650
	Reproduction cost	$58,900
Less	Contributory value of the improvements	– 50,650
Equals	Depreciation	$8,250
Depreciation Analysis		
	Curable physical	0
	Incurable physical, short-lived—motors and fans [$6,600 cost × (3-year age/10-year life)]	– 1,980
Balance of depreciation attributed to long-lived items		$6,270

The long-lived incurable physical deterioration divided by the reproduction cost of the long-lived items is the percentage of long-lived depreciation. The total percentage of long-lived depreciation divided by the effective age equals the annual rate of long-lived depreciation. The cost new of the short-lived items previously depreciated is deducted from total reproduction cost to avoid double-counting the depreciation. The calculations are summarized in Table 15.21.

Table 15.21 Property 1 Calculation Summary

Long-lived depreciation	$6,270
Long-lived cost ($58,900 reproduction cost – $6,600 short-lived cost)	$52,300
$6,270 divided by $52,300 is equal to 0.119885, or 12%	
12% divided by the effective age (3) is 4% per year	

Property 2

The analysis of Property 2 shown in Table 15.22 requires consideration of the curable deterioration resulting from damage to the side of the grain bin and analysis of short-lived items with the remaining deterioration attributable to incurable physical long-lived items. The deferred maintenance totals $2,000, of which $1,750 is included in reproduction cost and $250 is the excess cost needed to remove the old material prior to replacing the damaged panels. This grain bin is functional and does not represent an overimprovement. It is sufficient for one year's storage, therefore external obsolescence is not present.

As with Property 1, the depreciation balance (after deductions are made for deferred maintenance and incurable short-lived items) is divided by the reproduction cost of the long-lived items (after deductions for those portions of reproduction cost that have been depreciated as curable and incurable short-lived) to arrive at the percentage of long-lived depreciation. The calculations are summarized in Table 15.23.

Table 15.22 Analysis of Property 2

	Sale price	$700,000
Less	Land contribution	– 639,150
Equals	Contributory value of the improvements	$60,850
	Reproduction cost	$80,600
Less	Contributory value of the improvements	– 60,850
Equals	Depreciation	$19,750
Depreciation Analysis		
	Curable physical (in reproduction cost)	– 1,750
	Curable physical excess (not in reproduction cost)	– 250
	Incurable physical short-lived—motors and fans [$6,600 cost × (5-year age/10-year life)]	– 3,300
	Balance of depreciation attributed to long-lived items	$14,450

Table 15.23 Property 2 Calculation Summary

Long-lived depreciation	$14,450
Long-lived cost ($80,600 reproduction cost – $1,750 curable cost – $6,600 short-lived cost)	$72,250
$14,450 divided by $72,250 equals 0.20000, or 20%	
20% divided by the effective age (5) is 4% per year	

Property 3

The analysis of Property 3 illustrated in Table 15.24 requires the consideration of curable physical deterioration due to damage to the elevator leg and analysis of the short-lived items. Appraisers must also apply the incurable physical long-lived rate derived from Sales 1 and 2 to arrive at a residual applicable to the incurable functional obsolescence due to excess. The deferred maintenance totals $1,000, all of which is included in the reproduction cost estimate; there is no curable excess. The grain bin on this property exhibits incurable functional obsolescence due to the excess handling system. The grain bin itself is functional, but the market will not pay for the excess cost represented by the elevator leg, dump pit, boot pit, and U-trough unloading augers. Nevertheless, the bin is sufficient for one year's storage, so external obsolescence is not present.

Table 15.24 Analysis of Property 3

	Sale price	$547,775
Less	Land contribution	– 500,000
Equals	Contributory value of the improvements	$47,775
	Reproduction cost	$89,250
Less	Contributory value of the improvements	– 47,775
Equals	Depreciation	$41,475
Depreciation Analysis		
	Curable physical (in reproduction cost)	– 1,000
	Curable physical excess (not in reproduction cost)	0
	Incurable physical short-lived—motors and fans [$7,800 cost × (5-year age/10-year life)]	$3,900
	Incurable physical long-lived [$80,450 long-lived cost × (4%/year for 5 years)]	– 16,090
	Balance of depreciation attributed to incurable functional items	$20,485
	Total physical depreciation ($1,000 + $3,900 + $16,090)	$20,990
	Physically depreciated reproduction cost ($89,250 – $20,990 total physical depreciation)	$68,260
	Incurable functional obsolescence, superadequacy	$20,485

Incurable functional obsolescence due to a superadequacy is reported as a percentage of the reproduction cost after curable and incurable functional obsolescence due to a deficiency has been deducted. In this sale, there is no curable or incurable functional obsolescence due to a deficiency, so the reproduction cost base is the same as the physically depreciated reproduction cost. The estimated incurable functional obsolescence due to the superadequacy ($20,485) is divided by the physically depreciated reproduction cost ($68,260), indicating 30% depreciation.

Property 4

Due to the age and condition of Property 4, the short-lived items are fully depreciated and in need of replacing. Therefore, these items are classified as curable physical deterioration. There is additional curable deterioration due to deferred maintenance on the tanks. The total curable deterioration (both short-lived and damage) is $20,000, of which $17,500 is included in reproduction cost and $2,500 is excess cost to remove the old motors and fans and damaged panels prior to replacement. There is no incurable physical deterioration in short-lived items because all of the short-lived items were handled in the curable category. The grain bin on this property exhibits incurable functional obsolescence due to the excess handling system. The bin is sufficient for two years' storage. Because this sale occurred

Table 15.25 Analysis of Property 4

	Sale price	$716,400
Less	Land contribution	– 635,000
Equals	Contributory value of the improvements	$81,400
	Reproduction cost	$215,475
Less	Contributory value of the improvements	– 81,400
Equals	Depreciation	$134,075
Depreciation Analysis		
	Curable physical (in reproduction cost)	– 17,500
	Curable physical excess (not in reproduction cost)	– 2,500
	Incurable physical short-lived—motors and fans (Motors and fans included as curable)	0
	Incurable physical, long-lived [$197,975 long-lived cost × (4%/year for 10 years)]	– 79,190
	Balance of depreciation attributable to incurable functional obsolescence	$34,885
	Physically depreciated reproduction cost [$215,475 reproduction cost – $99,190 total physical depreciation]	$116,285
	Incurable functional obsolescence, superadequacy	$34,885

prior to the change in the government program, however, no external obsolescence is present.

The estimate of incurable functional obsolescence due to the superadequacy ($34,885) is divided by the physically depreciated reproduction cost ($116,285), indicating 30% depreciation. As with Property 3, this sale is indicating 30% incurable functional obsolescence due to the superadequate grain handling facility. Because this property sold prior to the change in the government program that reduced the need for on-farm storage, it exhibits no external obsolescence.

Property 5

The sale of what is identified as Property 5 is a resale of Property 4. The property has been well-maintained since the repair and replacement of short-lived items three years ago. Again, the grain bin exhibits functional obsolescence due to the excess handling system. The bin is sufficient for two years' storage and, since the change in government programs, it exhibits external obsolescence. Note that the reproduction cost has changed due to an increase in cost over the past three years.

The external obsolescence is reported as a percentage of the physically and functionally depreciated reproduction cost. The reproduction cost base is the same as the physically and functionally de-

Table 15.26 Analysis of Property 5

	Sale price	$683,000
Less	Land contribution	– 635,000
Equals	Contributory value of the improvements	$48,000
	Reproduction cost	$232,050
Less	Contributory value of the improvements	– 48,000
Equals	Depreciation	$184,050
Depreciation Analysis		
	Curable physical (in reproduction cost)	0
	Curable physical, excess (not in reproduction cost)	0
	Incurable physical, short-lived—motors and fans [$12,750 cost × (3-year age/10-year life)]	– 3,825
	Incurable physical, long-lived [$219,300 long-lived cost × (4%/year for 13 years)]	– 114,036
	Incurable functional obsolescence [$114,189 phys. dep. reproduction cost × 30%]	– $34,257
	Balance of depreciation attributed to external obsolescence	$31,932
	Physically and functionally depreciated reproduction cost [$232,050 reproduction cost – $117,861 total physical depreciation – $34,257 functional obsolescence]	$79,932
	External obsolescence	$31,932

preciated reproduction cost. The external obsolescence divided by the physically and functionally depreciated reproduction cost is $31,932 divided by $79,932, or 40%.

Property 6

There are no curable items for Property 6, so the appraiser analyzes the short-lived items and applies the incurable physical long-lived rate derived from Property 1 and Property 2 to arrive at a residual applicable to external obsolescence. The grain bin does not exhibit incurable functional obsolescence due to an excess handling system. It is sufficient for two years' storage, so external obsolescence is present.

Table 15.27 Analysis of Property 6

	Sale price	$532,900
Less	Land contribution	– 465,000
Equals	Contributory value of the improvements	$67,900
	Reproduction cost	$124,000
Less	Contributory value of the improvements	– 67,900
Equals	Depreciation	$56,100
Depreciation Analysis		
	Curable physical (in reproduction cost)	0
	Curable physical, excess (not in reproduction cost)	0
	Incurable physical, short-lived—motors and fans [$7,400 cost × (2-year age/10-year life)]	– 1,480
	Incurable physical, long-lived [$116,600 long-lived cost × (4%/year for 2 years)]	– 9,328
	Functional obsolescence	0
	Balance of depreciation attributed to external obsolescence	$45,292
	Physically and functionally depreciated reproduction cost [$124,000 reproduction cost – $10,808 total physical depreciation – $0 functional obsolescence]	$113,192
	External obsolescence	$45,292

The external obsolescence is reported as a percentage of the physically and functionally depreciated reproduction cost. The reproduction cost base is the same as the physically and functionally depreciated reproduction cost. The external obsolescence divided by the physically and functionally depreciated reproduction cost is $45,292 divided by $113,192, or 40%. This sale indicates that the external obsolescence is applicable to both flat- and cone-bottom storage.

Application of Depreciation Data to Subject

The subject property includes an 85,000-bushel grain bin. The bin has a concrete, hopper-bottom floor with a discharge auger. There is a 1,500-bushel-per-hour elevator leg with 9-in.-by-16-in. buckets. The elevator leg has a 50-ft. discharge height and two downspouts, one to the grain bin and one for loading trucks. The comparative-unit method is used to estimate the reproduction cost of the subject improvements. This method of estimating reproduction cost was used on the sales, and consistency in the analysis and application of depreciation rates dictates that the same method be used to estimate the reproduction cost for the subject. The reproduction cost of the subject's grain bin is $180,200.

The subject property's grain bin was constructed six years ago. Curable physical deterioration is present due to the four damaged metal panels. The estimated cost to replace these panels is $5,500, of which $5,000 is included in reproduction cost and $500 is the excess cost incurred in removing the old panels to replace them with new ones. There is curable functional obsolescence due to a deficiency because the electrical motor that powers the elevator leg is undersized and needs to be updated. The cost if included at the time of construction is $500. The cost to remove and update the motor now is $800, indicating curable functional obsolescence of $300. There is no salvage value for the motor. The new cost of the short-lived items is $7,800, and the projected life is 10 years. Thus, the incurable physical depreciation is $7,800 cost × 6-yr. age/10-yr. life = $4,680.

The sales of Property 1 and Property 2 provide the basis for estimating the incurable physical long-lived deterioration of 4% per year over a six-year effective age for a total incurable physical long-lived deterioration of 24%. This percentage will be applied to the cost new of the long-lived items to avoid double depreciation of the items previously depreciated. The reproduction cost to be depreciated as incurable physical long-lived deterioration is calculated in Table 15.28.

Table 15.28 Long-Lived Cost Calculation

Reproduction cost		$180,200
Curable cost in reproduction cost	–	5,000
Cost of short-lived items	–	7,800
Cost of long-lived items		$167,400
	×	24%
Incurable physical long-lived deterioration		$40,176

The sales of Property 3 and Property 4 provide the basis for estimating incurable functional obsolescence due to an excess. This depreciation reflects market resistance to the elaborate grain handling system, and the depreciation estimate is 30%. This percentage is applied to the physically depreciated cost new to avoid double depreciation of the items previously considered. The physically depreciated value is calculated as shown in Table 15.29.

Table 15.29 Physically Depreciated Value Calculation

Reproduction cost	$180,200
Total physical deterioration ($5,000 + $500 + $4,680 + $40,176)	– 50,356
Physically depreciated cost	$129,844
Functional incurable obsolescence ($129,844 × 0.30)	$38,953

The sales of Property 5 and Property 6 provide the basis for estimating external obsolescence due to an excess amount of storage. This depreciation is the result of a change in the government programs that govern farm grain storage. The percentage of depreciation indicated by these sales is 40%. This percentage is applied to the physically and functionally depreciated value, again to avoid double-counting the depreciation of the items previously considered.

Table 15.30 Functional Incurable Obsolescence Calculation

Reproduction cost	$180,200
Total physical deterioration ($5,000 + $500 + $4,680 + $40,176)	– 50,356
Functional obsolescence	– 38,953
Physically and functionally depreciated cost	$90,891
Functional incurable obsolescence ($90,891 × 0.40)	$36,356

Depreciation Estimate for the Subject Property

The case study concludes with the final estimate of depreciation in the subject property (see Table 15.31).

Table 15.31 Depreciated Cost of Subject Property

Reproduction Cost			$180,200
Depreciation			
	Physical deterioration		
	Curable (in reproduction cost)	$5,000	
	Curable (excess)	500	
	Incurable short-lived	4,680	
	Incurable long-lived	40,176	
	Total physical deterioration		$50,356
	Physically depreciated reproduction cost		$129,844
Functional obsolescence			
	Curable	300	
	Incurable	38,953	
	Total functional obsolescence		$39,253
Physically and functionally depreciated reproduction cost			$90,591
External obsolescence @ 40%			$36,236
Depreciated reproduction cost			$54,355

Chapter 16

The Valuation of Livestock Ranches

Over the last 30 years, US livestock production has shifted to much larger enterprises, in part because of economies of scale. For example, consider the production locus, which is the farm size (in annual sales of livestock) at which half of the total national production comes from larger farms and half comes from smaller farms. Between 1987 and 2002, the production locus increased by 60% for broilers, 100% for fed cattle, 240% for dairy farms, and 2,000% for hog production, according to the USDA.[1] And the trend seems to have continued since 2002. In 2016, cattle feedlots with less than 1,000 head of capacity comprised the vast majority of US feedlots, but those facilities accounted for a relatively small share of fed cattle sold. In contrast, lots with a capacity of 1,000 head or more made up less than 5% of the total number of feedlots, but those facilities marketed 80%-90% of fed cattle. At the upper end of the scale, feedlots with a capacity of 32,000 head or more marketed around 40% of fed cattle.[2]

Structural change has led to increased productivity and, through that, to lower commodity costs of production. For example, the dairy farms in the largest size category (1,000 cows or more) had average costs of $21.02 per hundredweight in 2015, which was 16% below the average for farms in the next largest size category (500-999 head) and 32% below the costs for farms with 100-199 head (estimated at $31.12 per hundredweight). Lower costs of production typically lead to lower wholesale and retail prices.

1. USDA Economic Research Service, *The Transformation of US Livestock Agriculture: Scale, Efficiency, and Risks* (Washington, DC: 2009), iii.
2. USDA Economic Research Service, www.ers.usda.gov/topics/animal-products/cattle-beef/background.aspx.

Figure 16.1 Fed Cattle Shift to Large Feedlots

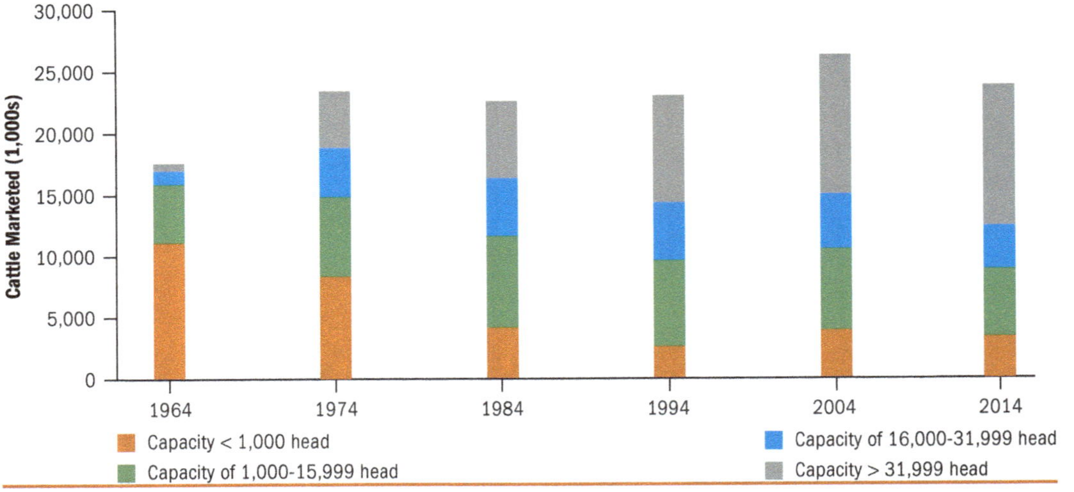

Source: USDA

Structural change in livestock agriculture has also had adverse effects. Livestock wastes are becoming more geographically concentrated in the United States, and excessive agricultural applications of the nutrients contained in manure pose risks to air and water resources. There is a clear association between farm size and the concentration of manure–larger operations are more likely to ship manure to other operations and apply manure to their own fields more intensively. However, the cost of removing manure to large farms is still modest in relation to their production cost advantages, and there are a variety of ways to mitigate the risks from the concentration of manure, such as reformulating feed to reduce the amount of nutrients (e.g., nitrogen) contained in the resultant animal waste.[3]

Types of Livestock Ranches

A distinguishing characteristic of livestock ranches is that forage grasses and crops grown on the land are the main source of feed for the livestock raised on the property. Although the term *ranch* is sometimes used synonymously with *farm* or *homestead* to describe a rural property, in this chapter the term refers to a livestock-producing property. Ranchland used for recreational purposes is a different type of rural property, with different valuation considerations than a livestock ranch.

Livestock ranch appraisals can be simple or very complex, like the livestock industry. Both natural and man-made features and

3. USDA Economic Research Service, *The Transformation of US Livestock Agriculture*, iii-iv.

relationships affect the value of livestock ranches. Livestock ranches in North America, Australia, Africa, South America, and the European continent vary greatly. Within the United States, a livestock ranch might be operated as a separate unit or in conjunction with various crop production or recreational endeavors. The appraisal of a livestock ranch may involve special considerations and knowledge depending on its location–e.g., the Pacific Northwest, the Sandhills of Nebraska, the desert areas of the Southwest, the fruit- and nut-producing areas on either coast, the Rocky Mountain region, the Corn Belt, or the high rainfall areas of the Gulf Coast. Livestock ranches are influenced by:

- climate (temperature and precipitation)
- topography
- vegetation
- availability of supplemental feed
- proximity to markets

These influences may also affect the operation and highest and best use of individual properties within a subregion.

The operation of a livestock ranch for the production of cattle, sheep, or goats is usually influenced by the types of forage available. Variations within or between regions may also have a significant influence on the ability of a property to achieve maximum productivity. A wide variety of factors typically affect the style and operation of a ranching unit and dictate the species or combination of species in that unit. Regional economics and lifestyles may require that the ranch have a primary function for recreation, with livestock production as a sideline. Many property owners have options, choosing to use the property for year-round cow/calf or farm flock production, seasonal grazing of yearling or mature livestock, commercial production, purebred production, or a combination of uses.

In appraising a livestock ranch, appraisers must become fully informed about the types of ranching operations in the market area and determine the flexibility or limitations in the operation of the subject property and comparable properties. Appraisers must also recognize the recreational amenities that relate to the subject property and comparable properties.

Land Base of Livestock Ranches

Throughout much of the United States and Canada, many livestock ranch operations are based entirely on deeded land held in fee simple title. Regardless of a ranch's geographical location, the operation of a ranch may include both owned and leased land. Ranching operations in the 17 western states often include privately owned land as a base unit to provide forage production or base water, combined with land leased from federal or state governments. Livestock operations throughout the United States include land leased from other entities such as municipalities, utilities, timber companies, and private owners. The terms of leases and permits vary from annual to multiyear. To properly analyze the use and value of either owned or leased lands, appraisers must research and understand all the real estate used for the livestock ranching operation.

In the 17 western states, ranching operations often include privately owned land combined with land leased from state or federal governments. The US federal government also owns and leases many millions of acres suitable for livestock ranching in these states.

State and Federal Leases

The US federal government gave deeded land to many state governments in order to stimulate development. States usually received specific sections within each township, such as sections 16 and 36. Although some states have sold this land to private individuals, many still retain ownership. Most state-owned lands are leased to private enterprises, often for agricultural purposes.

State ownership usually involves a wide variety of land use types, including dry rangeland, subirrigated pasture or meadow, forest, and cropland. Some states allow multiyear leases in which state-owned land becomes an integral part of a livestock ranching operation. Over time, this leased land may become extremely important to the successful operation of the associated owned land. Changing social and economic factors may heavily influence the utility and financial feasibility of state leases.

The federal government owns and leases many millions of acres suitable for livestock ranching, particularly in the 17 western states. Various federal government divisions may own the land, so a variety of lease or permit restrictions or opportunities may be involved. For example, lands within national forests are administered by the US Forest Service within the Department of Agriculture. Government-owned lands that fall under the purview of the Department of the Interior are administered by the Bureau of Land Management (BLM), and, to a lesser extent, the Bureau of Indian Affairs, the Bureau of Reclamation, and the US Fish and Wildlife Service. Those entities use a variety of lease terms, varying from annual to 7- to

10-year terms, and apply strong controls on the use of leased land. Leases and permits specify the species and number of livestock and the allowable grazing season.

Livestock grazing on either Forest Service or BLM lands is considered a privilege. Although this privilege is held by a private individual and associated with specific deeded land, permits and leases are not automatically conveyed with the sale of a ranch. Permittees must possess commensurate deeded land. That is, the permittee must have the ability to maintain the livestock on non-leased portions (i.e., government land cannot make up the total enterprise).

Land leased from the government often becomes an integral part of a ranch enterprise. Upon sale of the deeded land, the permittee must request a transfer of the grazing privileges to the buyer of the land. Although permits typically remain with the base property to which they have been attached for many years, a permit technically reverts to the Forest Service or BLM and does not have to be granted to the new owner. A second alternative is forest and BLM permits being transferred with the owned cattle. A permittee may sell cattle to a new owner and request that the federal permit be transferred with the cattle. Again, the new owner/permittee must possess commensurate deeded land.

Grazing permits are closely supervised. US Forest Service and BLM administrators have the right to set up strict pasture rotations, salting patterns, and periodic livestock counting. It is always sound appraisal practice to verify all details of a government grazing permit with the appropriate US Forest Service or BLM personnel. Other government entities such as military reservations may grant grazing leases to private operators.

In addition to state and federal grazing leases, many ranches today incorporate additional lands through private leases. Private leased land may be owned by individuals, families, grazing associations, investment companies, railroads, or municipalities. Railroad ownership in some western states is interspersed with private or state ownership of adjoining sections. Like federal lands, land held through private leases may become an integral part of a ranching operation. Conversely, the market value of the real estate included in these private leases may be controlled by the lessee due to the physical configuration of the owned and leased land. Regardless of its ownership characteristics, the feasibility and value of leased land within livestock ranches will vary with the requirements of the specific administrator.

Deeded Ranches

The livestock industry divides livestock ranches into specific classes or types of ranches. Although a significant portion of the national production of grazing livestock is associated with farm or timber operations, the characteristics of various types of ranches described below distinguish ranches from farm properties.

Ranch types within the 17 western states vary greatly. The term *mountain ranch* typically refers to properties at higher elevations and is often associated with land leased from the government. The base property may be included at the mountain location or located miles away. Cool-season grasses are predominantly found on mountain ranches. Forage production can vary from high feed production areas to irrigated and subirrigated lands to dry ranges with limited forage production.

On year-round mountain ranches, hay must be fed to the livestock during the five- to six-month winter period. When there is sufficient hay produced on the ranch to feed the livestock during the winter months, the ranch is in balance. If additional hay needs to be purchased, this suggests that the ranch is not in balance. In winter, the cost of hay may be an expensive item for a mountain ranch. Although pregnancy rates are fairly high, most mountain ranches do not have high weaning weights because of late calving and limited forage for summer grazing. A mountain ranch can also experience significant death losses from predators or poisonous plants.

Desert ranches can be found in several states. This type of ranch is characterized by very low forage production per acre and often a very small percentage of deeded land compared to leased land. In desert ranches, water sites are often part of privately owned land, while extensive acreages of government or state lands may have limited water. A desert ranch often uses specific terrain or forages for the nongrowing season. Cattle are adapted to harvesting browse, including various woody plants. This off-season grazing of browse may be supplemented by feeding the livestock high-protein energy supplements, known as *cake*. The large acreages required for desert ranches significantly reduce the ability to manage livestock. Pregnancy rates and even weaning weights can be significantly lower on desert ranches than they are on other types of ranches.

Ranches in the Gulf Coastal regions enjoy very high carrying capacities per acre due to high rainfall and a long growing season. The high forage production in these areas usually depends on the production of Bermuda grasses and requires large amounts of costly fertilization. This type of ranch may use hay or concentrate cake for the short winter period. Gulf Coast ranches allow for intensive management but have specialized requirements. High rainfall, heat, and other factors create excessive parasite problems. Routine treatment of livestock for both internal and external parasites including ticks and worms is required. Other types of disease and management problems are also common in this environment. In the operation of a Gulf Coast ranch, it is important to select cattle breeds that are resistant to parasites, disease, and heat-related problems.

The term *plains ranch* refers to properties in the central United States characterized by broad expanses of undulating to rolling terrain. From north to south, plains ranches vary in terms of rainfall and forage production. They can be operated as either (a) hay and cake or (b) range properties. The style of operation can vary with the availability of hay or the desires of the operator. From east to west, plains ranches also differ in terms of rainfall, soil type, and type of grass. Individual areas in the western, drier plains produce limited forage with drought-tolerant grasses such as buffalo and grama grass and may include extensive amounts of cactus. Tallgrass prairie ranches in the high rainfall areas of the eastern plains produce excessive forage per acre from big bluestem, switchgrass, and Indian grass. Intermixed are prairies of midsize grasses such as little bluestem, side oats grama, and wheat grass.

Livestock production in the plains can vary. Some ranches have intensive management and excellent pregnancy rates and weaning weights, while ranches in drier areas are subject to less favorable factors. Northern portions of the plains can experience severe winter storms, while southern areas require heat- and parasite-resistant cattle.

Obviously, many areas in the West include ranching operations with combinations of the previously described ranch characteristics. Small geographic areas may also possess widely different types of ranches. The coastal areas of California enjoy a reverse of seasons in which winter is the productive period for livestock grazing. Land leased from the government is typically more important to mountain and desert ranches. Recently, significant public pressure has mounted concerning government leases. In appraisal assignments, it is incumbent on appraisers to understand those factors that affect the value of the ranch property being appraised.

Function of a Ranch Property

An appraiser's analysis of highest and best use and the resulting estimate of value for a livestock ranch are typically based on the productivity and amenities provided by the ranch. For a typical livestock ranch, the basis of productivity, and therefore of market value, is the ability of the owner or operator to convert forage into income through the sale of livestock. In reality, a livestock operation is the "farming" of forage and "harvesting" by livestock. The value and highest and best use of a livestock ranch can be significantly influenced by amenities other than livestock production. Actual productivity may become a minor factor relative to the value of the property for scenic, recreational, or conservation purposes. A thorough study is needed to determine the amenities of a specific property and whether they contribute to value. Appraisers should realize that the highest and best use of a ranch property might not be its current use.

Converting Forage into Marketable Livestock

The economics of a ranch are directly related to forage production, usually measured as carrying capacity in animal units. Forage production, whether expressed directly through carrying capacity or indirectly on a per-acre basis, is the foundation that has to be established before a credible valuation can be completed. Most livestock ranches in the western United States are marketed based on carrying capacity. The normal basis in other parts of the country is forage production per acre. Analysis of ranches indicates that property price is directly related to forage production or carrying capacity.

Forage production for livestock use is typically expressed in animal units (AUs) or animal unit months (AUMs). The AUM measurement of productivity allows appraisers to relate and compare properties using different kinds of feed and feed sources on a standard basis. Thus, one property can be compared with another even if they do not have exactly the same kind of feed, timing of feed, or level of production per acre.

An AU is defined in this chapter as one mature cow weighing 1,000 pounds, either dry or with a nursing calf up to six months of age, or the equivalent. The cow is assumed to consume a certain amount of forage, typically 26 pounds of dry material per day. When related to forage, the AU is the amount of feed necessary to sustain the 1,000-pound cow or its equivalent for one year, which is 9,490 pounds of dry matter. The AU measurement provides a yardstick that can be related to other cattle or sheep. If, for example, a 1,000-pound cow equals 100% of 1 AU, then a 500-pound calf is equal to one-half or 50% of 1 AU. Each 100 pounds of live weight is equal to 0.1 AU.

Regardless of the definition that is prevalent in the area, reliable comparisons can be made if appraisers calculate the AU measurement consistently for the subject property and comparable ranch properties. Table 16.1 provides AU equivalents for various classes of livestock. Table 16.2 illustrates the calculation of AUs on a combined cattle-sheep operation when the number and class of livestock are known.

If the element of time is considered, a useful relationship can be described. The animal unit month, or AUM, is the quantity of feed needed for good growth or sustenance for one animal unit for one month (typically 26 pounds of dry matter feed per day).

animal unit (AU). The standard unit of measurement used to calculate the relative grazing impact of different classes of livestock, e.g., one animal unit is a 1,000-pound beef cow with or without a nursing calf with a daily dry matter forage requirement of 26 pounds per day.

animal unit month (AUM). The amount of forage to fulfill metabolic requirements by one animal unit for one month, e.g., one AUM is equal to 780 pounds (or 26 pounds per day) of dry matter forage.

carrying capacity. The maximum number of animals that can be grazed without damaging the forage production, quality, or soil of the range; expressed in AUs or AUMs.

forage. All browse and herbaceous food that is available to livestock or game animals; used for grazing or harvested for feeding.

Table 16.1 Animal Unit Equivalents

Animal	Animal Units
Mature cow*	1.00
Calf, four months to weaning	0.25
Weaner calf	0.50
Yearling	0.60
Two-year steer (long yearling)	0.80
Horse	1.25
Bull	1.25
Sheep	0.20
Goat	0.20

* Heifers are equivalent to 0.7 AU at weaning time and 1.0 AU at the time of pregnancy test, but they require extra feed for development, with some culled at an early date—summarized to 0.85 AU.

Table 16.2 Animal Units

Animal Class	Number of Head	Animal Unit Equivalent	Animal Units
Mature cows	500	1	500
Calves	475	0.25	119
Yearlings	80	0.6	48
Bulls	29	1.25	36
Horses	12	1.25	15
Sheep	1,000	0.2	200
Total	2,096		918

AUM is actually a formula:

$$A \times U \times M = AUM$$

where

A = the number of head of livestock

U = the size of the unit in decimals

M = the number of months

Using this formula, four animals that are about 500 pounds each (0.5 AU or one-half of a mature animal unit of 1.0) grazing for six months equal 12 AUMs:

$$4 \times 0.5 \times 6 = 12 \text{ AUMs}$$

Table 16.3 uses the information included in Table 16.2 and introduces the element of time in determining AUMs. Table 16.3 indicates that this mixture of 918 AUs requires 10,064 AUMs in feed annually. Appraisers must understand the distinction between the different methods of calculating AUMs. The relationship between the number of head, number of cows, and number of AUs often varies in different

Table 16.3 Animal Unit Months

Animal Class	Type or Number of Head	AU Equivalent	AUs	Time in Months	AUMs
Mature cows	500	1	500	12	6,000
Calves	475	0.25	119	4	476
Yearlings	80	0.6	48	12	576
Bulls	29	1.25	36	12	432
Horses	12	1.25	15	12	180
Sheep	1,000	0.2	200	12	2,400
Total	2,096		918		10,064

regions as do local terminology and methods of calculation. These differences must be understood to accurately estimate the forage production and value of livestock ranches.

Feed Requirements

As previously mentioned, the value of a livestock ranch is based on the conversion of forage into marketable livestock. Both operators and appraisers must have an understanding of the nutrient requirements of livestock. Appraisers do not need a degree in nutrition, but they should understand how feed requirements vary with the class and age of livestock and with the digestibility and utility of the growing or harvested forage supplies. Individual nutrient requirements–e.g., protein, total digestible nutrients (TDN)–increase significantly for growing animals compared to mature livestock. Additionally, feed requirements are much higher during periods of inclement weather, during the later stages of gestation, and throughout lactation.

The availability of feedstuffs in both summer and winter has a significant influence on the weaning weights, daily gain, and reproductive capabilities of both female and male livestock. All of these items affect the productivity and profitability of individual ranches. Highly qualified ranch appraisers have a thorough understanding of the feed requirements for livestock and know how to identify the availability of feed on a specific ranch property. They also understand how the natural elements of feed availability can be altered by management. Feed availability on a specific ranch is heavily influenced by elevation, growing season, precipitation, soils, and management. The availability and use of browse (tender shoots, twigs, and leaves of trees and shrubs used by animals for food), protein supplements, and harvested forage (hay or silage) will dictate the species of livestock, the season of use, and ranch balance.

Table 16.4 describes the feed requirements for livestock and is based on data from the National Research Council. Complete tables are available from most land grant universities. The table clearly illustrates that different livestock require different amounts of feed during different times of the year and production periods. Notice, for example, the "Cows nursing calves–first 3 months" class. The 1,100-lb. cow in the first row is considered to be an average milking animal, while the second 1,100-lb. cow is a superior milking cow.

When an appraiser is estimating AUs, AUMs, ranch capacity, and ranch balance, the actual amount of feed that is required by each livestock class must be correlated to the available forage production. Sources of information about the daily nutritional requirements of livestock are readily available online, often from university extension offices and government agencies.

The livestock ranch appraiser also needs reliable information as to the forage production on various grasses and range types. The best data

Table 16.4 Daily Nutritional Requirements

Species, Class	Body Weight*	Daily Gain	Min. Dry Matter	Total Protein	Digest. Protein	TDN	Vitamin A 1,000 IU
Growing steer calves/yearlings	440	1.5	12.6	1.34	0.86	7.9	13
		2.4	10.1	1.39	0.95	8.6	13
Pregnant yearling heifers–last trimester	880	0.9	15.4	0.97	0.46	8.7	21
		1.8	25.6	2.22	1.25	14	33
Dry pregnant mature cows–last trimester	880	0.9	15.4	0.97	0.046	8.7	21
	1,100	0.9	17.9	1.12	0.53	10	24
Cows nursing calves–first 3 months	1,100	21.6	1.98	1.17	11.7	24	
	1,100	26	2.84	1.67	14.8	38	
Bulls, growth and maintenance–moderate activity	1,100	1.5	26.9	2.35	1.36	16.5	48
	1,980		25.1	2.18	1.21	13.9	44
Ewes–last 6 weeks gestation	120	0.37	4.2	0.34	0.19	2.2	2.7
Ewes–first 8-10 weeks lactation	120	-0.07	5.1	0.42	0.23	2.9	2.7
Rams	120	0.2	4.2	0.32	0.18	2.1	1.5
	160	0.1	4.8	0.32	0.18	2.4	2

* All figures except last column are in pounds.

is obtained from local livestock operations. Using this data, appraisers can calculate actual forage production. Another excellent source for estimates of the forage availability for grazing of various grasses and range types is the USDA Natural Resource Conservation Service.

Various range lands and grass species can provide an assortment of available forage. Forage availability can vary from several AUMs per acre in high rainfall areas to less than 0.1 AUM per acre in desert areas. US Forest Service lands, BLM lands, and western ranges typically vary from 0.20-0.65 AUMs per acre. The forage that can reasonably be harvested per acre, measured in AUMs, is influenced by dry matter content, palatability, and management (rotational grazing, seasonal grazing, or intensive grazing practices). The availability of forage is also influenced by conservation practices and by livestock waste. Proper long-term grazing practices are a requirement in all types of livestock ranching operations.

Recreational Use of Rangeland

In addition to supporting domestic livestock, ranches may support a wide range of wildlife. In many areas of the Great Plains and western United States, fishing and hunting amenities may produce income equal to or exceeding the income derived from typical livestock operations. Therefore, the fish and wildlife population and habitat on a ranching unit are of utmost importance in estimating the ability of a ranch to produce income, which in turn affects property value and marketability.

Fishing leases are available on many ranches that include mountain streams, rivers, reservoirs, lakes, or earthen tanks. Fishing on mountain streams and rivers is typically on a per-day or per-week basis. Reservoir, lake, and earthen tank fishing is usually granted on a per-season basis.

Hunting leases are in effect in areas with upland game birds, waterfowl, and big game animals. The primary upland game bird species hunted in the plains are bobwhite quail, blue or scaled quail, and ringneck pheasant. Quail hunting is typically on a per-acre basis, whereas pheasant hunting is typically on a per-day basis. Waterfowl hunting includes wild ducks, geese, and sandhill cranes. They are hunted in grain fields, lakes, earthen tanks, and river bottoms. Hunting leases for big game animals such as deer, elk, bear, and mountain lions are found in all the intermountain regions as well as many areas of Texas and Oklahoma. Big game hunting leases come in various forms. One common hunting lease calls for a lump-sum amount to cover the property to be hunted and the time to hunt. Often the leases call for the number and size of the animals to be taken to protect the supply of wildlife. Another form of big game lease is a per-gun shared lease. The outfitter pays the landlord a set portion of the fee paid to him or her. The risk to the outfitter is less because, if some hunters cancel, the outfitter does not have to pay the full amount.

The amount of money an outfitter is willing to pay for a hunting lease is based on the known big game population, habitat, accessibility, and time of year the lease is negotiated. If a lease can be negotiated early in the year, the outfitter will pay more because there will be up to nine months to book hunters to pay for the lease.

Regional Data

When appraising livestock ranches, appraisers usually collect regional data first. This information is important to the ranching industry and ranch appraisers. Ranch appraisers study livestock markets because the distance to and accessibility of livestock markets are of major importance. A ranch that runs stocker cattle that will later be sold as feeders should be located in reasonable proximity to major suppliers to the high plains cattle feeding industry. However, not every producer or rancher wishes to market through the larger auctions. Therefore, it could be advantageous to be near a smaller local auction or be in an area where many order buyers and country traders are available for direct marketing.

Transportation is important to a ranching operation. The availability of trucks and, in some cases, access to the railroad should be considered on each ranch appraisal. Climate is also important. Relevant factors include the growing season, the normal high and low temperatures, average annual precipitation, and major deviations from the normal precipitation pattern.

Local customs should be considered. Some areas specialize in cow-calf operations or yearling operations. Some areas specialize in high-quality bred cattle, whereas other areas may specialize in quality crossbred cattle. Still other areas are known for handling plainer cattle. Ranches in some areas must use specialized breeds of cattle and sheep that have the necessary natural resistance to parasites, disease, and weather conditions.

Appraisers collect information about a particular community or area from a variety of sources, such as the Natural Resources and Conservation Service, the Farm Service Agency, the US Weather Bureau, local livestock auctioneers, livestock inspectors, and cattle and sheep buyers.

Property Data

Proper use and skilled interpretation of aerial photographs are needed to formulate a land inventory. An accurate map can indicate forage types, topographical features and elevations, water sources, fences and roads, livestock protection, and land use. An appraiser who can read maps and aerial photographs will have a good idea of what the property looks like prior to the site visit. Forage types can be determined by looking at an aerial map. For example, hard grass areas can be differentiated from sand grass areas on a high plains ranch by the topography. It is possible to differentiate between north and south slopes from an aerial map of a mountain ranch. The tree or brush canopy can also be estimated fairly accurately on both types of ranches.

Topographical features such as elevation, high points, deep canyons, and bottom land areas can also be determined. Water sources such as streams, rivers, earthen tanks, wells equipped with windmills, and irrigation ditches can be pinpointed from a map. Property boundary fences and cross fencing as well as exterior and interior roads can be plotted. Livestock protection such as canyons and timbered areas can be located. The land use may also be identified on the map as rangeland, pasture, or cropland. All these physical characteristics are critical to the ranch operation because they affect ease of operation and the available forage production.

Studying maps prior to visiting a property will give appraisers an idea of what is to be appraised. Landmarks, fences, roads, and watering facilities can be noted in advance and used as references during inspection of the property from the ground or air.

Aerial photographs can be obtained from the Farm Service Agency office that serves the area where the ranch is located. These maps are available online at www.fsa.usda.gov/programs-and-services/aerial-photography/imagery-programs/naip-imagery/. Alternatively, they may be ordered through the local FSA office or directly from the Cartographic Unit in Salt Lake City, Utah. The cost of these maps is nominal.

Maps obtained from the Natural Resource Conservation Service are essentially the same as those obtained from the FSA. Detailed maps of a particular ranch can be obtained from the Natural Resource Conservation Service with written permission from the landowner. These maps often show the fence layout, the acreage in each pasture or field, roads, livestock and irrigation water supplies, and other physical characteristics of the ranching operation. County assessor's maps are valuable in that they show the perimeter of the subject property as well as the boundaries of any comparable properties.

Topographical maps prepared by the US Geological Survey are fairly reliable. These maps show the elevation contour of an area as well as any wooded areas, streams, fencing, roads, or bodies of water. These maps are available from various providers in larger cities, or they can be downloaded directly from the US Geological Survey.

After obtaining good maps, a ranch appraiser may have to collect a significant amount of property data on ranch improvements. Ranch improvements may influence the productivity and value of individual comparable properties and the appraised property. As in any appraisal, the size, construction, age, and condition of the building improvements determine their utility. The presence or absence of residences, lambing or shearing barns, calving facilities, and goat protection can influence the value of a livestock ranch.

Improvements or developments to the real estate typically have a significant effect on the utility of livestock ranches, particularly in the western states. Perimeter and cross-fencing together with physical access influence the grazing patterns and feed use of ranges. The availability, reliability, and distribution of livestock water sources restrict or improve grazing patterns and therefore forage utility and ranch value. The establishment of improved grass species, irrigation development, or other management practices may increase the length of the grazing season or forage availability and improve ranch balance and value. In some geographic regions, management practices such as grass fertilization, range burning, sagebrush and willow control, and the control of pests and predators may be an important part of the standard management and ranch improvement scheme for livestock ranches.

Appraisers should obtain as much information as possible on the history of comparable properties and the subject property. Investigation into the number, species, class (e.g., cow, bull, calf, yearling), and age of the livestock produced on a ranch is key to the analysis and understanding of the property. However, operating history and claims by the owner may or may not be an accurate indication of future operations. Improvements to fencing and water or the development of improved grasses may increase carrying capacity. Productivity and value can also be altered by a change in water rights, reductions in BLM or forest AUMs, or an increase in weeds, predators, or pests. Past operators, the BLM, the US Forest Service, and state lease authorities can be contacted for information. Obviously, the probable future use including carrying capacity should be forecast. Factors such as the tons of hay produced, the grazing patterns used, and the amounts of saleable product (beef, mutton, wool, and mohair) are typically investigated. Care should be taken to analyze the livestock ranch based on typical management rather than the superior or deficient management skills of the current operator, which would include investigation of the feeding season, required supplements, bull-to-cow ratio, level of replacement heifers, and market availability.

The historic use of leased land can usually be determined. The historic and probable future use of the AUMs available from the BLM, forests, state leases or permits, and other leased land is important to the carrying capacity, balance, and future utility of the total ranching operation. Season and level of use are typically dictated by government leases. The right to renew or transfer the lease to a new owner is of special importance. The physical location and ease of operation of leased land relative to the base property should be analyzed for the comparable properties and the subject property. In addition to reductions in the permitted AUMs, leases of US Forest Service and BLM properties should be viewed in light of recent pressure for dual uses. That is, the public's demand for recreational uses such as hunting, fishing, and camping may alter the utility of a livestock ranch.

Forage Species

Forage species, their condition, and the percentage of the plant community that the various forage species occupy are key to ranch productivity. Some factors that should be considered are climax or key species, the season of growth, the season of use, the canopy or overgrowth, and invaders.

The *climax* or *key species* is the major grass species located on the rangeland. Rangeland can be described using a soil or group of soils that produce the same kind and amounts of forage if the range is in similar condition. Each major parcel of rangeland most likely has a different key species, and the key species is identified as one of the most desirable plants found on the rangeland that is capable of producing a substantial portion of the forage. This is the species to be managed by the ranch operator. For example, if blue grama makes up 20% of the plant population by weight on hard grass rangeland and this is the species that the operator thinks is the best grass on the rangeland, then the blue grama will be the key species even though another grass species might constitute 50% of the present plant community.

Forage species can be categorized into two basic seasons of growth: (1) the warm season and (2) the cool season. Warm season grasses typically grow in the late spring and into the fall months. Cool season grasses usually begin their growth in the fall and grow into the early part of the winter, at which time they are dormant. They begin to grow again in early spring and grow until hot weather returns and they become dormant again. A browse plant such as a live oak is neither a cool season nor a warm season forage plant. It might be green year-round, but its major growth periods are in the spring and early summer.

The season of use of a forage species depends on the ranching operation as well as the type of livestock or wildlife that are going to use the forage. Short grasses and midgrasses that are typically found on high plains hard grass ranches normally cure out–i.e., dry out to be preserved for future use–very well. They can be used as a standing hay crop or grazed during the growing season. The season of use would depend on the ranching operation. Conversely, tall grasses typically found on sand grass ranches do not cure out well and are not as desirable for a standing hay crop. The better season of use for tall grasses is during the growing season. Note that some grasses or forage plants may be desirable to livestock and wildlife only during certain seasons of the year. For example, a forage may be very desirable in the early spring and of little or no value during the dormant season. To analyze the seasons of use for forage species, a forage inventory should be developed for individual pastures. Then a grazing plan should be outlined based on the forage species available. This could indicate whether a ranch should be considered as a year-round, cow-calf operation or whether it would be better used as a stocker operation because the desirable forage is not available for two or three months of the year. An inexperienced ranch appraiser faced with this situation should seek assistance from a range conservationist with the Natural Resources Conservation Service.

The canopy or overgrowth is another consideration in the analysis of value. The canopy cover can be defined as the percentage of the soil surface that is shaded by invading brush species when the brush is in full foliage. *Overgrowth* and *canopy* are terms that can be used synonymously. The percentage of canopy cover affects the type and species of grasses and forbs that can be grown under the canopy and the amounts of forage that can be produced.

Land use maps and plant inventory lists also include invaders. Invaders are generally undesirable plants. They can compete with plants in the climax community for moisture, nutrients, and light. Invaders come in and grow along with increasers–which are species of plants that increase in amount as a result of management practices or biotic or abiotic influences–after the climax vegetation has been reduced by close grazing or natural causes. Many invaders are weeds and shrubs of little or no grazing value, like snakeweed.

With the forage inventory in hand, the rural appraiser needs to check for poisonous plants such as larkspur. This plant is easy to identify in early spring when it grows above the grasses and is easily taken by the grazing cattle. Pastures containing a high percentage of larkspur can only be used for grazing in the late fall, preferably after the first killing frost.

The property map should be correlated to the forage inventory list to help identify problems with rangeland deterioration due to wind or water erosion, rodents, and brush encroachment. Correlating the forage inventory and property maps will also help appraisers recognize the acceptability of the livestock distribution. The forage around primary water sources is often overgrazed. The livestock must be forced to seek the readily available forage away from the water. This can be accomplished with cross fences, additional water distribution, baiting with salt or feed supplements, and herding.

Over the years, many ranches have been built up, split off, and built up again. This has improved the convenience of operation on some ranches and made other ranches very difficult to operate. A good example would be a ranch with an assured grazing privilege that is a great distance from the base property and headquarters unit. Government control can also limit the convenience of operation. In some instances, the managing government body dictates the number of animals that can be grazed on the deeded base property as well as the public land. Similarly, some common grazing allotments and associations dictate the kind and class of livestock that can be grazed.

In summary, the analysis of forage species is one of the more critical parts of a ranch appraisal. All of the factors mentioned–i.e., the key species, the seasons of growth, the season of use, the canopy cover, invaders, poisonous plants, rodents, livestock distribution, and convenience

Snakeweed, pictured above, is known as an invader. It has little grazing value and competes with more useful plants for moisture, nutrients, and light on rangeland.

climax (or key species). The major grass species located on a piece of rangeland.

invaders. Plants, animals, and microbes not native to a region that, when introduced either accidentally or intentionally, cause economic or environmental harm or harm to human health.

overgrowth (or canopy). The percentage of the soil surface that is shaded by invading brush species when the brush s in full foliage.

Larkspur, shown above, is poisonous to cattle. Pastures with a high percentage of this plant can only be used for grazing after the first killing frost of autumn.

of operation–are primary factors to consider in estimating the carrying capacity for a ranch. These same factors are also important in determining the highest and best use of the ranch and the comparable properties. It is recommended that appraisers seek assistance from range management specialists qualified in the area of the ranch being appraised.

Ranch Capacity

Although carrying capacity is much less important in the valuation of a recreational ranch or one with a nonagricultural use, a study of ranch capacity should be completed for the subject property and each comparable property.

In the analysis of carrying capacity, appraisers estimate the amount of forage available on a ranch. Units of measure and comparison vary in different geographical areas. The following units of comparison may be used:

- head
- cow unit
- animal unit (AU)
- animal unit month (AUM)
- cow year long (CYL)
- yearling year long

These different units representing total forage or carrying capacity are not synonymous. The per-head unit represents the total number of head run, which would include cows, bulls, replacements, and other livestock normally measured in a fall count. The cow unit represents the number of cows run in a normal year on a particular ranch. The cow unit measure also includes the necessary bulls, replacement heifers, and weaners for operation of that ranch. In contrast, the AU uses the equivalent number of mature cows as a basis of measurement related to carrying capacity. This has created a significant amount of confusion within the industry. All of the different units, if used and represented correctly, will result in the same final estimates of forage production.

For example, "1,000 AUs," "1,200 head," and "800 mother cows" could represent the same amount of forage in certain localities. The amount of forage per head or per mother cow can vary by area based on local climate conditions and grazing patterns, and adjustments would be required for each. The AU is simply annualized AUMs. That is, it measures the total forage available on a specific property divided by 12 months based on the feed requirements for a mature 1,000-pound cow. It is recommended that carrying capacity be based on the total amount of forage available, normally expressed in AUMs. Ranch balance, economy of feed, and other factors should be addressed separately.

The first step in determining carrying capacity is to estimate the AUMs available from individual feed sources and then summarize all the sources. Table 16.5 relates to a ranch unit with both owned land and assured lease land.

Table 16.5 Ranch Capacity Analysis

Type of Crop or Pasture	Acres	Season of Use	Crop Yield per Acre	AUMs per Acre	Total AUMs Capacity
Owned land					
Alfalfa hay	130	Winter	3.50	8.85	1,151
Meadow hay	415	Winter	2.00	5.06	2,099
Hay aftermath*	545	Fall	0.00	0.33	180
Native pasture	4,853	4/1 to 11/30	0.00	0.40	1,941
Crested wheat pasture	2,000	4/1 to 4/30	0.00	0.25	500
Site and roads	0				
Total					5,871
Classification	**Acres**	**Season of Use**	**% Public Use Land**	**Number of Head**	**Total AUMs**
Assured grazing lands					
BLM #73062	6,066	5/1 to 6/30	100	645	1,290
BLM #73063	3,034	10/1/ to 11/30	80	406	650
National forest	11,300	7/1 to 9/30	100	408	1,224
Total					3,164

Summary

Total AUM capacity	9,035
Percent capacity on fee-owned land	64.98%
Percent capacity on assured leases and permits	35.02%
Total AUs = AUMs divided by 12 months	753

* Hay aftermath must be deducted from the acreage to prevent double-counting for acres.

Managers with the US Forest Service and the BLM dictate the season of use and the number of animals that can be placed on a particular grazing privilege. On BLM permits, appraisers must consider the percentage of public land use. Sometimes portions of the deeded land are not fenced separately from the BLM land, and the BLM controls the stocking rates of both types of land. For example, consider the mathematical calculation for the BLM permits from Table 16.5:

Percentage of Public Use Land	Season of Use	Number of Head	Total AUMs
100%	2 months	645	1,290
80%	2 months	406	650*
Total			1,940

* 2 months × 406 head = 812 AUMs × 80%

Ranch appraisers should be familiar with the length of the feeding season and the typical amount of supplemental feed, such as hay and protein supplements, required in the area where the subject property and comparable properties are located. Mathematically, it is easy to support the AUMs per acre on measurable feed such as hay. It is slightly more difficult to support the AUMs per acre on native or seeded pasture tracts. This information can be obtained from Natural Resource Conservation Service range scientists and county soil books. When using the soil books, the total amount of dry matter produced is recorded. Adjustments are required for nonpalatable forage, range conservation, and waste. Local range professionals generally use rules of thumb for forage adjustments that they can scientifically support.

A scientific, detailed, mathematical ranch capacity analysis is a logical and supportable method for measuring the forage production of ranch property. This system of forage measurement has checks and balances. The projected forage production or the ranch capacity analysis should correlate with the actual forage production that is obtained from knowledgeable, unbiased ranch operators in the area who provide "typical" management. The historical capacity of the subject property should be reviewed, but appraisers should not let past operations bias the ranch capacity analysis. If there is a sizable difference between the historical capacity and the more scientific capacity analysis, the difference should be addressed by appraisers.

The information typically provided by local operators is not the actual AUMs per acre, but the actual number of cattle that were allowed to graze a certain amount of pasture for a specific amount of time. Often this is referred to *acres per cow* or *acres per animal unit*. This information can be easily converted into AUMs per acre. For example, in the ranch capacity chart, the native pasture was estimated at 0.40 AUMs per acre. In other words, 30 acres of pasture is required per AU:

$$1 \text{ month}/0.40 \text{ AUMs per acre} = 2.5 \text{ acres per AUM}$$

$$2.5 \text{ acres per AUM} \times 12 \text{ months} = 30 \text{ acres per AU}$$

Ranch Balance

The culmination of an appraiser's analysis of a livestock ranch's physical characteristics is the determination of *ranch balance*, which is the combination of sufficient feed and forage resources in each season to promote the continuous satisfactory maintenance and production of livestock. The study of forage types, productivity levels, lease use, and wintering requirements and the calculation of carrying capacity are precursors to the determination of ranch balance. Appraisers study ranch balance to determine the carrying capacity on a year-round basis. The ranch may be correctly balanced for seasonal feed requirements, or it may be deficient or possess excess amounts of individual feed items. Ranch balance is important to the income-producing ability of a ranch property and to the application of the sales comparison approach on a per-unit basis.

The timing of feedstuff availability is critical to the feeding of livestock, so an analysis of carrying capacity should be presented in a spreadsheet or graph. This format shows the available feed supply on a month-by-month basis. Table 16.6 indicates the available feed supply by month, broken down by feed source. The spreadsheet was developed using the AUM information obtained from the previous ranch capacity analysis.

The ranch balance table indicates that the subject ranch is well-balanced at 750 AUs. To support this conclusion, appraisers must be confident that the subject ranch possesses the necessary physical characteristics to be operated as projected. A good point for further explanation would be the lack of fencing separating the hay ground from the pasture. If the hay ground is not fenced separately, then the pasture could not be used until after the crop-growing season, which could influence annual ranch balance.

The total 1,779 AUMs of native pasture in this analysis is different from the 1,941 found in the ranch capacity analysis (Table 16.5).

Table 16.6 Ranch Balance Analysis

Month	BLM Pasture 73062	BLM Pasture 73063	National Forest	Native Pasture	Crested Wheat	Hay Aftermath Pasture	Meadow Hay	Total AUMs	Diff. AUMs
January	-	-	-	-	-	-	750	750	-
February	-	-	-	-	-	-	750	750	-
March	-	-	-	-	-	-	750	750	-
April	-	-	-	250	500	-	-	750	-
May	645	-	-	105	-	-	-	750	-
June	645	-	-	105	-	-	-	750	-
July	-	-	408	342	-	-	-	750	-
August	-	-	408	342	-	-	-	750	-
September	-	-	408	342	-	-	-	750	-
October	-	406	-	222	-	122	-	750	-
November	-	406	-	71	-	58	250	785	35.00
December	-	-	-	-	-	-	750	750	-
Total	1,290	812	1,224	1,779	500	180	3,250	9,035	35.00
Summer pasture									-
Excess fall pasture									-
Excess hay in AUMs									42.00
Excess hay in tons									16.61

Inclusion of deeded land within lands controlled by the BLM creates this apparent discrepancy. The difference of 162 AUMs is actually listed in the total 812 BLM capacity. Because deeded and BLM lands are fenced into the same pasture, appraisers must be careful to avoid a double-counting of the capacity.

The ranch balance table illustrates how land use, type of forage, and season of use can be correlated to determine ranch balance. Appraisal assignments are not textbook cases, however, and ranch appraisers will encounter different scenarios related to ranch balance. Although a ranch may be out of balance, appraisers are still required to analyze and estimate carrying capacity. For example, consider a slight change in the land use described in Table 16.5, the previous ranch capacity analysis. In Table 16.7, it is assumed that only 100 acres of native pasture is converted to alfalfa hay.

Considering the slight acreage change, the total available forage production changes from 750 to 826 AUs. Based on the results in the first example, a total of 826 AUs would not be considered balanced. Thus, the ranch balance analysis must also be changed.

Based on this ranch balance analysis, now the ranch has 913 AUMs of excess hay in the month of April (see Table 16.8). It is most logical to consider the ranch balanced at 750 AUs with excess hay base. Appraisers must determine from the marketplace if the excess hay base (per AU or acre) contributes the same proportional value as the units required in the ranch balance. Generally, excess hay base sells for the same amount because of the market demand for hay, and additional income may be generated. In the example above, the 913 AUMs of excess hay is equivalent to 361 tons of alfalfa hay:

$$913 \times 26 \text{ pounds per day} \times 30 \text{ days (one month)} = 712{,}140 \text{ pounds of hay}$$

$$712{,}140 \text{ pounds of hay}/2{,}000 \text{ pounds per ton} = 356 \text{ tons}$$

When completing adjustments for the different degrees of ranch balance, appraisers must recognize that the different types of land do not contribute the same value as they do in forage production. For example, rangeland that produces 0.4 AUMs per acre sells for $125 per acre. An AU of rangeland of this productivity will require 30 acres, or $3,750 per AU (30 × $125). Alfalfa hay land producing 8.85 AUMs per acre sells for $1,500 per acre, equal to $170 per AUM, or $2,040 per AU.

When a ranch is out of balance, a complete study of alternative uses should be conducted. For example, if a ranch is short on winter feed, is hay or supplemental pasture readily available? If hay is not available in the marketplace, are term cattle or rental cattle available to use the extra supply of summer or winter forage? An obvious conclusion is that the highest and best use of a ranch property is directly related to ranch balance.

Table 16.7 Ranch Capacity Analysis

Type of Crop or Pasture	Acres	Season of Use	Crop Yield per Acre	AUMs per Acre	Total AUMs Capacity
Fee-owned land					
Alfalfa hay	230.00	Winter	3.50	8.85	2,036
Meadow hay	415.00	Winter	2.00	5.06	2,099
Hay aftermath*	645.00	Fall	0.00	0.33	213
Native pasture	4,753.00	4/1 to 11/30	0.00	0.40	1,901
Crested wheat pasture	2,000.00	4/1 to 4/30	0.00	0.25	500
Site and roads	0.00				0
Total					6,749
Classification	**Acres**	**Season of Use**	**% Public Use Land**	**Number of Head**	**Total AUMs**
Assured grazing lands					
BLM 73062	6,066.00	5/1 to 6/30	100	645	1,290
BLM 73063	3,034.00	10/1 to 11/30	80	406	650
National forest	11,300.00	7/1 to 9/30	100	408	1,224
Total					3,164

Summary

Total AUMs capacity	9,913
Percent capacity on fee-owned land	68.08%
Percent capacity on assured leases and permits	31.92%
Total AUs = AUMs divided by 12 months	826

* Hay aftermath must be deducted from the acreage to prevent double-counting for acres.

Cow Unit

The analysis of a property as a balanced, 750-AU ranch does not indicate 750 producing cows. Replacement heifers and breeding bulls must be included in the analysis of a ranch operation's carrying capacity on a cow/calf basis.

A producing cow unit is defined as the carrying capacity necessary to sustain a mature cow for a year and her nursing calf for six months, together with the necessary percentage of replacement heifers and breeding bulls. Death losses, culling rates, normal pregnancy rates, and required cow-to-bull ratios must be considered. These factors all vary with the geographical area and the style of the ranch operation.

For example, consider the following factors:

- 100 calving cows
- cow death loss and culling rate of 16%
- replacement heifer retention rate of 80%
- cow-to-bull ratio in herd count of 20:1

Table 16.8 Ranch Balance Analysis

Month	BLM Pasture 73062	BLM Pasture 73063	National Forest	Native Pasture	Crested Wheat	Hay Aftermath Pasture	Meadow Hay	Total AUMs	Diff. AUMs
January	-	-	-	-	-	-	750	750	-
February	-	-	-	-	-	-	750	750	-
March	-	-	-	-	-	-	750	750	-
April	-	-	-	250	500	-	913	1,663	913
May	645	-	-	105	-	-	-	750	-
June	645	-	-	105	-	-	-	750	-
July	-	-	408	342	-	-	-	750	-
August	-	-	408	342	-	-	-	750	-
September	-	-	408	342	-	-	-	750	-
October	-	406	-	182	-	162	-	750	-
November	-	406	-	71	-	51	222	750	-
December	-	-	-	-	-	-	750	750	-
Total	1,290	812	1,224	1,739	500	213	4,135	9,913	913
Summer pasture									-
Excess fall pasture									-
Excess hay in AUMs									913
Excess hay in tons									361.02

Because of a combination of death loss, culling for physical problems, and infertility, 16% of the mature cow herd must be replaced each year. Death loss and infertility factors also affect replacement heifers and the number being retained so that only 80% of those selected ever make it into the breeding herd. Therefore, 16 replacements divided by the 80% retention indicates that 20 heifers must be exposed during the breeding period to be able to replace 16 mother cows. Many factors influence the pregnancy rate. However, a generally accepted management practice is to provide 5 bulls for each 100 cows. These numbers could represent bulls actually placed in the breeding pasture or could account for a percentage held in reserve.

Measured at the time of pregnancy testing and culling of the mature cow herd, the following calculations result in the required AU carrying capacity:

Animal Type	AUs per Head	Total AUs
100 calving cows (as defined)	1	100
20 replacement heifers (as calculated)	0.85*	17
6 breeding bulls (as defined)	1.25	7.5
126 head count		124.5

* Heifers are equivalent to 0.7 AU at weaning time and 1.0 AU at the time of pregnancy test, but they require extra feed for development, with some culled at an early date—summarized to 0.85 AU.

These calculations indicate a required 1.25 AU carrying capacity for every mature cow to calve in the herd. Thus, a 100-cow unit ranch requires 125 AUs of feed availability.

Various assumptions should be considered when analyzing the number of head of mature cows, replacement heifers, and bulls:

- cow culling rate
- heifer replacement rate
- cow-to-bull ratio
- AU classification
- death loss

Those assumptions are specific to the appropriate ranching area. For example, the anticipated death loss is likely to be higher in hotter and drier ranching regions. Ranch managers in some markets may use rules of thumb for heifer replacement rates, cow-to-bull ratios, and other measures of the makeup of a profitable herd. The more an appraiser knows about the expectations of ranch operators in a specific market, the better that appraiser can judge the appropriate assumptions for analyzing the number of AUs that the rangeland being appraised is likely to support.

Valuation Procedures

All three approaches to value are applicable to the valuation of livestock ranches. Livestock ranches are inherently land-intensive, with building improvements making up a nominal portion of the overall value. Units of comparison for the appraisal of livestock ranches vary with locality. The AU is normally accepted and understood everywhere. The AU can be refined to reflect local units of comparison such as cow units, head, and yearlings year long. Various other marketing units can be used for appraisal analysis, including price per acre for deeded land and price per AUM for government leases.

Cost Approach

The cost approach can be applied to a ranch property (land and buildings) in the same way it is applied to any other type of agricultural property (consisting of land and buildings). These basic procedures are covered in Chapter 15. Real estate improvements on ranches, such as livestock fences, irrigation systems, corrals, and stock water facilities, are found on nearly all of the comparable properties and will be included in the unit price. Care should be exercised, however, because irrigation wells (not irrigation equipment) may contribute a significant portion of the irrigated land value. These improvements are not found on most ranch properties and can vary considerably, so they need to be addressed separately in the cost approach.

The cost approach must also address the building improvements. The utility of a ranch improvement provides significant contributory value. Ranch properties often include a wide assortment of buildings that appear to have some remaining physical life, but are no longer functional for a modern ranch operation. Appraisers need to understand the purpose and utility of each building and how the market is recognizing its contribution to value. The presence or absence of a specific building may or may not influence the market value of the subject property.

Applying the cost approach to the real estate portion of a ranch typically includes developing incremental values by land type (e.g., irrigated meadow, dry rangeland, subirrigated pasture, sand grass, hard grass) and by lease type (e.g., state AUM lease, BLM AUM lease, US Forest Service AUM lease). When applying the cost approach for a property with various components, it is critical to support the premise that the sum of the values of the parts is equal to the value of the whole.

Land Valuation

A land value opinion is generally developed in conventional fashion using unimproved ranch sales. Both the owned land and the leased acreage in the comparable properties should be allocated by land type (e.g., irrigated crop and dry cropland, permanent or improved pasture, native range). For example, Table 16.9 identifies six properties to be used in the land valuation process. Like the subject property, all of these properties are of native rangeland, so no additional allocation between the land classes is required.

Once the sales of comparable properties have been identified and researched, they must be adjusted for market conditions (time) and financing, as applicable. The market conditions adjustments are extracted in Table 16.10. All the sales in this data set were cash or cash equivalent and required no adjustment for financing. Table 16.10 demonstrates the pairing of sales, which is used to extract the market conditions adjustment indicated in the subject market area.

Once the applicable market conditions and financing adjustments are estimated, the land valuation process can proceed as shown in Table 16.11. Note that the subject property and comparable properties are all native rangeland. Therefore, the carrying capacities of the various sites are used to allocate the subject component values. When multiple land classes are present, the valuation process should be patterned after the processes described in Chapter 14.

Table 16.12 shows the allocation of the value conclusion to the subject property. The estimated price per AU is applied to the total AUs supported by the subject property after analyzing the carrying capacity of the site's various types of rangeland. That is, the amount of acreage for each type of rangeland on the site can be converted into an estimate of AUs and then summed.

Table 16.9 Analysis of Sales of Unimproved Properties

Record No.	4405	4404	4410	14356	14355	14357
Sale no.	U-1	U-2	U-3	U-4	U-5	U-6
Grantee	Haley	Cromera	Consolidated	Alpha Cattle	Gordon	Ellison
Sale date	11 Months	11 Months	6 Months	Current	Current	11 Months
Size (in acres)	11,726	5,760	5,158	10,431	18,910	7,431
Sale price	$3,870,000	$1,380,000	$1,392,500	$2,655,000	$5,551,000	$2,400,000
Acre price	$330	$240	$270	$255	$294	$323
AU price	$9,579	$9,583	$9,473	$9,316	$8,811	$7,430
AUs	404	144	147	285	630	323
Acres/AU	29.0	40.0	35.0	36.6	30.0	23.0
Land type	Panhandle	R Plains	R Plains	R Plains	Panhandle	R Plains
Improvements	None	None	None	None	None	None

Table 16.10 Market Conditions Adjustment Analysis

Sale	**U-1**	**U-3**
Sale U-1 vs. U-3		
Date	11 Months	6 Months
Price per AU	$9,579	$9,473
Compound change/month	-0.22%	
Sale	**U-1**	**U-4**
Sale U-1 vs. U-4		
Date	11 Months	Current
Price per AU	$9,579	$9,316
Compound change/month	-0.25%	
Sale	**U-2**	**U-3**
Sales U-2 vs. U-3		
Date	11 Months	6 Months
Price per AU	$9,583	$9,473
Compound change/month	-0.23%	
Sale	**U-2**	**U-4**
Sale U-2 vs. U-4		
Date	11 Months	Current
Price per AU	$9,583	$9,316
Compound change/month	-0.26%	
Sale	**U-3**	**U-4**
Sale U-3 vs. U-4		
Date	6 Months	Current
Price per AU	$9,473	$9,316
Compound change/month	-0.28%	
Conclude	-0.25% per month	

Table 16.11 Market Conditions and Financing Adjustments to Sales of Unimproved Properties

Record No.	4405	4404	4410	14356	14355	14357
Sale no.	Sale U-1	Sale U-2	Sale U-3	Sale U-4	Sale U-5	Sale U-6
Grantee	Haley	Cromera	Consolidated	Alpha Cattle	Gordon	Ellison
Sale date	11 Months	11 Months	6 Months	Current	Current	11 Months
Size (in acres)	11,726	5,760	5,158	10,431	18,910	7,431
Sale price	$3,870,000	$1,380,000	$1,392,500	$2,655,000	$5,551,000	$2,400,000
Acre price	$330	$240	$270	$255	$294	$323
AU price	$9,579	$9,583	$9,473	$9,316	$8,811	$7,430
AUs	404	144	147	285	630	323
Acres/AU	29.0	40.0	35.0	36.6	30.0	23.0
Land type	Panhandle	R Plains	R Plains	R Plains	Panhandle	R Plains
Improvements	None	None	None	None	None	None
Financing	Cash	Cash	Cash	Cash	Cash	Cash
Sale price/AU	$9,579	$9,583	$9,473	$9,316	$8,811	$7,430
Adjustments						
Financing	0	0	0	0	0	0
Price adj. for financing	$9,579	$9,583	$9,473	$9,316	$8,811	$7,430
Market conditions adjustment						
Months	11	11	6	0	0	11
% adjustment	-2.75%	-2.75%	-1.50%	0.00%	0.00%	-2.75%
$ adjustment	-$263	-$264	-$142	$0	$0	-$204
Price adj. for market conditions and financing	$9,315	$9,320	$9,330	$9,316	$8,811	$7,226
Conclude	$9,300 per AU					

Table 16.12 Land Valuation of Subject Property

Type of Rangeland	Acres/AU	Acres	AUs
Clay loam	30	9,440	314.7
Very shallow	40	8,110	202.75
Shallow clay	37	7,600	205.4
Sandy land	40	2,780	69.5
Gypland	50	1,908	38.2
Loamy bottomland	20	441	22.05
Total		30,279	852.6
	Estimated subject land value via cost approach	$7,929,180	

Building Residual and Depreciation Analysis

The next step in the cost approach is to analyze the improved property sales to extract the depreciation rates prevailing in the market area. Analysis of the improved property sales must include building residual analysis to estimate how much the improvements contribute to the value of each improved sale. Table 16.13 illustrates the improved sales analysis to allocate the value to the land and improvements. The resulting indication of contributory land value for each sale is then deducted from the sale price, resulting in the indicated value that the improvements contribute to the improved sale.

After the total building contribution of the sale is estimated, the rate of depreciation can be estimated improvement by improvement or on an overall basis. Table 16.14 details the process of estimating the annual depreciation based on the allocated building values developed by sales comparison.

The depreciation analysis shown in Table 16.14 produces overall annual depreciation rates for each improvement class. Based on these indications, an appropriate rate can be selected for application to the subject improvements. In this analysis, depreciation has been treated on an overall basis. A complete breakdown allocating the various forms of depreciation to each class of improvement was not conducted. If market properties demonstrate significant functional or external obsolescence, the breakdown method as described in Chapter 15 would be preferred. Whether an overall depreciation estimate or the breakdown method is used, the results of this analysis are ultimately applied to the subject improvements based on an appraiser's estimate of the subject improvements' effective age. Table 16.15 shows the application of the estimated rates of depreciation to the subject property's improvements, resulting in the estimated value of the improvements. Land value is then added to arrive at the indicated value of the subject property through the application of the cost approach.

Sales Comparison Approach

Nearly all of the analytical techniques covered in Chapter 12–including direct comparison, paired sales, and statistical analysis–can be used to develop an opinion of the value of a ranch using the sales comparison approach. The available market evidence will determine which technique is most applicable and which unit of comparison is most appropriate. Possible units of comparison for this approach include animal units, cow units, yearling units, and acres.

If the properties are compared based on AUs, appraisers should note that the value per AU very often decreases as the percentage capacity on assured leases increases. The value per AU may also decrease as the percentage capacity on irrigated or cultivated land increases. Adjustments may be required for the physical characteristics

Table 16.13 Improvement Residual Analysis

Improved Sale I-1 15 Months				**Ranch Type: Panhandle**		
Improved Sale Price	**Improved Sale Land**	**Improved Sale Acres**	**Improved AUs**	**Unimproved Sale U-1 Contribution (Acre or AU)**	**Improved Sale Land Contribution**	**Improved Sale Improvement Contribution**
	Range land	33,000	1,320	$9,666	$12,759,120	
	Imp. range	0	0	0	0	
	Cropland	0	0	0	0	
	Leased land	0	0	0	0	
$12,900,000	Total	33,000			$12,759,120	$140,880
Market Conditions Adjustment of Unimproved Sale to Improved Sale's Date of Sale						
	Sale U-1	11 months		Market conditions adjustment		
	Sale I-1	15 months		-0.25% per month		
	Difference	-4 months		Sale U-1 adjusted price		$9,666
Improved Sale I-2 11 Months				**Ranch Type: Rolling Plains**		
Improved Sale Price	**Improved Sale Land**	**Improved Sale Acres**	**Improved AUs**	**Unimproved Sale U-1 Contribution (Acre or AU)**	**Improved Sale Land Contribution**	**Improved Sale Improvement Contribution**
	Range land	8,548	389	$7,428	$2,889,492	
	Imp. range	0	0	0	0	
	Cropland	0	0	0	0	
	Leased land	0	0	0	0	
$3,077,250	Total	8,548			$2,889,492	$187,758
Market Conditions Adjustment of Unimproved Sale to Improved Sale's Date of Sale						
	Sale U-6	11 months		Market conditions adjustment		
	Sale I-2	11 months		-0.25% per month		
	Difference	0 months		Sale U-6 adjusted price		$7,428
Improved Sale I-3 6 Months				**Ranch Type: Panhandle**		
Improved Sale Price	**Improved Sale Land**	**Improved Sale Acres**	**Improved AUs**	**Unimproved Sale U-1 Contribution (Acre or AU)**	**Improved Sale Land Contribution**	**Improved Sale Improvement Contribution**
	Range land	43,585	1,362	$9,450	$12,870,900	
	Imp. range	0	0	0	0	
	Cropland	0	0	0	0	
	Leased land	0	0	0	0	
$13,596,231	Total	43,585			$12,870,900	$725,331
Market Conditions Adjustment of Unimproved Sale to Improved Sale's Date of Sale						
	Sale U-1	11 months		Market conditions adjustment		
	Sale I-3	6 months		-0.25% per month		
	Difference	5 months		Sale U-1 adjusted price		$9,450
Improved Sale I-4 2 Months				**Ranch Type: Rolling Plains**		
Improved Sale Price	**Improved Sale Land**	**Improved Sale Acres**	**Improved AUs**	**Unimproved Sale U-1 Contribution (Acre or AU)**	**Improved Sale Land Contribution**	**Improved Sale Improvement Contribution**
	Range land	8,916	637	$9,363	$5,964,231	
	Imp. range	0	0	0	0	
	Cropland	0	0	0	0	
	Leased land	0	0	0	0	
$6,300,000	Total	8,916			$5,964,231	$335,769
Time Adjustment of Unimproved Sale to Improved Sale's Date of Sale						
	Sale U-4	0 months		Time adjustment	-0.25% per month	
	Sale I-4	2 months				
	Difference	-2 months		Sale U-4 adjusted price	$9,363	

Note that when adjusting the price of the land sale to the improved sale for market conditions produces a negative adjustment, the adjusted price will be higher in a declining market.

Table 16.14 Market Depreciation Extraction Analysis

	Sale I-1	Sale I-2	Sale I-3	Sale I-4
Sale price	$12,900,000	$3,077,250	$13,596,231	$6,300,000
Land contribution	$12,759,120	$2,889,492	$12,870,900	$5,964,231
Building contribution	$140,880	$187,758	$725,331	$335,769
Replacement cost of dwelling	$147,000		$896,400	$405,000
Depreciation	$84,000		$371,229	$180,000
Dwelling contribution	$63,000		$525,171	$225,000
% Depreciated	57.1%		41.4%	44.4%
Effective age	25		25	20
% Dep. per year	2.3%		1.7%	2.2%
Replacement cost of tenant dwelling		$157,500	$310,800	$157,500
Depreciation		$67,500	$177,600	$87,750
Dwelling contribution		$90,000	$133,200	$69,750
% Depreciated		42.9%	57.1%	55.7%
Effective age		20	25	25
% Dep. per year		2.1%	2.3%	2.2%
Replacement cost of bunk/cook	$192,000			
Depreciation	$142,080			
Bunk/cook contribution	$49,920			
% Depreciated	74.0%			
Effective age	20			
% Dep. per year	3.7%			
Replacement cost of barn(s)	$90,000	$408,000	$216,000	$144,000
Depreciation	$62,040	$310,242	$151,200	$102,981
Barn(s) contribution	$27,960	$97,758	$64,800	$41,019
% Depreciated	68.9%	76.0%	70.0%	71.5%
Effective age	20	25	20	25
% Dep. per year	3.4%	3.0%	3.5%	2.9%
Replacement cost of shed			$8,100	
Depreciation			$5,940	
Shed contribution			$2,160	
% Depreciated			73.3%	
Effective age			20	
% Dep. per year			3.7%	

Table 16.15 Cost Approach Estimate of Subject Property

	Units	Unit Cost	Amount
Replacement cost—owner's home and garage	2,136	$226.50	$483,804
Less depreciation			
% Depreciation	22.00%		
Depreciation amount			– $106,437
Improvement contribution			**$377,367**
Replacement cost—foreman's home and garage	2,016	$127.50	$257,040
Less depreciation			
% Depreciation	26.40%		
Depreciation amount			– $67,860
Improvement contribution			**$189,180**
Replacement cost—cook shack	1,216	$96.00	$116,736
Less depreciation			
% Depreciation	55.50%		
Depreciation amount			– $63,948
Improvement contribution			**$52,788**
Replacement cost—hunting lodge (mobile home)	1,600	$78.60	$125,760
Less depreciation			
% Depreciation	40.00%		
Depreciation amount			– $50,304
Improvement contribution			**$75,456**
Replacement cost—utility barn	3,600	$34.50	$124,200
Less depreciation			
% Depreciation	60.00%		
Depreciation amount			– $74,520
Improvement contribution			**$49,680**
Replacement cost—corrugated metal barn	2,400	$22.50	$54,000
Less depreciation			
% Depreciation	52.50%		
Depreciation amount			– $28,350
Improvement contribution			$25,650
Replacement cost—hay barn	3,600	$13.50	$48,600
Less depreciation			
% Depreciation	14.80%		
Depreciation amount			– $7,194
Improvement contribution			$41,406
Replacement cost—3-bay overhead cake bin	60	$480.00	$28,800
Less depreciation			
% Depreciation	16.00%		
Depreciation amount			– $4,608
Improvement contribution			**$24,192**
Total improvement contribution			**$834,879**
Plus land value via unimproved sales			$7,929,180
Estimated subject value via cost approach			$8,764,059
Rounded to			$8,765,000

listed in Figure 16.2. It is important to develop a list of physical characteristics that are recognized by buyers of ranch properties in the subject property's area. Needless to say, it is also important that adjustments be derived from market evidence.

When applying the sales comparison approach to ranch properties, appraisers should follow the methodology described in Chapter 12. The specific adaptation of the methodology to ranch properties is demonstrated in Tables 16.16 through 16.19.

As previously discussed, certain nuances must be considered in appraisals of hay-based ranches in contrast to appraisals of Great Plains ranches. Hay-based ranches derive a portion of their productive capability from the growing of hay in a cultivated or uncultivated environment and harvesting and storing the production to use for feeding later in the year. A typical Great Plains operation uses the livestock to consume the forage produced on the property.

When analyzing hay-based ranches and using AUs or AUMs as the unit of comparison, the carrying capacity of the properties must be converted to a consistent unit of comparison. This may be accomplished by converting the carrying capacity of the comparable properties and subject property to "grass equivalent" AUs or AUMs. This technique recognizes that hay-based ranches usually consist of rangeland in addition to irrigated crop/hay and pasture. Irrigated lands are more management-intensive than rangeland, and the costs associated with harvesting the feed, which include labor and machinery, must be considered in the analysis. To equalize the data, conversion is appropriate. The basis for conversion is the typical cropshare arrangement, in which the owner provides the land and another individual provides the labor and machinery for irrigation and harvesting. Depending on the area, the share split may vary slightly, but it is generally close to a 50-50 split. To recognize the contribution of the hay/crop land while keeping the unit of comparison constant, the number of AUs from the hay base is reduced by 50% in the case of a 50-50 split on both the comparable properties and the subject property. The corresponding "grass equivalent" AUs, therefore, account for any difference in capacity derived from the hay-based lands.

As in any valuation, the unit of comparison selected for the property must be the unit recognized by the market in the subject area. In most ranching areas, the preferred unit is the AU or AUM. However, in areas where the market is driven by nonagricultural uses such as

Figure 16.2 Adjustments for Physical Characteristics

- Land use, which may include assured leases
- Water, both livestock and irrigation
- Forage productivity
- Balance and convenience of operation
- Site improvements
- Building improvements
- Size
- Percentage of leased land
- Location/access
- Special amenities/recreation

recreation, commercial hunting, or fishing, a per-acre unit of comparison may be indicated by the market. Table 16.16 summarizes the sales to be analyzed and adjusted in the sales comparison approach.

Table 16.16 Improved Sales

Record No.	**4411**	**12809**	**4407**	**12808**
Sale no.	Sale I-1	Sale I-2	Sale I-3	Sale I-4
Grantee	Willard	Porter	TBP Group	Williams
Sale date	15 months	11 months	6 months	2 months
Size (in acres)	33,000.00	8,548.08	43,584.58	8,916.04
Sale price	$12,900,000	$3,077,250	$14,250,000	$6,300,000
Acre price	$390.90	$360.00	$326.94	$706.59
AU price	$9,774	$7,920	$10,464	$9,891
AUs	1,320	389	1,362	637
Acres/AU (rounded)	25	22	32	14
Land type	Panhandle	Rolling plains	Panhandle	Rolling plains
Improvements	Yes	Yes	Yes	Yes
Oil field activity	None	Extensive	None	Moderate

The improved sales must be analyzed to properly allocate the contribution of the improvements using residual analysis, as described in Chapter 12. The underlying analysis of carrying capacity is critical to the process and must be completed before processing of the approach can begin. The building contributions can be allocated for application in the sales comparison approach as shown in Table 16.17.

The improvement and land adjustment process can proceed once the building allocation analysis is complete. This analysis allows direct comparison of the comparable property's improvements with the subject property's improvements. In both processes, appraisers must exercise informed judgment in making comparisons and deriving adjustments. Tables 16.18 through 16.21 demonstrates the development of the improvement adjustments, which will be used in the sales comparison adjustment grid.

The final step in the process is the direct comparison of the sale properties to the subject property. In this example, only one type of land–native rangeland–was present. When multiple land types are present, a land adjustment must be developed based on the sales of improved properties as detailed in Chapter 12. Hay-based ranches generally require development of a "grass equivalent" analysis as well as categories for land subject to various types of federal or state grazing permits or leases. Once the pairings and adjustment estimation process is complete, the sales adjustment grid can be constructed. Table 16.22 incorporates the adjustments developed above.

Table 16.17 Building Allocation

Sale No.	Sale I-1		Sale I-2		Sale I-3		Sale I-4	
Improvement	**Cond./Util.* (Size†)**	**Contrb. $ ($/Unit)**	**Cond./Util.* (Size)**	**Contrb. $ ($/Unit)**	**Cond./Util.* (Size)**	**Contrb. $ ($/Unit)**	**Cond./Util.* (Size)**	**Contrb. $ ($/Unit)**
Main dwelling	F/F	$63,000			G/G	$525,171	F/F	$225,000
	1,400	$45.00			4,980	$105.45	3,000	$75.00
Tenant dwelling			A/F	$90,000	F/F	$66,600	F/F	$69,750
			1,500	$60.00	1,480	$45.00	1,550	$45.00
Tenant dwelling					F/F	$66,600		
					1,480	$45.00		
Bunk/cook house	F/F	$49,920						
	2,560	$19.50						
Barn(s)	F/F	$27,960	P/F	$97,758	A/F	$64,800	F/F	$41,019
	3,000	$9.33	13,600	$7.20	7,200	$9.00	4,800	$8.55
Shed(s)					A/F	$2,160		
					360	$6.00		
Total building contribution		$140,880		$187,758		$725,331		$335,769

* Condition/utility ratings: E = excellent; G = good; A = average; F = fair; P = poor

† Size in square feet

Table 16.18 Building Comparison (I-1) to Subject Property

Improvements	Size	Condition/ Utility	$/Sq. Ft.	Contrib. Value
Sale I-1				
Dwelling				
Tenant dwelling	1,400	F/F	$45	$63,000
Tenant dwelling				$0
Tenant dwelling				$0
Mobile home				$0
Bunk/cook	2,560	F/F	$19.50	$49,920
Barn(s)	3,000	F/F	$9.33	$27,990
Shed(s)				$0
Bins/etc.				$0
Total improvement value				$140,910
Price per AU@ 1,320 AUs				$106.74

Improvements	Size	Condition/ Utility	$/Sq. Ft.	Contrib. Value
Subject				
Owner's house	2,136	G/E	$180	$384,480
Foreman's house	2,016	G/G	$90	$181,440
Hunting lodge	1,600	F/A	$45	$72,000
Cook shack	1,216	F/G	$36	$43,776
Barns (2)	6,000	F/F	$9	$54,000
Hay barn/shed	3,600	G/G	$6	$21,600
Cake bins	60	G/G	$405	$24,300
Subject improvement value via Sale I-1				$781,596
Price per AU @ 850 AUs				$919.52
Indicated adjustment				$812.78

Table 16.19 Building Comparison (I-2) to Subject Property

Improvements	Size	Condition/ Utility	$/Sq. Ft.	Contrib. Value	Improvements	Size	Condition/ Utility	$/Sq. Ft.	Contrib. Value
Sale I-2					Subject				
Dwelling					Owner's house	2,136	G/E	$180	$384,480
Tenant dwelling	1,500	A/F	$60	$90,000	Foreman's house	2,016	G/G	$90	$181,440
Tenant dwelling				$0	Hunting lodge	1,600	F/A	$45	$72,000
Tenant dwelling				$0	Cook shack	1,216	F/G	$36	$43,776
Mobile home				$0	Barns 2	6,000	F/F	$9	$54,000
Bunk/cook				$0	Hay barn/shed	3,600	G/G	$6	$21,600
Barn(s)	13,600	P/F	$7.20	$97,920	Cake bins	60	G/G	$405	$24,300
Shed(s)				$0					
Bins/etc.				$0					
Total improvement value				$187,920	Subject improvement value via Sale I-1				$781,596
Price per AU @ 389 AUs				$488.09	Price per AU @ 850 AUs				$919.52
					Indicated adjustment				$431.43

Table 16.20 Building Comparison (I-3) to Subject Property

Improvements	Size	Condition/ Utility	$/Sq. Ft.	Contrib. Value	Improvements	Size	Condition/ Utility	$/Sq. Ft.	Contrib. Value
Sale I-3					Subject				
Dwelling	4,980	F/G	$105.45	$525,141	Owner's house	2,136	G/E	$180	$384,480
Tenant dwelling				$0	Foreman's house	2,016	G/G	$90	$181,440
Tenant dwelling	1,480	F/F	$45	$66,600	Hunting lodge	1,600	F/A	$45	$72,000
Tenant dwelling	1,480	F/F	$40	$59,200	Cook shack	1,216	F/G	$36	$43,776
Mobile home				$0	Barns 2	6,000	F/F	$9	$54,000
Bunk/cook				$0	Hay barn/shed	3,600	G/G	$6	$21,600
Barn(s)	7,200	A/F	$9.10	$65,520	Cake bins	60	G/G	$405	$24,300
Shed(s)	360	A/F	$5.95	$2,142					
Bins/etc.				$0					
Total improvement value				$718,603	Subject improvement value via Sale I-1				$781,596
Price per AU@ 1,362 AUs				$527.61	Price per AU @ 850 AUs				$919.52
					Indicated adjustment				$391.92

Table 16.21 Building Comparison (I-4) to Subject Property

Improvements	Size	Condition/ Utility	$/Sq. Ft.	Contrib. Value	Improvements	Size	Condition/ Utility	$/Sq. Ft.	Contrib. Value
Sale I-4					**Subject**				
Dwelling	3,000	F/F	$75	$225,000	Owner's house	2,136	G/E	$180	$384,480
Tenant dwelling				$0	Foreman's house	2,016	G/G	$90	$181,440
Tenant dwelling				$0	Hunting lodge	1,600	F/A	$45	$72,000
Tenant dwelling	1,550	F/F	$40	$62,000	Cook shack	1,216	F/G	$36	$43,776
Mobile home				$0	Barns 2	6,000	F/F	$9	$54,000
Bunk/cook				$0	Hay barn/shed	3,600	G/G	$6	$21,600
Barn(s)	4,800	F/F	$9	$43,200	Cake bins	60	G/G	$405	$24,300
Shed(s)				$0					
Bins/etc.				$0					
Total improvement value				$330,200	Subject improvement value via Sale I-1				$781,596
Price per AU @ 637 AUs				$518.37	Price per AU @ 850 AUs				$919.52
					Indicated adjustment				$401.15

The adjustment process shown in Table 16.22 required downward adjustments for market conditions based on an average annual decline of 3%, which was derived from paired data analysis. The market conditions adjustments were based on compound change. Straight-line or compound-change calculations are certainly acceptable, but using straight-line change requires extra analysis if sales both before and after the date of valuation are analyzed. Assuming compound change eliminates this extra step and produces the same result.

The only other adjustments required, based on the sales analysis, were the improvement adjustments detailed in the building adjustment analysis and adjustments for oil field activity, which were estimated through paired data analysis.

Income Capitalization Approach

Nearly all of the income and expense information and methods of capitalization discussed in Chapter 13 can be applied to ranch properties. The most common technique employed is direct capitalization, in which a single year's income estimate is converted into an indication of value in one step, either by dividing the income estimate by an appropriate rate or multiplying it by an appropriate factor.

Deriving capitalization rates from comparable sales is essential. Data on the sale price, income, expenses, and market condition of each property at the time of sale is needed. In addition, appraisers must be certain that the net operating income of the subject property and each comparable property is calculated and estimated in the same manner.

Table 16.22 Improved Sales Adjustment Grid

Record No.		4411	12809	4407	12808
Sale no.	Subject	Sale I-1	Sale I-2	Sale I-3	Sale I-4
Grantee		Willard	Porter	TBP Group	Williams
Sale date	Current	15 months	11 months	6 months	2 months
Size (in acres)	30,279.00	33,000.00	8,548.08	43,584.58	8,916.04
Sale price		$12,900,000	$3,077,250	$14,250,000	$6,300,000
Acre price		$390.90	$360.00	$326.94	$706.59
AU price		$9,774	$7,920	$10,461	$9,891
AUs	850	1,320	389	1,362	637
Acres/AU (rounded)	36	25	22	32	14
Land type	Rolling plains	Panhandle	Rolling plains	Panhandle	Rolling plains
Improvements	Yes	Yes	Yes	Yes	Yes
Oil field activity	Slight	None	Extensive	None	Moderate
Financing Adjustment					
Financing		Cash	Cash	Liberal	Cash
Sale price/acre		$390.90	$360.00	$326.94	$706.59
Adjustments					
Financing adjustment		$0.00	$0.00	-$15.00	$0.00
Financing-adjusted price/acre		$390.90	$360.00	$311.94	$706.59
Financing-adjusted price/AU		$9,744	$7,920	$9,981	$9,891
Financing-adjusted price		$12,900,000	$3,077,250	$13,596,231	$6,300,000
Market Conditions Adjustments					
Months		15	11	6	2
Market conditions adjustment		-$360	-$216	-$150	-$16
Market conditions-adjusted price/AU		$9,414	$7,704	$9,831	$9,843
Other Adjustments					
Improvements		$813	$431	$392	$401
Improvement adjusted price		$10,227	$8,135	$10,223	$10,244
Oil field activity		$0	$2,085	$0	$0
Oil field-adjusted price		$10,227	$10,220	$10,223	$10,244
Location		0	0	0	0
Indicated subject value/AU		$10,227	$10,220	$10,223	$10,244
Estimated subject value via sales comparison approach	850 AUs @ $10,225 = $8,691,250				

There are three generally accepted methods for analyzing the income and expenses of ranch properties:

- cash lease
- cropshare rental
- owner-operator

A ranch lease can be structured to fit any situation. Most ranch properties are leased for a flat rental rate, based on a per-AU or AUM rate. In areas where hunting or recreation is the primary motivating factor in the market, income is often quoted on a per-acre basis. Many current leases tend to be short-term. Appraisers may treat these short-term leases with a vacancy loss, if applicable, depending on how vacancy is perceived in the market. Caution should be exercised in determining when a ranch property becomes vacant.

For ranch properties, if a tenant does not renew a lease in October after the growing season and forage has been harvested, the land does not become vacant until the spring when the next growing season begins. Considering the physical characteristics of ranch properties and agricultural operations, the supply and demand relationship, and local economic conditions, a vacancy loss rate for a flat-fee, short-term lease is generally not required. Appraisers should survey the local market to determine if a vacancy estimate is necessary. As long as the vacancy rate is treated consistently (whether used or not used) in analyzing the comparable properties and the subject property, the appraisal will be market-supported.

There are many different types of leases written on ranch properties. For example, most ranches on the eastern plains rent on a per-acre basis, usually with limits on the carrying capacity. Ranches in the intermountain region tend to rent on the basis of carrying capacity per day, AUM, or AUs. Many different kinds of leases can be found in each area, and it is important to find the best method of lease analysis to correlate the different lease information into a single, common unit. For example, a $14.40 per acre rent for pasture in an area that requires 30 acres per AU is the same rental rate as $36 per AUM, or $432 per AU.

Although information on intensive farming areas may be easy to find, it is often very difficult to find ideal rental information on ranch properties. Consider ranches that have a high percentage of their carrying capacity from assured lease land. Special arrangements must be made with the US Forest Service and the BLM to allow nonowner livestock to be grazed. As the spread or percentage capacity from assured lease land changes, it causes a large variance in the capitalization rates. This makes it somewhat difficult to compare and analyze these rates.

When rental information is not adequate or consistent, an owner-operator statement is an alternative method of capitalization. Owner-operator analysis is a detailed procedure, and appraisers who apply it must be intimately familiar with the area and the ranching industry to be able to allocate income between real property and other assets of the business. Income and expense information is available from most agricultural colleges and local extension services.

The income capitalization approach is applied to ranch properties with the same procedures used for all other classes of property. The only difference is the unit of comparison used, which should be the unit prevalent in the subject property's market area. The units of comparison most commonly used are AU, AUM, and acre. Table 16.23 demonstrates the development of capitalization rate analysis based on the sales of improved properties analyzed in the sales comparison approach.

Table 16.23 Capitalization Rate Analysis

Sale No.	Subject	Sale I-1	Sale I-2	Sale I-3	Sale I-4
Sale price		$12,900,000	$3,077,250	$14,250,000	$6,300,000
Acres	30,279	33,000	8,548.08	43,584.58	8,916.04
Date	Current	15 months	11 months	6 months	2 months
Land type	Rolling plains	Panhandle	Rolling plains	Panhandle	Rolling plains
Acre price		$390.90	$360.00	$326.94	$706.59
Price per AU		$9,774	$7,920	$10,461	$9,891
Total AUs	850	1,320	389	1,362	637
Estimated gross income					
Rangeland		$356,400	$104,907	$367,746	$171,951
Improved range		$0	$0	$0	$0
Cropland		$0	$0	$0	$0
Hunting (per acre @ $0.50)		$47,025	$0	$62,109	$12,705
Total gross income		$403,425	$104,907	$429,855	$184,656
Total gross income/acre		$12.24	$12.27	$9.87	$20.70
Less operating expenses					
Taxes		$49,500	$10,770	$65,376	$11,235
Maintenance		$40,343	$10,490	$42,987	$18,465
Management		$16,137	$4,197	$17,139	$7,386
Total operating expenses		$105,980	$25,457	$125,556	$37,086
Net operating income		$297,445	$79,450	$304,299	$147,570
Net operating income/acre		$9.01	$9.29	$6.99	$16.56
Net operating income/AU		$225.34	$204.24	$223.41	$231.66
Capitalization rate		2.31%	2.58%	2.14%	2.34%
Estimated capitalization rate	2.35%				

The analysis of income data on the comparable properties indicated that the current market rent for grazing in the subject area was $270 per AU. The hunting income estimate was based on prevailing lease rates in the subject area. Sale 1 and Sale 3 were leased for hunting at $0.50 per acre based on 95% of the total acreage. The headquarters and improvement areas were excluded from the hunting lease area.

Taxes were based on actual taxes at the date of sale. Maintenance and management fees were based on typical rates prevailing in the market area for ranch properties.

The indicated capitalization rates developed from the sales of comparable properties range from 2.14% to 2.58%. In the final analysis, a rate of 2.35% was adopted as reasonable for the subject property.

The application of the income capitalization approach to the subject property is detailed in Table 16.24 based on a landlord-tenant arrangement.

The pro forma analysis in Table 16.24 uses data derived from the analysis of properties in the market area. In this example, no privilege or improvement income was included because all of the leases in the market area included the use of improvements at the lease rate. In some jurisdictions, separate line items for improvements may be applicable depending on local custom and market activity.

Table 16.24 Pro Forma Operating Statement for the Subject Property

Income Estimate	**Units**	**Price/Unit**	**Total**	
Grazing (per AU)	850	$270 per AU	$229,500	
Hunting (acres)	28,765	$1.50 per acre	$43,148	
Total gross income				$272,648
Less operating expenses				
Ad valorem taxes (land)	$1.26	$1.26 per acre	$38,151	
Management (% of gross)	4%		$10,905	
Maintenance (% of gross)	5%		$13,632	
Total operating expense				– $62,688
Net operating income				$209,960
Capitalization	$209,960	@ 2.35%		$8,934,468
Estimated value via the income capitalization approach				$8,934,468
Rounded to				$8,935,000

Chapter 17

The Valuation of Dairy Farms and Facilities

American consumers spend between 10% and 12% of their food dollars on milk and other dairy products. Total sales of dairy products amounted to $102 billion in 2014, and the per capita consumption of milk products for that year was 614 pounds.

According to data from the US Department of Agriculture, the number of individual dairy farms in the United States has decreased over time through consolidation, but the total number of milk cows has remained relatively stable, with herd size, milk production per cow, and total milk production all increasing. Total milk production for 2015 was 208.3 billion pounds, from 9.3 million cows, with an average of 22,365 pounds of milk per cow.[1]

Although dairy cattle are raised in all 50 states, the major concentrations of dairy farms are in the Northeast, the Southwest, Florida, and the states bordering the Great Lakes. The top five states in total dairy production are Wisconsin, California, Idaho, New York, and Texas.

The long-term trend of consolidation in US agriculture is evident in the dairy industry. In the United States, the number of dairy farms decreased from 139,670 in 1995 to 49,331 in 2012, according to the Census of Agriculture. The *Progressive Dairyman* estimated that the number of dairy herds dropped 3% in 2015, to 43,584.[2] However, productivity continues to increase steadily because of economies of scale and the ongoing evolution of dairy management, as illustrated in Figure 17.1.

1. USDA, *Livestock, Dairy, and Poultry Outlook*, www.ers.usda.gov/publications/ldpm-livestock,-dairy,-and-poultry-outlook.aspx.
2. 2015 US Dairy Statistics, www.progressivepublish.com/downloads/2016/general/2015_pd_stats_highres.pdf.

Figure 17.1 US Dairy Farm Productivity (Milk per Cow)

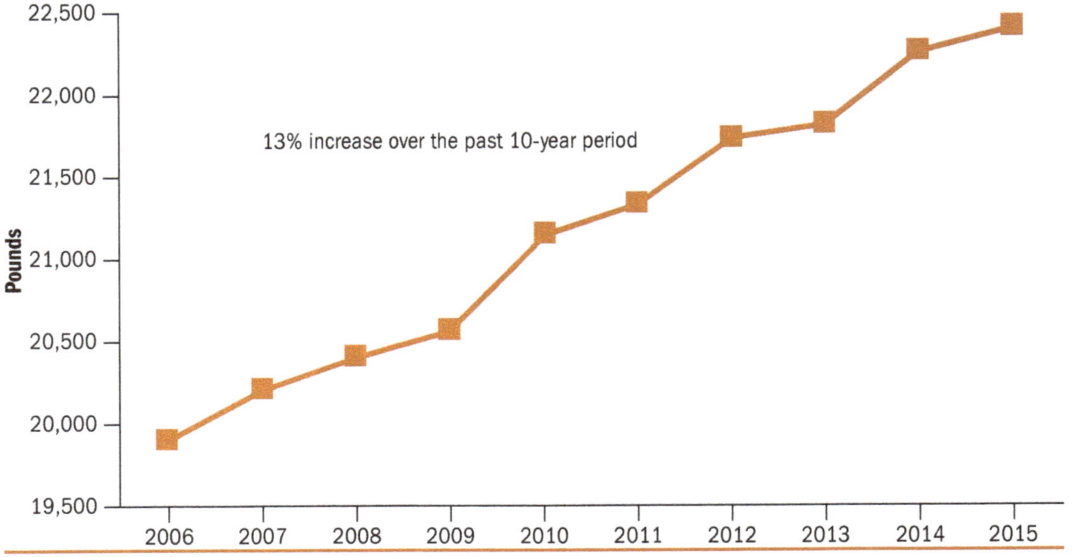

USDA-NASS; 02-19-2016; Last Modified: 10/19/2016

Glossary

pasture. Land devoted to the production of domesticated or native forage used for grazing.

dry lot. An area where livestock are confined, fed, and milked; usually equipped with feed alleys and automatic waterers; also called an *open lot*. Sometimes, dry lots include pen shades and wind breaks.

milk house. A dairy building used exclusively for the storage of milk.

milking parlor. A dairy building used exclusively for milking cows.

free stalls. Resting cubicles or "beds" in which dairy cows are free to enter and leave, as opposed to being confined in stanchions or pens.

tie-stall barn. A traditional barn, used for both housing and milking, designed to give each cow her own stall with bedding and free access to food and water in front of her in a manger. The cow is milked in her individual stall, alongside her herd mates. Also known as a *stanchion barn.*

Historical Background

Throughout history, people have kept domestic cows as a convenient source of milk and have turned milk into other, longer-lasting dairy products such as butter, cheese, ice cream, and yogurt. Scientific developments have made storing and transporting milk easier and more hygienic. These innovations include the condensing and evaporating of milk, pasteurizing milk to improve preservation and destroy harmful bacteria, and homogenizing milk to render it more appealing and easily digestible. In addition, improvements in refrigeration and the speed of transportation have allowed milk producers to distribute more milk over greater distances and preserve it longer on the shelves of grocers. New techniques are continuously being explored, including the development of longer shelf lives and unrefrigerated storage.

Veterinary science has contributed to the improved health and breeding of dairy cattle. Vaccines are now used to prevent such diseases as brucellosis, leptospirosis, and blackleg. Effective medical treatments for dairy cattle have decreased losses due to milk fever, mastitis, and other diseases. Careful selective breeding using production monitoring and component testing has greatly increased the productivity of dairy cows.

Genetic improvements have been facilitated by the use of artificial insemination. Semen from proven bulls can be frozen and shipped anywhere in the world. This has helped both small and large dairy enterprises to upgrade desirable genetic traits. New techniques, such as superovulation and embryo transplants, also help produce more profitable offspring for future dairy herds.

Dairy Management

A dairy farm owner or manager assesses the resources available (farm real estate and facilities) and adapts management practices that will result in maximum profits. Any appraisal of a dairy farm should assess whether its available resources will make it difficult for the manager to adopt the technology and practices required by current industry economics.

For example, an operator may realize that the dairy farm to be operated has shallow and poorly drained soils. To adapt the soils to a roughage program that will support milk production, the operator may need to establish a pasture feeding system rather than an intensive crop production program. In some instances, the manager may have to purchase concentrates and roughage because feed cannot be grown on the property. Marginal soils may be too much of a handicap. If so, the highest and best use of the land may change from use as the site of a dairy to the site of some other farming enterprise, residential use, or recreational use.

When appraising a property to be sold, an appraiser may not know who the manager or purchaser of the dairy farm will be. In that case, the appraiser would have to assume that the property will be competently managed and operated in order to realize maximum profitability. In most market value definitions, there is an assumption of sale. As such, in almost all cases the appraiser will need to assume competent management. If the client is also requesting some form of analysis under current management, then the appraiser will have to account for poor management or potential super management as well.

Today's dairy industry is highly competitive, and below-average management produces marginal returns. The industry provides the highest level of reward to the highest levels of management, much like other agricultural enterprises. In many parts of the United States, dairy farms have ceased operations because the available resources are not adaptable. The highest and best uses of the properties in these areas have changed to something other than use as the site of a dairy, regardless of the level and ability of management. It is incumbent on the appraiser to evaluate the trends in the market area thoroughly. Key questions to ask are

- What is the trend for dairy units in terms of size, capacity, and quality?
- What is changing in the infrastructure supporting the industry, including processors, suppliers, and equipment dealers?
- What factors are influencing these changes–e.g., milk pricing, transportation and marketing costs, changing consumption habits?
- What is the regulatory environment–e.g., waste and nutrient management?

The size and capacity of a dairy is more important in the contemporary market than it was as recently as 1980. In 2012, the midpoint herd size–at which half of the total number of cows are in larger herds and half are in smaller herds–was 900 head, according to the Census of Agriculture. In contrast, the mean herd size at the same time was much lower, at 144 head.[3]

Dairies with fewer than 100 cows are typically operated with a family labor force supplemented with part-time help. Properties of 200-300 acres with facilities can adequately support a herd of 50-100 cows. Farms of this size generally have adequate land resources for pasturing the animals and raising a significant amount of the herd's roughage requirements. These properties do not usually have additional housing because they rely on a family labor force. The income stream produced on smaller farms is often subsidized by the off-farm income of one or more family members. Properties of this size may be purchased for use as a rural residence with rural subsistence as a secondary objective. The farms are generally located in transitional areas and have alternative uses for other livestock farming. The buildings on these dairy operations generally exhibit some functional obsolescence.

Dairy farms with 100-500 cows have utility if they are profitable at their current capacity or can be expanded through combination with smaller farms located in the immediate area. They can be profitable if

3. James M. MacDonald, Jerry Cessna, and Roberto Mosheim, *Changing Structure, Financial Risks, and Governmental Policy for the US Dairy Industry* (Washington, DC: USDA Economic Research Service, 2016), 7.

Figure 17.2 Midpoint and Mean Herd Size of US Dairy Farms (1987-2012)

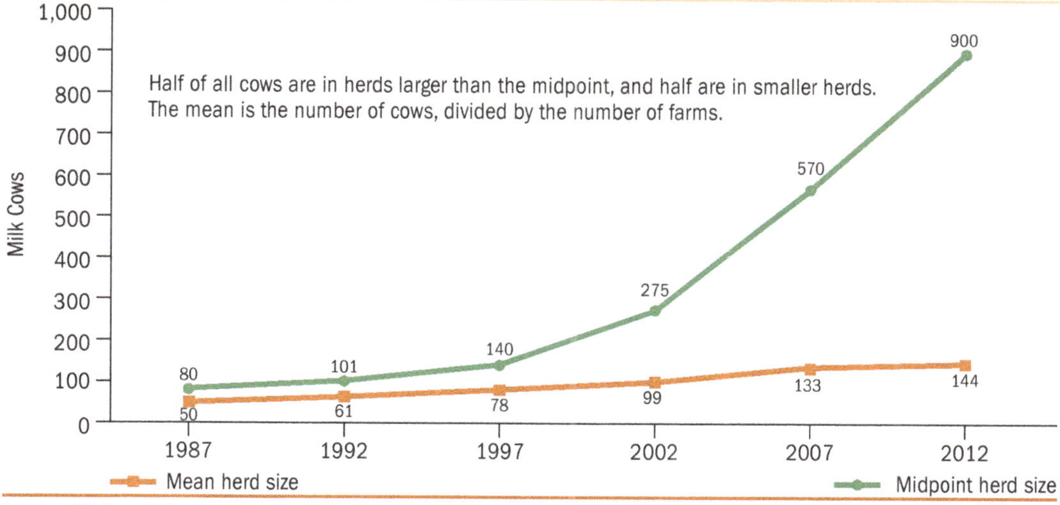

Source: USDA Economic Research Service (ERS) using data from the Census of Agriculture. Means are from published census estimates; midpoints are ERS estimates from unpublished census records. Graph from *Changing Structure, Financial Risks, and Governmental Policy for the US Dairy Industry* (2016).

there are custom crop growers in the area from whom operators can purchase feed for their facilities at a reasonable price. It is difficult to make small changes to these farms to increase their efficiency of operation. When appraising a dairy of this size, the appraiser must assess whether the facilities have the capacity to handle crops and cows efficiently. If they do not, the unit becomes less valuable. When the land resources and production facility are out of balance, major facility changes may be required to maximize profits. Or specialized management may be required to correct the imbalance and maintain profitability. For example, the operator may have to emphasize herd management rather than crop management if the land base is deficient and feed must be purchased to support the dairy operation.

For dairy farms that have more than 500 cows, a significant land base is required for waste and nutrient management to balance profitable production and environmental concerns. Management of animal wastes is of particular concern in cold climates, where it is difficult to move manure off the farm during winter months. With large concentrations of livestock, the storage, processing, and distribution of wastes can be a significant concern. Generally, a land base of 1.5-2.0 acres per animal unit is required. By this measure, a farm with 600 cows on 600 acres is out of balance, and it would have a lower value than a farm with similar capacity but 900-1,000 acres of cropland. Cropland requirements can be met either by ownership or by leases. Long-term leases may be executed to provide sufficient acreage to support the dairy.

If the capacity of the farm buildings exceeds the land base by a significant amount, an appraiser should identify and measure any functional obsolescence resulting from this superadequacy. It is not uncommon to find a dairy in this size range that has limited land resources, inadequate waste disposal or storage facilities, and a great amount of physical deterioration as well.

Dairies of more than 2,000 cows are typically found in California, the Southwest, and parts of the Southeast. These dairies are characterized by high-density confinement of dairy animals, limited capital investment in dairy housing, limited land resources, and almost exclusive dependence on purchased feed. These dairies may have 5,000 or more head at one location with no land base on which to grow feed. After 2008-2009, when milk prices were significantly below the break-even point and prices of alfalfa hay and silage (green fodder that is chopped or shredded and preserved in a silo or pit) were at record highs, dairy farmers began to rethink this position. Where land has been available, dairy farmers have been increasing their landholdings. Corn production caught up to ethanol demand, hence the dramatically lower corn prices seen over the last few years as compared to the peak price immediately following the ethanol mandate.

Herd production is typically high, as is the cull rate in the herd. Management of animal wastes is also of great concern on these farms. Many farms have either contracts or leases that allow for the disposal of animal wastes. Management must pay careful attention to nutrient management on the lands where wastes are disposed of. Sometimes plant varieties that consume high levels of nitrates, nitrites, phosphorous, and trace elements must be grown to avoid oversaturation of the soil with nutrients.

Physical, Legal, and Economic Characteristics

Dairy facilities, including feedlot and milking parlor operations, need to be labor-efficient in their design and operation. In general, the more animals that can be managed by one individual, the more efficient the unit.

In warm climates, cow comfort is critical during stressful periods of high temperatures and humidity. Facilities that have been modernized to cope with heat stress have significant advantages. Shades for cattle in feed lots, misting systems in holding areas, and automatic cow preparation washes in the holding area are representative of features that contribute value to the improvements.

The overall support system in the area–i.e., infrastructure–is critical. An ample supply of labor is needed for milking, cleaning the facility, and performing other tasks around the farm. Skilled and unskilled workers for the milking crew must be available near the farming unit.

The farm infrastructure also includes veterinarian and feed service personnel to assist with the nutrition and health needs of the herd. Active dairy extension programs administered through state land grant colleges are generally found in states where dairies are an important part of the agricultural economy. Other services that must be available are

- feed and nutritional supplement suppliers
- equipment dealers and repair facilities
- processors
- a transportation system to move raw products such as feed and fertilizer to the farm and milk from the farm to the processor

Feed must be readily available at a reasonable cost. Feed makes up a significant portion of the total cost of dairy production in a warm climate area, so an adequate supply of feed must be available on an ongoing basis.

The environmental regulations in the area should be researched by appraisers. Local zoning, health, and animal density requirements must be reviewed. For example, appraisers may need to know if test wells are required and need to be monitored. Likewise, appraisers may need to be aware of whether manure can be disposed of off site or if it must be stored and treated on site.

Cold climate areas tend to have a large number of dairies that are sometimes referred to as "picture book" or "traditional" farms. These are well-kept farming operations with upright silos. The cows are usually housed in tie-stall barns, milked with a pipeline milker, and fed inside the barn. Some smaller farms may have free-stall barns and milking parlors. The common trait for all of these farms is herd size, which is generally 150 animals or less. These farms may be competitive in an area as entry-level, or starter, farms.

Small Family Operations

Small farms are not equipped to handle a large number of cows per worker, nor do they typically have high levels of production per worker. Buyers of these properties are not interested in milking large herds, do not want a large workforce, and will generally supply a majority of the farm labor themselves. To be financially viable, these farms require specific management skills. The farming unit is often situated in a transitional market. The farm may change in use from full-time dairy to part-time dairy/rural residential to another livestock use. Some of the factors to be evaluated on these farms include

- animal housing

- ventilation
- feeding
- feed storage
- young stock

Small dairies may use tie stalls or free stalls. The size of the stall is critical. Most milk cows on modern dairies are large, often weighing up to 2,000 pounds. The stalls should be large enough to accommodate these animals. Some barns are designed for Jersey, Guernsey, or other cow breeds that are considerably smaller than the more commonly used large Holstein, Ayrshire, and Brown Swiss breeds. Stalls and facilities designed for the smaller breeds may not be adaptable to larger animals.

Barns should have adequate ventilation systems that force outside dry air in and moist air out. A significant amount of research has been done on the need for ventilation in dairies. The moist air in the building must be removed and replaced with dry air from the outdoors. Failure to ventilate a barn adequately may result in herd health problems, deterioration of the buildings due to moisture concentration, or both.

The profitability of a dairy farm is partly a result of feeding. The system on a small farm must be able to feed as quickly as possible, as efficiently as possible, and as often as demanded. The system can be either an automatic feeding system located in the barn or a feedbunk located outside the barn or in the common areas of a free-stall barn. An automated feeding system that relies on belts, augers, and motors can result in additional operating and maintenance expense. The appraiser must be aware of the efficiency, or lack of efficiency, provided by the feeding system on the property.

On small dairy farms, feed storage is a critical issue. When evaluating upright silos, the appraiser must assess the age of the silo, the condition of the unloaders, if present, and whether maintenance and repair will be costly. Many upright silos offer sealed storage capabilities. The appraiser must assess the utility of these units and determine the market's acceptance of the structures.

Most small farms have housing for young stock that is separate from the housing for the milking herd. These facilities must be well-ventilated, allow for easy, efficient feeding, and have adequate capacity. Many smaller farms maintain a 100% replacement program for the milking cow population. If young stock facilities are too small or overcrowded, they may become obsolete. The appraiser needs to consider whether other barn complexes in the area are available for use as housing for young stock. These facilities may be readily available in a transition area where some dairy farms have ceased operation. One alternative to on-farm housing of young stock is the use of a custom heifer grower. In some areas, this practice is commonplace.

Large Dairies

The objective of a large dairy is to operate at a low cost of production. This requires maximum efficiency. Efficiency is required in all facets of the operation:

- moving feed from the field to the cows
- grouping cows according to production, stage of lactation, and other characteristics
- operating milking systems

The standard criteria in a large dairy in a cold climate is to handle at least 50 cows per worker and produce a million pounds of milk per worker. Herd averages are typically above 18,000 pounds of milk per cow and can be as high as 25,000 pounds of milk per cow or more. With such high levels of production, facilities must be designed for maximum efficiency without sacrificing herd health and comfort. The criteria to be evaluated in large dairy units include

- milking parlors
- ventilation
- animal grouping
- waste
- feed storage

Free-stall barns with milking parlors are the standard for large dairy units. Milking parlors are generally operated by one person and feature automatic crowd gates. The cows can be changed with a minimum of movement on the part of the operator. For example, a double 32 parlor is generally constructed so that all 32 cows on one side can be moved in and out at one time. A modern dairy barn with an older parlor can have a significant amount of physical and functional obsolescence. It is expensive to replace or update a parlor.

Free-stall barns must be well-ventilated. Warm free stalls with solid sidewalls should have a fan system for use during summer months. In some moderate climates, the ventilation systems are used year round. Cold free stalls, which have open or partially open sidewalls, should have a system that moves the air rapidly. The eave height on many older free-stall barns is 12 feet. However, barns constructed since 1995 have an eave height of 14 feet or higher to accommodate modern equipment.

Large free-stall units should be designed so that cows can be segregated into groups. Large commercial dairy herds typically are divided into as many as five groups based on factors such as stage of lactation, production, size, and age. If the housing facility does not allow for the grouping of milking animals, the appraiser should consider the potential for functional obsolescence.

Waste disposal and management are becoming more critical. Spreading of manure is still the most common method of waste management and disposal. Solids are separated by a "separator" or an evaporation pond, then scooped up and spread or hauled off-site. The liquid goes to the lagoons and is pumped onto the fields, typically by center pivot irrigation. However, new environmental regulations may make it difficult to apply manure to nearby fields on a daily basis year round. Complementary waste management facilities and equipment are becoming more common on large dairies. Many farms have a strategically placed lagoon. Centrally located lagoons with easy access to fields are preferable. Alley scrapers and storage pits in the middle of the housing facility can be efficient methods of waste management.

Many local and regional authorities are regulating waste management through land use or zoning regulations. The density of livestock allowed on the acreage may be regulated through zoning. Wetland regulations may dictate what wastes can be disposed of and where they can be disposed of. When large concentrations of livestock are housed, there may be requirements to monitor aquifer water quality. The appraiser should thoroughly review all land use, environmental, and zoning regulations as a part of the appraisal process.

Cattle on large dairy farms consume large amounts of feed, and commodity storage is required on all farms. Most large dairy farms have a centrally located storage complex where silos (generally bunker silos), grain bins, and commodity storage form a feeding center. With this type of system, mixer wagons can be used efficiently. The wagons are used to collect the feed ingredients, mix the ration, and then distribute the ration to the animals at feedbunks located either in the free-stall barn, along the side of the barn, or on feedlanes next to an open/dry lot.

Characteristics Common to All Dairy Farms

Milk marketing is the most critical factor to be considered in the evaluation of a dairy. When assessing a market area, appraisers must determine the long-term viability of the milk market. A viable market has several handlers or dairy cooperatives, which provide a degree of competition and offer alternatives to the producer. In addition to the number of outlets, the access to these outlets must be considered. Modern milk haulers use large trucks, often tractor-trailers, to move the milk from the farm to the plant. A dairy farm located on roadways that cannot accommodate these large vehicles is at a disadvantage in the marketplace.

Infrastructure, the availability of support services and industry, and an adequate concentration of similar dairy farms are required for a successful dairy farm and industry. As herd size increases, farms become more equipment-intensive and farming practices become more specialized. Dairy farm operators are increasingly dependent on the services of nutritionists, feed suppliers who can provide specialized ingredients, dealers who can supply and service specialized equipment, and sufficient competing farms to attract and maintain the infrastructure. In the areas where feedlot dairies dominate, an adequate supply of high-quality roughage must also be readily available.

Dairy facility and farmstead layout have a significant influence on profitability. The more efficiently the site can be configured, the more cost-effective the operation becomes. Herd housing, feed storage, milking facilities, and milk storage must be designed with efficiency of movement in mind. As an example, five small milk barns might be used to milk 6,000 head, but two more efficient modern facilities (or even one well-configured facility) could easily accommodate the same number of cows.

Land Resources

The physical requirements for operating a dairy farm fall into two basic categories:

1. Buildings
2. Land

Characteristics of the building improvements vary, depending on environmental limitations, local requirements, and trends. (Dairy farm buildings will be discussed later in this chapter.) Land requirements remain constant regardless of where the farming activity is located. The basic land needs are for crop production, animal exercise, waste disposal, and water resources. Land use is generally classified as tillage or cropland, pasture or grazing, and building sites or farmstead. Some farms also have woodland, brushland, wetlands, and other miscellaneous land categories.

The building site (or sites) on dairy farms can represent a significant contribution to the property in terms of both value and utility. The following factors must be considered in the evaluation of building sites:

- location (relative to population centers, community needs, schools)
- access (roads, transportation arteries, driveways)
- frontage (exposure)
- topography
- shape (regular, divided by road, multiple sites)

Each of these characteristics can influence the utility of the site by either positively or negatively affecting the maneuverability of vehicles and equipment, the flow of operations between buildings, and the potential for expansion.

Tillage or cropland provides the primary land base for production of the roughage needed to feed the livestock. The quality of the soils that make up the tillable land is critical. On a dairy with good-quality soils, as little as one acre of cropland may be needed per cow. As mentioned earlier, 1.5-2.0 acres of cropland is typically needed per cow. However, if the soils are of poor quality, the requirement may be as high as five acres or more per cow. Factors that affect tillable land and its desirability for a dairy farm include

- location
- water rights, particularly in the Western states
- soil types (especially the availability of Class I and II soils)
- topography and drainage
- annual rainfall
- field shape, configuration, and size
- surface and subsurface drainage
- natural fertility and tillage
- ability of the soil to use nutrients
- conservation practices
- proximity of the cropland to the building site

Crop rotation, strip cropping, alternative cropping, and no-till or minimum tillage practices are relatively modern techniques that have enhanced crop yields and reduced soil erosion.

Pasture and grazing land is becoming more important on organic dairy operations. The confinement housing trends of the 1970s and 1980s are being reversed in some areas. Many dairy farmers are recognizing that they may need to provide a supplemental feed source and more exercise for the herd. Traditionally, pasture for mature animals has been located close to the barn and included minimal land area–just enough to permit some exercise. Pasture areas for herd replacements and dry cows were established farther from the building complex, mainly because these animals required less observation and management than mature, milking animals.

Grazing land is sometimes measured in two productivity units: (1) animal unit months, abbreviated as *AUMs*, or (2) cow year long or, alternatively, cattle year long, both abbreviated as *CYL*. These measures indicate the capacity of a particular plot to provide adequate nourishment for mature animals for one month or year. These terms are most commonly used in the western portions of the United States.

Most dairy farms have some woodland. Historically, the trees were used for lumber, timber sales, firewood, fence posts, sugar bush (in maple-producing areas), and fruit or nut production. Many of these uses are no longer relevant to modern farming needs. Therefore, the contribution of woodland to the value and utility of a dairy has been reduced.

Occasionally, wetlands may be included in the land base of a dairy farm. Because large acreages are associated with dairies, wetlands are often situated between and within the more productive tracts. While some wetland areas may once have been drained and converted into cropland, this practice is now restricted.

One of the most important uses of the land base of a dairy farm is to serve as a site for waste disposal. While manure can be used to supplement the fertility of the soil, there are limitations to the quantities that can be applied. Application of animal wastes on an ongoing basis may require intensive crop management. Crops that can better use the nutrients found in animal wastes may be planted, crop rotation becomes more important as a management tool, and the timing of waste applications may have to be adjusted. Recent legislation and public perceptions may further restrict where, how, and how much animal waste can be applied to the land.

Historically, dairy farms were balanced with all, or most of, the required land base included as part of the farm. While some acreage may have been rented, the farm unit was typically self-contained. The percentage of rented land used by a dairy farm tends to vary with business cycles. That is, dairy operators can expand the size of the herd without investing in new land when rental rates are affordable and may have to contract operations when land values rise. Rental agreements may include many different types of consideration, including payments in services rather than cash (e.g., garden plowing, driveway snow plowing), which must be considered when an appraiser evaluates the subject property's land base.

Certain dairy farms, ranging from small to very large units and including most feedlot dairies, may purchase all or most of their feed rather than raising it on the farm. These farming operations balance the costs of purchasing feed from other sources with the cost of owning large tracts of land and owning, maintaining, and operating equipment lines. These farms may be able to purchase the feed for less than it would cost to raise it.

When land is rented rather than owned, waste disposal is again a major consideration. Without a land base on which manure can be spread, alternative methods of waste handling and disposal must be available. The appraiser must assess market acceptance of these operations, their economic viability, and what, if any, effect they have on property value.

"Right to Farm" laws of various sorts have been adopted throughout the country in response to changing demographics. These laws may fall under the categories of transferable development rights

(TDR), purchase of development rights (PDR), agricultural districts, agricultural assessments, and others. Some governmental entities and not-for-profit land trusts such as the American Farmland Trust have been acquiring conservation easements on farmland in an attempt to preserve its agricultural utility and integrity. Many of these programs call for special or reduced assessments if the land is kept in agricultural production. The appraiser should be aware that these programs can also impose penalties if the land is converted to nonagricultural use.

The appraiser should investigate the existence of easements, deed covenants, and other legal use restrictions affecting the property being appraised. Easements are acquired for a variety of purposes, including conservation, fishing, hunting, trapping, recreational use, timber, development, mining, and scenic preservation. They can be short-term or remain in effect into perpetuity. Easements typically affect the value of the land, but the land's utility as the primary land base or support land for a dairy may not be affected. The appraiser must determine what, if any, influence legal restrictions have on the value of the farm being appraised.

Improvements

Typically, the value of improvements relative to the total value of a dairy is significant. The total value of the improvements may exceed the value of the land in many instances. To estimate the value of dairy farm buildings and improvements, the appraiser must understand the purpose of each structure as well as its overall utility and contribution to the whole enterprise.

Building improvements on a dairy typically fall into six general categories:

- herd housing
- milking facilities
- feed storage
- equipment, fixtures, and personal property
- machinery storage
- waste storage and handling

Herd Housing

The primary purpose of herd housing facilities is to protect the herd from the natural elements. Because climates vary, there is a wide variety of improvement styles. In the most extreme northern climates, shelter from wind, cold, snow, rain, and heat must be provided. Farm buildings are typically fully enclosed and of wood frame or pole construction. The design of buildings must provide for adequate ventilation and air movement in all seasons. The building must not be a source of stress for the animals. Styles commonly found in northern climates include stanchion or tie-stall barns (generally found on registered and purebred farms and older, small farms), stanchion or tie-stall barns that have been converted to loose housing or free-stall operation, and free-stall barns, which are designed for, and used by, most commercial farms. Some large dairy operators also use cross-ventilation barns.

Free-Stall Dairy Barn

Source: http://takingstock.asas.org/takingstock/wp-content/uploads/2014/02/Freestall-Dairy-Barn.jpg

In the moderate climates of the Southeast and Southwest, animals may only need shade and shelter from extreme sun and heat. In these areas, the housing facilities may be open on one or more sides and have large fans to facilitate air movement during very hot periods. Some facilities may even have cooling equipment or sprinkler systems that provide constant temperature control. Again, stress on the animals must be minimized and adequate ventilation must be provided.

Tie-Stall Barn

Source: GEA Group, www.gea.com/global/en/binaries/Dairy Farming_Tie-Stall%20Barn_1_tcm11-14302.jpg

Regardless of the style of housing provided, certain common attributes are required. Efficient movement of the herd is essential. The location of feeding areas, the proximity of housing to milking facilities, and the presence of hospital and calving facilities and areas for breeding, clipping, hoof trimming, and waste removal should be considered in evaluating a building's efficiency and utility. Stall size must be sufficient to accommodate modern dairy animals. As mentioned previously, breeds such as Guernsey and Jersey cows have a smaller stall space requirement. Larger animals such as Holsteins can weigh up to 2,000 pounds and need significantly more stall space to ensure their comfort and protect against udder injury.

Cross-Ventilation Barn

Source: http://50.28.41.185/~primebld/wp-content/uploads/2012/07/crossVentFreestall.png

Concrete and paved walkways must be designed without steep grades or slippery surfaces. Many exercise yards and barn alleys are grooved. Older facilities are typically regrooved on a regular basis.

Unless the farm operator is purchasing all replacements (first calf heifers), housing for young stock (calves and yearlings) and replacement animals is required. Nursery facilities for newborn calves, housing for young stock, and buildings to house bred heifers and dry cows are needed on most dairy units. Young stock may be housed in individual calf hutches, which are mobile units that can be moved around the farmstead, or their housing may be included in the design of a free-stall barn. Sometimes the housing of young stock is provided through the use of low-cost, aluminum hoop structures similar to greenhouses. Efficient feeding, cleaning, and animal care should be considered in the evaluation of the young stock barns. Ventilation and temperature control are requirements in all young stock and replacement housing.

Milking Facilities

Most modern milking facilities consist of two components: (1) the milking area and (2) the milk room or milkhouse, where milk storage, the cleaning of milking equipment, office operations, and other support functions take place. In some barns, milking is conducted where the animals are housed. Typically, these are stall barns with milk pipelines or milk transfer systems. Most free-stall barns have a parlor/milkhouse attached to the housing unit. The animals are transferred from the housing area to the milking area without leaving the main building. Some dairies have separate milking centers. Animals leave the housing area by walkways or holding areas that may or may not be enclosed, and the animals are milked in a parlor that is detached from the herd housing unit.

Milking parlors are designed to accommodate the milking equipment they contain. Parlor styles range from in-line systems and herringbone styles to rotary parlors, trigon parlors, or polygon parlors. The type of parlor generally influences the design and construction of the milking center. Size, layout, location (relative to housing), cow traffic patterns, milking routes, and the amount of mechanization required are all related directly to the type and style of the milking system. Other factors influencing milking facilities include herd expansion plans, available labor, available capital, available milking time, herd production levels, and personal preference on the part of the farmer.

The support portion of a milking center should include the milkhouse and other supplemental areas. The design of the milking center should allow adequate space for bulk storage tanks, compressor units, washing systems, a farm office, general utility rooms, lavatories, and general storage. Because this area houses the milk storage facilities, it is imperative that the building be conveniently located within the farmstead. Bulk hauling trucks, generally semi-trailers, must be able to get to the building with a minimum of effort. If trucks must back into the facility through a maze of other buildings, the placement of the structure is not efficient. Ideally, the building should be situated so that a large truck can drive into the area of the milk-house, turn around conveniently to load the milk, and exit smoothly from the farmstead.

A lavatory, and preferably a shower and locker facility on large farms, should be conveniently located in the milking center. Because farm records are increasingly being kept electronically, the farm office should be located in or near this building. Herd production records, feeding records, breeding records, and financial records can all be consolidated in this one location.

Feed Storage

Feed for the dairy should be stored as close to the herd as possible. Feed may include roughage in the form of hay, silage, concentrate, feed ingredients, or feed supplements. As the size of dairy farms increases, the traditional upright, or tower, silo seems to be losing its acceptance. While sealed storage (usually steel silos with glass linings and bottom unloading capabilities) may be found on many dairies, thorough analysis of the properties is required to determine the level of market acceptance for them. Silage and ingredients may be stored in bunker silos, which are also called *pit* or *trench silos*. These structures typically have concrete sides and floors, although occasionally they may have asphalt floors, treated lumber sides, or even dirt sides dug into a hillside. Bunker silos have sides approximately 8-10 feet high and no roofs. Commodity barns are similar in construction to bunker silos but are covered.

Bunker Silos

Commodity Barn

Source: Hanson Silo Company, www.hansonsilo.com/images/commodity-storage-7.jpg

Buildings used for ingredient storage typically have roofs and are located near the silage and roughage storage facilities. Vitamin and mineral supplements may be stored in elevated metal bins. This configuration allows for the efficient mixing of rations using mixer wagons. Feed storage should be located on the farmstead in an area that allows adequate space for ingress and egress for filling the storage units and removing the feed. Access for large trucks, bucket loaders, tractors, and other equipment must be considered. The capacity of the feed storage must be sufficient to store feed for an extended period of time. In areas such as the Northeast, dairies require sufficient silage storage for a 12-month period. Many farmers in that area prefer to have excess storage in case crop yields are below normal.

Machinery Storage

Most dairy farms require a considerable amount of machinery and equipment for day-to-day operations. Storage and repair facilities for the equipment should be located at the farmstead or near the farm unit. In evaluating machinery storage, eave height is a primary consideration. A 12-ft. to 16-ft. eave height is adequate for most farm equipment. The width of access doors is critical, especially if large, articulated tractors and harvesting equipment capable of multirow harvesting are used.

Machinery storage should include provisions for a farm shop. It is preferable to perform tractor and equipment repair indoors and out of the elements. The area's climate will dictate the degree of insulation required in the building, and whether heating or cooling facilities must be installed. The shop should be located in an area that is accessible to all equipment and close enough to the milking and housing facilities that the equipment in those buildings can be repaired efficiently.

Utilities must be available at the shop site. Three-phase electrical service may be required for some of the large welding equipment found in farm shops. Provisions for discarding or recycling waste materials such as oils and paints must be considered. Fuels and lubricants are usually stored near the shop. The method of storage must comply with local, state, and federal requirements. Permits and registration certificates for the storage facilities should be obtained and reviewed by the appraiser.

Provisions must be made to store agricultural chemicals used in crop production in and around the machinery storage and shop facilities. These storage units must meet local and regional standards as well as state and federal regulations. The appraiser is responsible for reviewing the suitability of the storage facilities.

Waste Storage and Handling

Increased scrutiny of agriculture and the environmental impact of large concentrations of livestock have made the storage and disposal of manure and other types of waste an important consideration in dairy farming. The typical dairy generates wastes from the milking parlor, the herd housing units, the shop facilities, and the treatment of ill animals. All of these wastes must be carefully stored and disposed of.

Many jurisdictions strictly regulate waste removal and storage functions. Local, regional, or state law may regulate the maximum number of animals on tracts. Disposal of animal waste may be regulated with requirements that the waste be "turned under," i.e., plowed into the soil within a certain period of time after application.

Mechanization of waste disposal is a necessity. Waste may be pushed by tractor or directed through large, sewer-like pipelines from the dairy barns and parlors to a lagoon specifically built to hold waste. Many parlors and milking centers have separate septic facilities that must be pumped out at regular intervals. It is not uncommon to find injection or irrigation systems that use the liquid effluent or excess water from lagoons and storage units. Some farms may even incorporate advanced technology for composting solid wastes, separating liquid and solid wastes for ease of handling, irrigation systems using waste effluent, or methane generation of electricity from by-products and wastes.

The disposal of used paints, lubricants, medical syringes, medications, and other specialized waste material used in the normal course of business is closely regulated. Many areas require that these items be collected separately and taken to specialized collection and recycling facilities. However, on-farm storage for short periods of time may still be required.

To properly review the adequacy and utility of a dairy's waste handling and storage capacity, the appraiser should be familiar with local, regional, state, and federal environmental requirements. While the appraiser need not be an expert in this field, a general knowledge of the requirements and constraints imposed on the industry is required. A lack of familiarity with regulations could result in a misleading value conclusion. If strict waste control is required and facilities or provisions for this function are not present on the farm, this fact could be inadvertently ignored or overlooked by the appraiser.

Equipment, Fixtures, and Personal Property

Equipment for milking, milk storage, feeding, and manure movement is critical. It is the appraiser's responsibility to differentiate between the equipment that is considered personalty and the items that are considered fixtures required for the ongoing farm operation. The market will indicate which items are required to maximize the value of the dairy improvements. Then, the appraiser must account for any and all equipment included in the total value opinion in the final valuation. The identification should describe the item and assess its value contribution. Furthermore, if livestock is included, the value of the livestock must be identified and quantified.

Intangible Business Assets

The operation of a dairy is a business enterprise that involves more than occupying or leasing the real estate. Depending on the scope of work, appraisers may be required to quantify the component of value attributable to intangible assets in addition to the value of the real property, allocating value between the various forms of property.

Highest and Best Use

To meet the test of *legally permissible* uses, a dairy must conform to local zoning and land planning requirements. The location of improvements, especially those that are "to be built," must conform to local building and zoning laws. The use of the land and building improvements must be allowed under the same regulations and comply with any regional or statewide requirements for waste management and animal density and housing.

To be a *physically possible* use of the land, the infrastructure required to support dairying must be considered. The distances to, and availability of, equipment dealers, dairy suppliers (of specialized ingredients, tools, and equipment), feed suppliers and feed mills, large animal veterinarians, and inseminators should be evaluated. The questions to be answered include

- What is the state of the infrastructure in the area?
- Is the infrastructure expanding, contracting, or static?
- Does the local dairy farmer have to travel far to find services such as an implement dealer, a source of concentrates, or a commercial hoof trimmer, or are these services available nearby?

The land base must be evaluated for its ability to support a dairy unit. For example, can organic wastes be disposed of on the soils without the threat of contaminating water supplies? Similar consideration must be given to the location of the herd housing and waste storage facilities. Soil resources must be available in sufficient quantities to provide adequate feed for the size of the facility that exists or is planned. The land required to sustain an animal varies from region to region, but sufficient land must be available through ownership or rental to sustain a profitable operation. Some dairy farms purchase all of their feed rather than grow it. In this case, the physical considerations center on waste storage and disposal and the availability of sufficient facilities to store feed.

The *financial feasibility* of a dairy and the *maximally productive* use of the land and building complex require careful analysis and consideration. Appraisers are charged with determining property value under "typical" or "average" management. Neither good nor poor management are to be considered in the valuation process. However, the success and profitability of a dairy depend on the management skills employed in its operation.

An appraiser needs to collect as much information as possible about the dairy's income and expense characteristics. The type and size of the dairy farm being appraised will determine the information required. If the dairy being valued is in an intensive dairy area and is similar to most other local farms, the economic data required to evaluate financial feasibility should be relatively easy to gather. An appraiser who regularly works in the area should have a significant amount of data on hand. If, however, the farm being appraised is much larger than what is typical in the area, the scope of information-gathering may have to be expanded. Rather than a small area of consideration, investigation of the market may cross regional or state lines.

Considering trends in dairy economics is a helpful part of the highest and best use analysis. The industry has a cyclical nature, so placing the date of the appraisal within the observable business cycle may assist the appraiser in evaluating feasibility.

If financial data is insufficient or unreliable, an alternative method of analysis may be considered. The four points discussed below, which relate to financial feasibility and the maximization of income potential, may be investigated:

- Recent developments in the area. Are dairy farms being used as dairies after their sale or are they being put to alternative uses? In some areas, the number of dairy farms may be decreasing, but the number of cows being milked may remain constant or be increasing. This could suggest that small units are being combined into larger, more efficient units. Typically, the changing concentration of a particular use, in this case dairy farming, is a direct reflection of the financial feasibility of that use of the land.
- The supply of and demand for dairy farms as evidenced by the number of property transfers (sales of dairies for use as dairies), the number of dairy farms offered for sale (listings), or the occupancy rates of existing units. If it is feasible to operate a dairy farm, the available units will be in operation. If profit margins are declining, the supply of farms for sale and the number of vacant units may rise.
- The purpose that dairy farms are put to after their most recent transfer. This observation provides an indication of whether or not use as a dairy is maximally productive. A dairy may generate positive cash flow (and thus be financially feasible) but not generate enough cash flow to maximize net income. In such a case, the farm may become absorbed by a larger dairy farm unit or may be put to another, unrelated agricultural or nonagricultural use.
- Comparison of the (vacant) site value to the value of the improved property less building (improvement) demolition costs. If

the site would be more valuable after the improvements are demolished, then the improvements make a negligible contribution to the value of the whole property. In other words, the improved property is legally permissible, physically possible, and financially feasible, but fails the test of maximum productivity.

Valuation Procedures

After a thorough investigation of the market area and the trends influencing value in the industry, the appraiser must identify the problem to be solved: What is being appraised? Usually it is the real property, which consists of the land, site improvements, buildings, and dairy fixtures that are attached or built into the buildings (such as bulk tanks, compressors, automatic washing equipment, feeding equipment, and waterers). Appraisers must have a clear understanding of the ownership of the specific fixtures and equipment.

Because improvements contribute significantly to the value of a dairy farm and tend to be highly specialized, the appraisal report usually presents a detailed description of all improvements. The typical report also includes a discussion of soils, property balance, and the unit's overall utility. The appraisal report should also include

- soil maps
- plats showing boundaries, drainage, and different uses of acreage
- photographs and sketches with measurements of all the significant improvements

In an appraisal, the choice and application of the approaches to value depend on the purpose and intended use of the appraisal and the available data. The sales comparison approach has traditionally been the most reliable and understandable method for valuing dairy farms, which are still primarily owned and operated by individuals. When the number of comparable sales is limited, particularly for sales of larger dairy properties, the income capitalization approach can be applicable. While leases of dairy farms do exist, they are usually lease-purchase options. During the last dairy downturn, many dairy operators either walked away from the lease-purchase option or renegotiated those arrangements. Most dairy operators have differing views on management practices (e.g., the choice of Jersey or Holstein cattle, the balance of cost and quality for feed), which complicates the analysis of comparable properties in both the sales comparison approach and the income capitalization approach. The cost approach is recognized as a viable valuation technique because of the extensive improvements on dairy farms. It is especially appropriate when the subject property has relatively new building improvements. Depreciation can be difficult to measure, but it can be verified by applying market extraction techniques to the sales of other dairy farms. This approach tends to result in the valuation of the underlying real property components only when that is the purpose of the appraisal.

Sales Comparison Approach

In the application of the sales comparison approach, the comparable dairy farms that are analyzed must be similar to the property being valued in terms of size, capacity, use, adaptability, and utility. A small dairy with a stanchion or tie-stall barn that has a limited tillable land base should not be appraised using sales of large farms with free-stall barns and excess land.

The comparability of the properties should extend to

- size (both herd capacity and size in acres of total land)
- housing facilities
- method of feed supply (purchased feed or grown feed)
- physical location
- type of milk marketing (fluid use, manufacturing use, a combination of the two)
- soil quality
- quality of improvements
- overall farm balance
- property rights conveyed

Appraisers must verify many different elements of a dairy farm sale. Information on the improvements, including their quality, number, utility, age, configuration, capacity, ownership, and category (personal property, real property, leasehold, etc.) must be gathered. The balance of the unit must be identified, and the present and intended uses of the farm should be evaluated. When gathering data on a dairy farm, attention should be directed to its income characteristics as well as to the costs of the different building improvements on the property. Figure 17.3 illustrates a format for data collection.

Improvements on a dairy farm generally make a major contribution to the overall sale price and, hence, to the overall value of the property being appraised. The appraiser must have sufficient information to adequately analyze the sales of improved properties that have occurred in the market. This necessitates the collection of sales of vacant land suited for use as a dairy farm, sales of farms with few improvements, and sales of farms with many improvements. By estimating the contributing market value of the land on each improved property and subtracting that value from the total sale price, the remainder of the purchase price will necessarily represent the depreciated value of the improvements.

Figure 17.3 Data Collection

Sale no. ____________________

Grantor: ____________________ Volume (book): ____________________

Address: ____________________

Grantee: ____________________ Page: ____________________

Address: ____________________

Deed date: ____________________ Transfer tax: ____________________ Financing: ____________________

Recording date: ____________________

Sale verified with: ____________________ Sale verified by: ____________________

Purchase price	**Total $**	**$/Cow**	Property rights conveyed:
Total			
Personal property			
Net to real estate			

Land	**Acres**	**$/Acre**	**$/Cow**	**Total Contribution**
Total				
Tillable				
Pasture				
Woods/waste/etc.				
Total				
Improvements				

Description	**Est. Eff. Age**	**Est. Econ. Life**	**Replacement Cost**	**Total Contribution**	**Accrued Depreciation**
Dwelling(s)					
Dairy barns					
Milking parlors					
Heifer/young stock housing					
Machinery/equipment sheds/shop					
Equipment/fixtures					
Feed storage					
Waste storage/ handling					
Total					
Total per cow					

Facility capacity: ____________________ Comments: ____________________

Estimated gross income: $ income/cow: $ ____________________

Estimated expenses: $ ____________________ Expenses/cow: $ ____________________

Estimated *NOI*: $ ____________________ *NOI*/cow: $ ____________________

Analyzing the properties in order of the degree of improvement allows the appraiser to determine the value relationships of the different components. In some instances, mathematical relationships may emerge. For example, housing facilities for herds, heifers, and young stock may demonstrate a relationship that can be expressed in ratios. These ratios, which are determined through the analysis of sales of properties with only a few improvements, can then be applied to highly improved properties to create order out of data that initially seems to be chaos. When a large array of dairy sales is analyzed, the appraiser develops an understanding of the specialized improvements' contribution to the total sales prices of operating dairy farms that have recently sold in the marketplace.

The comparison process requires identification of both elements of comparison and units of comparison. An element of comparison is a factor recognized by the market as either adding to, or subtracting from, total value. The elements of comparison for most dairy farm appraisals include

- land balance and mix
- improvement balance and mix
- property rights conveyed
- financing terms and conditions
- conditions of sale
- market conditions
- location to market and within a region
- size of farm in total acres as well as capacity, expressed in total cows or total animal units
- quality of the dairy unit
- any other physical or economic characteristics that the market may recognize

Some appraisers of rural property may consider only acres as a unit of comparison. However, a particular market may or may not recognize this unit. Before determining the appropriate unit to apply, an appraiser should analyze the array of sales data to determine what unit of comparison the market is using to evaluate property. For example, the properties can be compared to one another using not only acres but also capacity of milking animals, number of wet cows, total animal capacity, mature animal equivalent capacity, unit carrying capacity, total building capacity, and number of stalls.

Generally, if a market is buying and selling dairy farms based on per-cow units, recent sales will reflect that fact. The better properties have higher per-unit sale prices expressed in dollars per cow, and the poorer properties have lower per-cow-unit sale prices. Comparing the sales to each other on a per-cow basis will help the appraiser extract appropriate elements of comparison as well as the specific amounts of adjustment that may be required.

Table 17.1 illustrates the presentation of unadjusted sales to be used in the valuation process. After an appropriate unit of comparison is selected, sales of improved properties that are similar to the property being valued are compared on an individual basis with the subject property. Land is compared to land in terms of balance and mix as well as quality and utility. Improvements are compared in much the same way, although the utility of the buildings relative to the total farm is addressed in a separate adjustment. Herd housing on the comparable property is compared directly to herd housing on the subject property. Other improvements are compared in the same way.

Table 17.2 illustrates the process for deriving adjustments based on analysis of paired sales. Since no difference in financing terms exists in these sales, no adjustment is warranted. If financing terms did affect the value of any sale, however, this adjustment should be addressed.

The location adjustment reflects the transportation distance to markets for Sales 1 and 3, which are located 125 miles south of the subject area. The two comparisons indicated adjustments of $115 and $135 per cow, respectively, for this difference. In this instance, $125 per cow is selected as the appropriate adjustment.

An adjustment must be considered for the overall utility of the farm unit and the unit's condition and quality, as reflected in the market's acceptance or lack of acceptance. In this example, Sale 2 has a milking parlor that is in good condition and functions well but is out of balance with the housing capacity of the farm. The parlor is much smaller than the size of the unit demands, and an upward adjustment is required. Sale 1 is adjusted downward slightly because its overall condition and utility is excellent, a trait not typically found in the market.

Cost Approach

The cost approach is seldom used as the sole method for valuing dairy farms. It can, however, provide support for the sales comparison and income capitalization approaches. Use of the cost approach may be

Table 17.1 Unadjusted Sales Chart

Sale No.	Sale Date	Sale Price	Size (Ac.) Total/Crop	Capacity (Cows)	Location	Condition/ Utility	Effective Age of Facility	Price per Acre	Price per Cow
1	3 years ago	$2,050,000	675/520	1,025	Fair	Good	3 years	$3,037	$2,000
2	Current	$2,145,000	950/800	1,400	Good	Fair	10 years	$2,258	$1,532
3	2 years ago	$2,566,000	850/800	2,000	Fair	Good	1 year	$3,019	$1,283
4	Current	$2,675,000	900/830	1,450	Good	Good	2 years	$2,972	$1,845
Subject	Current		1,000/850	1,200	Good	Good	2 years		

Table 17.2 Sales Comparison Approach Adjustment Grid

	Sale 1	Sale 2	Sale 3	Sale 4
Sale date	3 years ago	Current	2 years ago	Current
Acres	675	950	850	900
Capacity (cows)	1,025	1,400	2,000	1,450
Financing terms	Market	Market	Market	Market
Location	Fair	Good	Fair	Good
	125 miles SW	25 miles NE	125 miles S	5 miles S
Condition/utility	Good	Fair	Good	Good
Price/acre	$3,037	$2,258	$3,019	$2,972
Price/cow	$2,000	$1,532	$1,283	$1,845
Units of comparison: $ per cow				
Land adjustment	251	83	159	111
Improvements adjustment	-556	-159	233	-146
Adjusted $ per cow	$1,695	$1,456	$1,675	$1,810
Location				
	Compare # 1	To # 4	$115/cow	Conclude
	Compare # 3	To # 4	$135/cow	$125/Cow
Adjustment	$125	0	$125	0
Adjusted $ per cow	$1,820	$1,456	$1,800	$1,810
Condition/utility				
	Compare # 1	To # 2	$364/cow	Conclude
	Compare # 3	To # 2	$344/cow	$350/Cow
	Compare # 4	To # 2	$354/cow	
Adjustment	0	$350	0	0
Adjusted $ per cow	$1,820	$1,806	$1,800	$1,810
Subject capacity (cows)	1,200	1,200	1,200	1,200
Indicated value of subject	$2,184,000	$2,167,200	$2,160,000	$2,172,000

preferred when valuing new or relatively new buildings, to appraise a proposed development, or for allocating value between various improvements (e.g., eminent domain or insurance valuation). The cost approach is also useful in quantifying non-realty value components.

In the cost approach, an opinion of the property value is reached by

- developing an opinion of land value
- calculating reproduction cost
- deducting depreciation as of the date of valuation

Refer to Chapter 15 for a thorough discussion of the cost approach.

The value of the land portion of a dairy property is estimated using undeveloped land sales that have a similar highest and best use. Good-quality farmland can be adapted to a variety of uses, including forage production for dairy facilities. The market recognizes that properties differ in their quality characteristics and suitability for a wide range of farming uses, ranging from pasture and grazing land to high-value crop production. Typically, a dairy operation must compete against these alternative uses for land.

The ownership risk of land varies with its use. Unimproved farmland is generally subject to relatively little risk, as reflected in low capitalization rates. The more intense and specialized the use, the higher the associated risk and the higher the capitalization rate.

The land is valued in its vacant state and as if available for development to its highest and best use. There are six ways to value land:

- sales comparison
- capitalization of land rent
- allocation
- extraction
- land residual
- subdivision development

In rural appraisal, land values are typically developed using sales comparison techniques, capitalization of ground rent, or extraction.

Sales comparison requires sales of similar properties that can be compared to the subject property. Extraction is a variation of allocation in which land value is extracted from the sale price of an improved property by deducting the value contribution of the improvements. Ground rents can be capitalized at an appropriate, market-derived rate to indicate the value of the subject site. Sales comparison techniques are most commonly used to determine the value of land for dairy farms. Table 17.3 displays land sales that could be used to value a site. By analyzing verified sales data, the appraiser can develop an opinion of land value for use in the cost approach to value.

An appraiser may use cost guides, local contractors, or market extraction to obtain costs. If cost guides are used, the provided costs should be compared to local costs to ensure that the multipliers supplied in the guide are accurate for the property's area. In areas with an established dairy industry, cost figures can be obtained from contractors specializing in the construction of buildings for this type of farm. The appraiser is encouraged to establish relationships with these contractors so that the cost estimates used are as accurate and current as possible. An example of the compilation of construction costs is shown in Table 17.4.

The estimation of depreciation depends on the availability of sufficient, verified sales data. To estimate depreciation on dairies, most appraisers use market extraction of depreciation rates. Use of this method requires proper sales data analysis, including the estimation of the economic age of, and curable obsolescence in, the farmstead and improvements. Table 17.5 illustrates the application of this method in analyzing the sale of an improved dairy farm.

A similar analysis can be conducted using the economic age-life method. Applying this method to dairy farms is complicated since these farms usually include a complex of several buildings. Either each building must be treated separately or the entire complex must be analyzed as a unit. Market abstraction of depreciation allows the appraiser to develop a range of depreciation rates for different types of improvements. These rates are then applied to the improvements on the property being valued. The market-derived rates provide empirical support for the rates that the appraiser uses in valuing the subject property components.

In the breakdown method, all types of depreciation are considered separately. Curable and incurable physical deterioration (short-lived and long-lived), curable and incurable functional obsolescence, and external/economic obsolescence are all considered in sequence when the breakdown method is used.

Table 17.3 Unadjusted Land Sales

Sale	Sale Date	Sale Price	Total Acres	Total $/Acre	Crop Acres	Crop $/Acre	Other Acres	Other $/Acre
1	1 year ago	$670,000	800	$838	725	$900	75	$233
2	Current	$605,000	750	$807	700	$850	50	$200
3	1 year ago	$455,000	650	$700	550	$800	100	$150
4	Current	$660,000	800	$825	775	$850	25	$50
5	Current	$655,000	800	$819	675	$935	125	$191

Table 17.4 Replacement Cost of Improvements

Description	Buildings	Equipment	
Milking parlor/milk house	$378,000		
Milking equipment		$750,000	
Free-stall barn	775,680		
Lockups		104,000	
Feed lanes	202,680		
Pens/corrals/gates		190,000	
Commodity/feed barn	121,500		
Feeding machinery		50,000	
Site improvements	260,000		
Shop building	42,750		
Total	$1,780,610	– $1,094,000	= $2,874,610

Table 17.5 Market Extraction of Depreciation Rates

Comparable Dairy Facility	Total	Cost/Cow
Replacement cost		
Buildings	$1,232,000	$1,202
Equipment	$1,200,000	$1,171
Total	$2,432,000	$2,373
Indicated contributory value	– 1,900,000	$1,854
Total indicated depreciation	$532,000	$519
Estimated physical curable deterioration	– 50,000	$49
Estimated physical incurable short-lived deterioration	– 200,000	$195
Remaining physical incurable long-lived deterioration	$282,000	
Estimated % physical incurable long-lived deterioration ($282,000/$1,182,000)	23.9%	
Estimated effective age	4 years	
Estimated annual rate of depreciation	6.0%	
Estimated economic life (1 ÷ 0.06)	17 years (rounded)	

While the physical deterioration and most forms of functional obsolescence found on a dairy farm can be estimated with relative ease, the estimation of external/economic obsolescence can be somewhat difficult. Market value is influenced by external forces including the condition of the general economy, the status of the dairy industry, the number of dairy farms available for purchase, highway networks, and distances to metropolitan centers, suppliers, and markets. In addition, changing environmental regulations exert external pressures on dairy properties.

The easiest way to estimate external/economic obsolescence is to analyze sales of properties with new improvements or improvements that are nearly new and exhibit minimal functional obsolescence. Table 17.6 illustrates the analysis of external obsolescence in a property improved with a free-stall barn and milking facility, feed storage, and shop/machinery storage. After depreciation rates have been derived from the market, these rates are applied to the cost of the subject property. Table 17.7 summarizes the results.

Income Capitalization Approach

The income capitalization approach measures the present value of anticipated future benefits. The application of direct capitalization (as illustrated in Chapter 13) is appropriate for most dairy farms, especially those that are well-established and similar to other dairies in the area. The appraiser must use care in estimating the income and expenses for these farms because there may be income related to farming other than milk production. Family living expenses and

Table 17.6 External Obsolescence Estimate

Sale price		$1,750,000
Less: land value		– 240,000
Contribution of building improvements		$1,510,000
Replacement cost new of building improvements	$2,187,381	
Less: improvement contribution	– 1,510,000	
Total depreciation		$677,381
Curable physical deterioration	$15,000	
Incurable physical deterioration–short-lived	$120,000	
Replacement cost new of improvements less curable physical and cost of short-lived items ($2,187,381 – $15,000 – $1,079,300 = $1,093,081)	$1,093,081	
Effective age of improvements	3 years	
Annual depreciation rate from sales	6%	
Physical deterioration–long-lived ($1,093,081 × 18%)	$196,755	
Total physical deterioration	331,755	
Physically depreciated replacement cost new		$1,855,626
Less functional obsolescence	0	
Physically and functionally depreciated replacement cost new		$1,855,626
Total depreciation	$677,381	
Identified physical and functional	$331,755	
Excess identified as external obsolescence	$345,626	
Percent of external obsolescence ($345,626/$1,855,626)		19%

Table 17.7 Appraisal of Subject Property

Replacement cost new		$2,874,610
Curable physical	$60,000	
Incurable physical–short-lived	100,000	
Incurable physical–long-lived ([$2,874,610 – $60,000 – $1,094,000 = $1,720,610] × 12%)	206,473	
Total physical deterioration		– 366,473
Physically depreciated replacement cost new		$2,508,137
Functional obsolescence	$0	
Physically and functionally depreciated replacement cost new		$2,508,137
External obsolescence ($2,508,137 × 19%)		– 476,546
Depreciated improvement cost		$2,031,591
Plus land value		250,000
Total		2,281,591
Indicated value by cost approach (rounded)		$2,300,000

the breeding and sale of high-quality cattle, registered cattle, embryo transplants, and 4H animals all affect the income stream of a dairy farm. All dairy farm appraisals should include consideration of the income capitalization approach because a dairy farm is an income-producing entity. The appraisal report must discuss the reasons for excluding the approach if it is not used.

Dairy farms are typically owner-operated businesses. As such, the estimation of income and expenses is critical because cash rental data can rarely be obtained. The historical income and expenses of the subject property are important data for the income capitalization approach. Income and expense information for the dairy industry is available from many sources, including accounting services, lenders, dairy cooperatives, and county extension agents supported by land grant agricultural colleges.

Figures 17.4 and 17.5 list revenue sources and expenses that appraisers commonly encounter when analyzing dairy farms. Determining which revenue sources relate to the operation of the business rather than the real property is a critical consideration in the application of the income capitalization approach. For example, the value of chattel such as dairy cows could be estimated using income capitalization, but the value of the chattel would have to be reported separately from the value of the real property.

On smaller farming operations, farm operators and other family members may also work outside the farm. This non-farm income is not included in the evaluation of the farm's net income potential.

The development of overall capitalization rates requires the analysis of income and expenses for comparable sale properties. Rates can also be developed using income capitalization techniques such as the band of investment developed for appraising other rural properties.

Rental and lease information may be available for smaller farms. From this information, the appraiser can extract overall rates. How-

Figure 17.4 Typical Revenue Sources for Dairy Farms

- Milk sales
- Calf sales
- Cull cow sales (i.e., removal of older or less desirable cows from a herd)
- Government payments and fuel tax refunds
- Patronage dividends
- Fuel tax refunds
- Utility sales (i.e., the sale of power resulting from methane generation)
- Sales tax and insurance refunds
- Crop insurance proceeds
- Excess feed sales

Figure 17.5 Typical Expenses for Dairy Farms

- Purchased feed
- Chemicals
- Custom hire
- Fertilizer and lime
- Freight and trucking
- Fuel costs
- Insurance
- Repairs
- Lease payments
- Seeds and plants
- Supplies
- Farm portion of taxes and utilities
- Miscellaneous expenses such as (a) purchase of replacement livestock, (b) building, machinery, and equipment replacement, and (c) entrepreneurial risk and management changes

ever, the appraiser must be careful to select rates from homogeneous data–i.e., rates can be extracted from cash rents or from cropshare rents, but not from both. Typically, the overall rates abstracted from cash rentals are lower than those abstracted from cropshare rentals, which in turn are lower than those abstracted from owner-operator data. The overall rates reflect the level of risk inherent in each mode of operation.

Chapter 18

The Valuation of Permanent Plantings

In 2014 American consumers used 264.8 pounds per capita of fresh and processed fruit, including 27.9 pounds of bananas, 18.8 pounds of fresh apples, and 56.7 pounds of orange juice and fresh oranges, according to USDA data. In the 2014 season, growers produced 9.4 million tons of citrus fruit valued at $3.7 billion, 19 million tons of noncitrus fruit valued at $16.3 billion, and 2.6 million tons of tree nuts valued at $10 billion, for a total value of $30 billion in US fresh fruit and nut production.[1]

Fruit and nuts are classified according to the kind of trees–deciduous or evergreen–that bear them. Deciduous fruit plants lose their leaves, usually in the winter, and have a period of apparent dormancy. They produce the following:

- pome fruit such as apples and pears
- stone fruit such as peaches, plums, apricots, and cherries
- berries or small fruit such as grapes
- nut crops such as almonds, chestnuts, pecans, pistachios, and walnuts
- miscellaneous crops such as figs, kiwifruit, mulberries, and persimmons

Evergreen fruit-bearers retain their leaves throughout the year, even when dormant, and produce the following:

- citrus fruit, including oranges, lemons, grapefruit, tangerines, and tangelos

1. US Department of Agriculture, *Fruit and Tree Nut Yearbook 2015*, http://usda.mannlib.cornell.edu/usda/ers/89022/2015/FruitandTreeNutYearbook2015.pdf.

- other fruits such as olives and avocados

Orchards and vineyards vary in size from small family-owned and -operated enterprises to large orchards of more than 50,000 acres. According to the USDA, 4,296,800 acres in the United States were improved with permanent crops in 2014, consisting of:

- 777,700 acres of citrus fruit crops
- 1,738,400 acres of major deciduous fruit crops, including apples, peaches, pears, grapes, cherries, plums, and apricots
- 353,700 acres of minor fruit crops, including nectarines, figs, olives, avocados, and dates
- 1,427,000 acres of nut tree crops[2]

Special Considerations for Valuation

A permanent planting such as an orchard or vineyard is a specialized property that presents a unique valuation situation. Appraisers must focus on the specific horticultural crop produced by a permanent planting, which requires in-depth knowledge of the particular variety of tree, vine, or bush and its characteristics, production, and requirements. If an appraiser is not knowledgeable about the variety of produce being cultivated on a property being appraised, a horticultural consultant should be engaged. Professional standards require appraisers to disclose any lack of competence or assistance required from other professionals.

Permanent plants require a number of years to reach maturity, and these plantings commit the land to that particular use for the economic life of the trees, vines, or bushes. Specialized agricultural properties such as orchards, vineyards, and groves usually have a higher value per acre than open land in the same locale.[3] The development and maintenance of permanent plantings requires specialized skills to ensure that the plants reach and maintain their genetic potential. Poor management can cause stress due to parasites or lack of water. The resulting damage may adversely affect the value of the property.

Development Time and Income-Producing Life

Orchards, vineyards, and groves are unique agricultural properties. There is significant variation in permanent crops, even within small geographic regions such as the Napa and Sonoma Valleys of California, where more than 100 varieties of wine grapes are grown.

2. Ibid.
3. There are some exceptions, such as highly specialized vegetable land in the rich coastal valley areas of California.

Appraisers must be knowledgeable of the land's potential capacity to produce different varieties of permanent crops as well as all costs associated with their development and production.

In addition to the normal ground preparation, planting, and fertilizing, it usually takes several years before permanent plantings generate a positive cash flow. The amount of time required to establish a permanent planting typically varies from 3 to 12 years, as compared to coastal vegetable land in California that is capable of producing two to three crops per year. The life cycle of permanent plantings moves from infancy to the point of establishment, when annual income exceeds preharvest and harvest costs, through a period of mature production, and ultimately to decline. For example, one variety of apple orchard may take 5 to 7 years to reach mature production, produce at a stable level for 30 years, and decline after 40 years. Certain peach varieties may reach mature production after 7 years but begin to decline after 20 years. Some pear varieties do not reach mature production until 12 years, but only start to decline at 45 to 50 years.

The life cycles of permanent crops vary in timing and duration, changing with different crops and even different varieties of the same crop. Figure 18.1 illustrates the projected life of a typical almond orchard in the southern San Joaquin Valley of California, with establishment occurring in the sixth year and mature production occurring in Years 8 through 19.

An understanding of bioengineering, genetics, and plant development has become increasingly important for appraisers, helping them judge where a permanent planting is in its economic life. Scientists can isolate genes that control plant development and growth,

Figure 18.1 Economic Life Cycle of an Almond Orchard

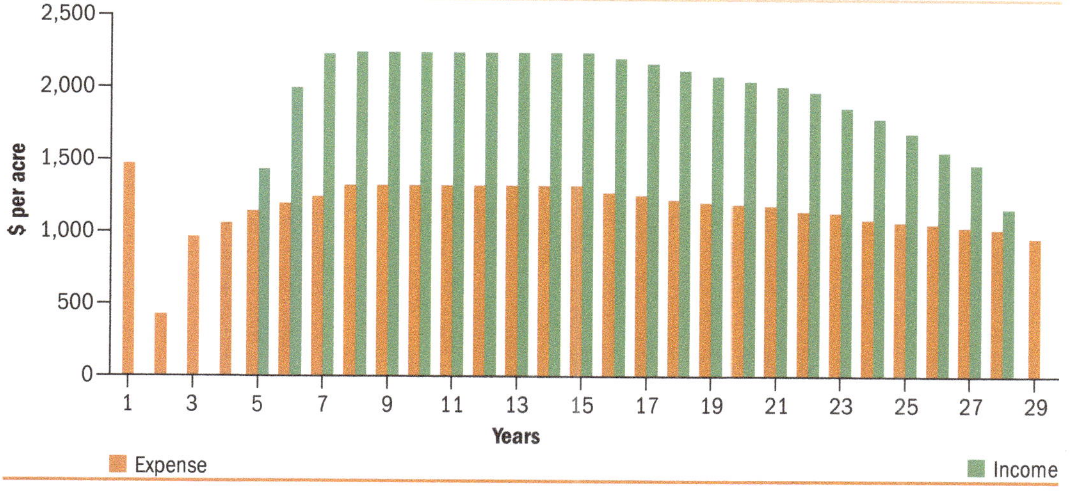

and laboratory techniques based on plant physiology and biochemistry may accelerate the process of creating, identifying, and incorporating new and useful plant traits.

Table grapes offer one example. For many years, the Thompson seedless variety, which is a white grape, was the only seedless grape. In 1973, the Agricultural Research Station in Fresno, California–jointly funded by the USDA and a University of California cooperative extension–introduced the Flame seedless variety, a flavorful red table grape that was well accepted by consumers. Since 1973, several other red seedless table grape varieties have been developed, market tested, and accepted by consumers. A review of recently released varieties indicates that approximately 12 to 15 years are spent researching and developing improved varieties of trees and vines. However, so much research is being conducted by corporations, individual farming operations, and universities, and so many new varieties are being brought to market, that the market advantage of a new successful variety may only last for a short time (such as 5 to 7 years for some types of tree fruit). Appraisers of permanent crops should remember that trees, vines, and bushes are wasting, or depreciating, assets. Therefore, their age, condition, and expected life are crucial factors.

Stages of Development

Permanent plantings are typically found in one of three stages:

- developing or immature
- mature
- decline

Each of these three stages requires a different appraisal emphasis.

The permanent planting must represent the highest and best use of the land as improved. During investigation of a developing or immature vineyard or orchard, it may be discovered that the property cannot be expected to develop into a successful operation, and its value for alternative uses should be considered. The property value, in this case, could be less than the vacant land value, after deducting the cost of removing the trees or vines and preparing the land for alternative uses.

The maximum value of a permanent planting is realized and sustained when the following conditions are met:

1. The soil is appropriate for the plant species and varieties cultivated.
2. The locale, including climatic conditions and water availability and quality, are appropriate for this crop.
3. Markets for the crop are well-established and accessible.
4. The permanent planting is effectively managed. Effective management includes the consistent replacement of damaged or dead

plants, the reduction of climatic hazards through irrigation, frost protection, and drainage facilities, and the use of appropriate and cost-effective pest control measures.

Soil Conditions

While it is generally accepted that superior soil quality is advantageous to field, row, and permanent crop production, poor quality soils are compatible with certain crops. For example, the presence of wine grape vineyards on the hillsides of California's Napa Valley have shown that Class IV and Class V rocky soils can, with good climatic conditions and proper management, produce good yields, superior quality fruit, and higher incomes than vineyards on Class I soils in other, less compatible areas of California. Similarly, citrus groves on soils of lesser quality but located in more temperate zones may be more productive than groves on higher-quality soils in colder locations. In these cases, climatic conditions rather than soil quality are the dominant factor.

Local Growing Conditions

Climate not only limits what kinds of crops can be grown but also determines the degree of success and the quality of the crop. If the climate is appropriate for a particular variety, the fruit can be produced more economically and with less risk. Temperate-zone trees have a complex set of requirements. For example, winter temperatures must not kill or seriously damage the trees, but the temperatures must be cold enough to satisfy the dormancy requirements of the trees. The alternative is delayed foliation, which reduces the crop production in that year. The area temperature and light must be adequate for the particular variety of crop to mature during its growing season, and commercial production is usually concentrated in appropriate climates. Climate also affects the time of harvest and the time window during which the crops mature. Harvest time has become crucial to producers, both domestic and foreign, who compete for the high prices commanded by the first fruits of the season.

In most orchards, vineyards, and groves, irrigation is provided to ensure adequate water at crucial times. Insufficient water for growing fruit, for example, may result in fruit that is too small to be commercially marketable. Appraisers must explore the adequacy, quality, dependability, source, terms, water rights, and cost of the irrigation water supply. In the United States, new and expanding water issues are complicating appraisers' research and understanding of water supply and quality problems in permanent crop production. The central issues are (1) limited water supplies accompanied by increasing agricultural water demand, (2) decreasing groundwater quality,

(3) increasing water costs, and (4) environmental and social attitudes that support reallocation of traditional water supplies for urban expansion, wildlife habitat, and fisheries. The costs to pump groundwater for irrigation have increased in recent years due to higher energy costs and declining groundwater levels.

Established Markets

Fruit can be produced for either (a) the fresh market or (b) the processing market, which includes processing for juice, freezing, canning, and dehydrating. The demands of each market are specific and distinct. For example, the fresh market generally equates bigger with better. Therefore, a grower aiming for the fresh market may thin the fruit to increase its size (i.e., remove a certain amount of fruit from a tree so that the remaining fruit can grow to the desired size). This type of grower is usually more quality-conscious and has higher operating expenses than the grower who aims for the processing market. Fresh market prices are usually higher than those obtainable in the processing market. Harvest costs are usually lower for fruit grown exclusively for processing.

The fresh market often depends on cold storage or controlled-atmosphere storage facilities. The availability of nearby storage facilities is important because field heat must be removed from the fruit as soon as possible to extend its shelf life. Some growers have their own on-site storage facilities, and some fresh fruit operations are completely integrated with storage facilities, packing lines, and shipping facilities.

Controlled-atmosphere storage can extend the storage life of some fruit varieties. The fruit can be harvested when ready and stored in these facilities without deterioration until demand warrants marketing it. Effective controlled-atmosphere storage necessitates an air-tight chamber in which the air has been replaced by an inert gas (usually nitrogen) refrigerated at a constant temperature. When a sealed room is opened, the entire contents should be marketed rapidly.

Fruit processing facilities should also be located relatively close to the producer. Soft fruits are particularly susceptible to damage from long and rough hauling, which causes them to lose grade. Some fruit producers operate orchards that produce both fresh and processed fruit. Production may be graded by quality or size for combination use. For example, a pear orchard may grow Bartlett pears for processing and Bosc and D'Anjou pears primarily for the fresh winter pear market. Bosc and D'Anjou pears that fail to reach market quality can usually be sold for processing.

A few growers in suburban locations operate successful "mom-and-pop" pick-your-own orchards. These are usually not considered commercial operations because they account for a minuscule portion of the permanent crop industry. Prices for the fruit still on the trees may be higher, lower, or approximately equal to contract prices. But, by avoiding harvest costs, a grower can usually sell fruit to be picked by the consumer for less than the contract price.

When appraising an orchard, appraisers should note the size and quality of the fruit produced, the average production (preferably for the last 5 to 10 years), the method of marketing, and significant varieties. The desirable size is affected by the characteristics of the variety and the consumer's preferences. For example, California navel oranges are typically richly flavored, seedless, deeply colored, and large. Good-quality, medium-sized navel oranges may bring the best prices, but all vineyards, orchards, and groves produce varied sizes of fruit. Packing house reports usually supply information on grower returns and the fruit's size, quality, and pricing.

Appraisers should also understand and evaluate the marketing strategy for the crop of the subject property and competing properties, which may vary from direct sales to the consumer to sales through sophisticated marketing cooperatives such as the Almond Board of California and the Raisin Bargaining Association.

Management

Even with a compatible combination of crop, soil, and climate, the profitability of a permanent planting is largely dependent on management decisions concerning

- pollination
- spacing
- replacement
- pruning
- frost protection
- pest eradication
- energy sources

Appraisers of permanent crops should develop the ability to identify and understand the influence and subtle implications of each significant aspect of the planting–from layout and design, through physical planting, care, and production, to consumer recognition and marketing strategies.

Many varieties of trees cannot pollinate themselves. Even for self-fertile trees, better fruit is usually produced through cross-pollination. The pollen from one variety of fruit is used to fertilize another variety, so the proper variety, number, and arrangement of pollinating trees is necessary for the success of an orchard. The sweet cherry, for example, can present particular difficulty because most varieties cannot fertilize themselves, and a high percentage of fertilized blooms is needed for a full crop of these small fruits. A typical solution is to have equal numbers of the pollinizer and the main variety, both good marketable cherry types, even though each type of cherry usually needs to be harvested and marketed separately. The pollen of some varieties is carried by the wind, while other pollen is transferred by domestic bees or bees delivered to agricultural areas by migratory beekeepers during the pollination season for particular crops. Management decisions such as arranging an orchard in a pattern that takes advantage of wind currents and the normal flying behavior of pollinating insects can affect the effectiveness of pollination by native bees. The recent phenomenon of colony collapse disorder–the disappearance of the majority of a bee colony's worker bees, causing the hive to fail–has created instability in the supply of pollinators, which may result in additional costs for orchard operators reliant on migratory beekeepers. For example, almond pollination fees rose sharply after 2004, according to the California State Beekeeping Association (see Figure 18.2).

One of the objectives of orchard management is to develop maximum yields of high-quality fruit per acre with minimum time and cost. The variety of rootstock planted, the size of the trees, and their pattern of arrangement directly affect this goal. The original planting pattern should be suitable for the vigor of the stock and the ultimate size of the trees. It should facilitate pest control, cultivation, harvesting, and efficient land use, regardless of whether dwarf trees, standard trees, or high-density plantings are used. In high-density plantings, the excess trees may eventually be removed or formed into a hedgerow to give the mature trees adequate room and sunlight. The economic soundness of this decision will be determined by future production.

Figure 18.2 Average Almond Pollination Fee (1995-2015)

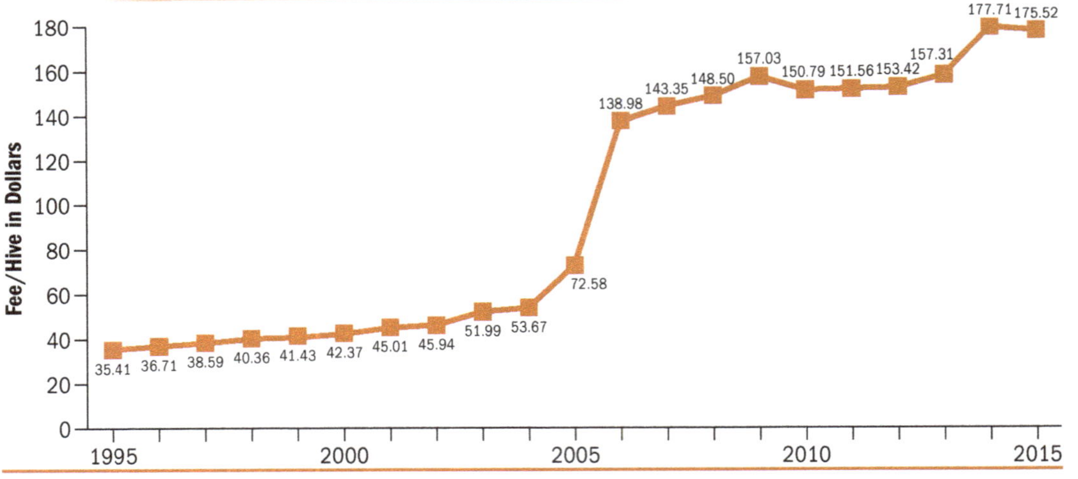

Planting density should be determined by variety, rootstock, and soil fertility. The development of dwarf and semidwarf rootstocks, particularly of apples and pears, has resulted in very high density plantings, which can be trellised like vines with special training and pruning. Trellis, or plant support, systems may represent a significant part of the development costs for many permanent plantings and may be a relevant element of comparison. Many types of plantings are developed on sophisticated, costly support systems. In grape and kiwifruit vineyards, for example, the trellis system may be the most costly item in the development process.

Sometimes growers try to renew old orchards by interplanting young trees alternately with old trees. The plan is to establish the young trees, gradually prune back the older ones, and then remove the old ones entirely. This plan has disadvantages, however, because the young trees are often shaded by the older trees, choked by weeds, damaged by equipment, or not properly irrigated.

Most permanent plantings require pruning, which can be accomplished either by maintenance pruning or by training or cutting young plants into certain forms or shapes. Some standard-sized fruit trees are trained to a certain form, such as modified leader, central leader, or open center (see Figure 18.3). Dwarf apple and pear trees are often trained in hedgerow systems, forming tree walls designed to provide a maximum number of trees per acre and to facilitate mechanical cultivation and harvesting. Maintenance pruning is done to maintain fruitfulness and to aid orchard management, harvesting, and pest control. To be successful, pruning should be adapted to the vigor of the particular rootstock and to the growth, flowering, and fruiting habits of the variety.

Prudent control of damaging vine and tree pests is an essential part of good management, and a wide range of options are available. For example, the traditional choice has been timely applications of sufficient commercial pesticides. In recent years, however, research and consumer agencies have focused attention on the possible risk of many of these chemicals to the consumer, and the public conversation has evolved to focus on the costs and benefits of organic farming and the effect of genetically modified organisms (GMOs) on the people and animals who consume those agricultural products. Appraisers of permanent crops need to monitor government policies on the regulation and registration of pesticides and how they affect pest control and preharvest costs.

Figure 18.3 Forms of Standard-Sized Fruit Trees

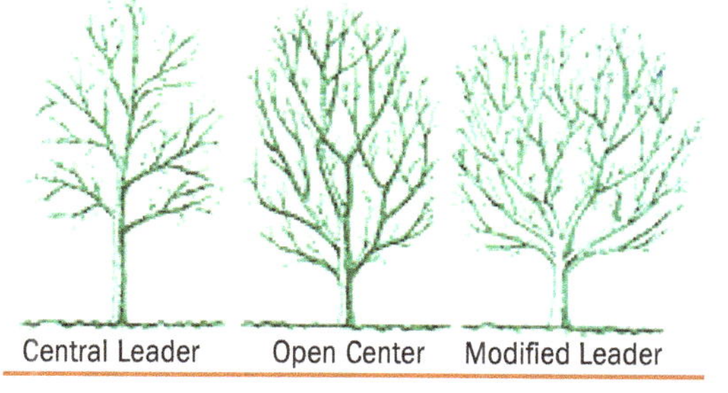

Another management choice that can affect the economic production of a permanent planting is the selection of electricity, natural gas, petroleum, or, more recently, solar and wind power to operate equipment and machinery.

Trend Analysis

Proper analysis of market trends is important in the valuation of permanent plantings and can be performed on most types of plantings. As previously stated, the development of an orchard or vineyard commits the land to the production of that crop for a finite economic life. The user of the land has thus lost the flexibility typical of open land. The owner-operator can no longer adjust to changes in the marketplace by switching crops in response to changes in the market. He or she is now committed to the production of the single crop chosen for development. As a result, the future viability of the market for the crop produced by that orchard or vineyard is critical to the value of the asset.

It is incumbent on appraisers to understand the structure of the market for the permanent crop under analysis. Markets for most permanent crops are typically regulated or controlled in some manner by marketing orders, commissions, boards, or other commodity groups. Membership in these marketing groups is often voluntary. Growers have the option of joining or complying with specific orders or regulations, and in return they benefit from the group's marketing program. In other cases, participation is mandated by federal regulations, as in the case of the raisin industry. These entities may control both the production and sale of the crop, either through volume shipment controls or production diversion programs. The controlling entity may also establish size or quality standards for product shipment, adding another element of control to the market. Crop shipments into the market may be controlled by industry reserve requirements. Under this arrangement, predetermined portions of a crop are held in reserve for carryover into subsequent years or for shipment to export markets.

Payments for production controlled by these entities may also be nontraditional. Crop proceeds may be distributed by various methods that prolong payment over a significant period of time after the actual physical delivery of the crop. Returns to the grower, then, may not follow the typical pattern of cash payment upon delivery of the crop, which is assumed in most markets.

Marketing or regulatory commissions or entities may also be effective sources of data for market analyses. They may be able to offer appraisers historical information on relevant factors such as bearing and nonbearing acres, past production, prices, and yields. Some groups track the sales of tree or vine nursery stock for plantings, which can allow appraisers to monitor trends in recent plantings that are not yet productive.

Varietal preferences for many crops can shift rapidly, as new varieties with superior physical characteristics are developed, introduced, and gain popularity over older, established varieties. A new variety of plum, for example, may have a deeper, brighter color, mature a few weeks earlier, or have a longer shelf life and thus be preferred over an older variety. Appraisers must keep up with market preferences. Varietal preferences may be quantified by tracking market share or by surveying packers or fruit brokers to determine whether the current planting on the subject property remains viable in the market.

In the analysis of market trends, appraisers should attempt to reduce yield and price information down to the net return per acre. Prices and yields may fluctuate dramatically in response to changes in supply and demand, so merely tracking unit prices may provide an inaccurate picture of true conditions. For example, a severe frost or hailstorm in a particular year may greatly reduce yields, which could trigger an unusually high unit price for that crop. Reporting the high unit price for that one year would give an unrealistic view of the market conditions for that crop. A more accurate method of measuring these trends would be to track the total gross revenue per acre or develop a weighted average of the commodity price. Years of low yields and high prices thus reflect a stabilized return, with the higher price offset by the lower yield. Conversely, the lower price is offset by the greater level of production for years of low prices and high yields.

For example, an appraiser might use the mathematical average of the simple price per pound over the past five years as an indicator of stabilized income to track the historical price of almonds, and the result could be considered a reflection of the commodity's recent market health. However, another appraiser might calculate the weighted average–that is, the total value of all five years' crops divided by the same five years' total crop production–and arrive at a more realistic vision of the stabilized income for the particular almond orchard and overall conditions in the industry.

Table 18.1 presents data on California almond production from 1995 to 2015. The inordinately high prices in 1995, 2005, and 2014 reflect significantly shorter crops in those years resulting from inclement weather conditions. These prices tend to distort the data, elevating the average price per pound derived as a simple arithmetical average. Developing a weighted average, however, smooths the effect of these high prices by compensating for the very low volume of nuts sold at that price. Again, merely viewing the high prices in 1995, 2005, and 2014 would provide a distorted view of market conditions. The data from Table 18.1 is shown in graph form in Figure 18.4.

For many permanent crops, marketing entities control the volume of product released to the market. For example, the Raisin Administrative Committee recommends to the Secretary of Agricul-

Table 18.1 California Almond Acreage, Production, and Value (1995-2015)

Year	Bearing Acres	Non-Bearing Acres	Yield per Acre (Pounds)	Production (Million Pounds)	Price per Pound (Dollars)	Value of Production (1,000 Dollars)
1995	418,000	65,700	890	370	2.48	880,896
1996	428,000	72,400	1,190	510	2.08	1,018,368
1997	442,000	63,000	1,720	759	1.56	1,160,640
1998	460,000	120,000	1,130	520	1.41	703,590
1999	485,000	115,000	1,720	833	0.86	687,742
2000	510,000	100,000	1,380	703	0.97	666,487
2001	530,000	75,000	1,570	830	0.91	740,012
2002	545,000	65,000	2,000	1,090	1.11	1,200,687
2003	550,000	60,000	1,890	1,040	1.57	1,600,144
2004	570,000	70,000	1,760	1,005	2.21	2,189,005
2005	590,000	110,000	1,550	915	2.81	2,525,909
2006	610,000	145,000	1,840	1,120	2.06	2,258,790
2007	640,000	125,000	2,170	1,390	1.75	2,401,875
2008	710,000	115,000	2,300	1,630	1.45	2,343,200
2009	750,000	90,000	1,880	1,410	1.65	2,293,500
2010	770,000	85,000	2,130	1,640	1.79	2,903,380
2011	800,000	75,000	2,540	2,030	1.99	4,007,860
2012	820,000	110,000	2,300	1,890	2.58	4,816,860
2013	850,000	120,000	2,360	2,010	3.21	6,384,690
2014	870,000	150,000	2,150	1,870	3.50	6,464,500
2015	890,000	n/a	2,080	1,850	n/a	n/a

SOURCE: USDA/NASS, California Field Office, www.nass.usda.gov/Statistics_by_State/California/Publications/Fruits_and_Nuts/201505almpd.pdf

Figure 18.4 Price per Pound of California Almonds

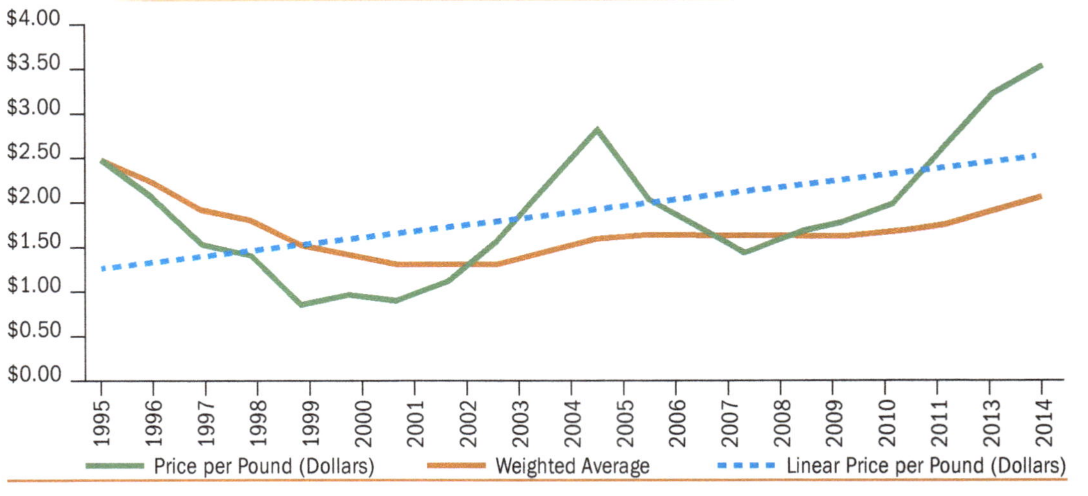

ture the total amount of the raisin crop to be designated as "free" tonnage. The remainder of the crop is then designated as "reserve" tonnage. The Raisin Bargaining Association negotiates the "field price" for free tonnage, and member growers are paid based on this price.

In years of large crops and soft markets, free tonnage may be as low as 50% of the total crop. While the reported field price may still reflect higher values, the grower may be receiving this price on only 50% or 60% of the total crop. The reserve is held in storage, at the grower's expense, to be sold at a later date or in other markets. Typically, returns to the grower on this reserve portion of the crop are significantly lower. Appraisers must be aware of the complexities within the market for the crop in question and recognize the resulting effect of these peculiarities on the incomes generated by these crop-producing properties.

Emerging Markets

In the United States, new geographical areas are emerging as sites of permanent plantings, such as vineyards east of the Mississippi and in Texas. For example, Texas winemakers have seen significant recent growth, producing 36% more cases of wine in 2014 than in 2010, according to Wines Vines Analytics.

Global competition is a growing concern because agricultural products are commonly traded internationally. New suppliers continue to emerge from the developing world, and new demand for a variety of agricultural products is shifting with the increasing purchasing power of Chinese and South Asian consumers. For example, agricultural exports from the United States to China grew on average 13% annually from 2005 to 2015, with exports of horticultural products continuing to grow despite an overall decline in exports of all other types of agricultural products.

Sales Comparison Approach

When applying the sales comparison approach, appraisers identify market-recognized factors of comparability and their influence on the prices paid for a commodity. An appraiser usually compares each variety of produce on a per-acre basis. The elements of comparison can be different, but some appear to be universal, such as time, size, age, production, and profitability. This approach is considered to produce a good indication of the market value of permanent plantings when sufficient recent sales of comparable properties are available for verification and analysis. Appraisers of permanent plantings must have adequate knowledge of the specific crop to identify and quantify significant factors of difference, which vary for different crops. For instance, appraisers should know the advantages and disadvantages of specific varieties, planted rootstocks, and budwood, irrigation water requirements, water sources, water costs, types of irrigation distribution systems, typical cultivation practices, and preharvest, harvest, and marketing costs.

Appraisers of permanent plantings throughout the United States use data from the cooperative extension services chartered under the land grant university system and funded by federal, state, and county governments. Wholesale nurseries can also provide appraisers with essential information on genetic capability, patents, and consumer response. Patented varieties–which are developed for the sole use of a private breeder, sold by license, or developed for public use and distributed without royalties–are subject to the forces of supply and demand that can affect value significantly. The values of permanent crops, even patented varieties, are also affected by the principle of change, in that geneticists are constantly trying to develop strains with improved vigor, size, color, shelf life, productivity, and harvesting dates. As mentioned previously, it may take 12 to 15 years to develop, test, patent, and market a new variety, while its market advantage may last only 5 to 7 years or until a newer variety is preferred by consumers.

Much information on comparable properties can be obtained through interviews with sellers and buyers. The physical data on specific properties should be supplemented with

- data from wholesale nurseries
- information from local farm advisors
- information from grower organizations and cooperatives
- statistical data from county and state agencies
- sale verification data, including property rights conveyed, buyer and seller motivations, financing terms, date of possession (i.e., will the buyer take possession at close of escrow or after harvest?), cultivation costs (i.e., were cultivation costs reimbursed as part of the sale or outside escrow?), holding period by seller, marketing period, and net operating income (for analyses of the overall capitalization rate and internal rate of return)

Wholesale nurseries are often a source of valuable information on new plant varieties that have not been market-tested or varieties for which data is not available due to the limited acreage planted. Data can include special soil requirements, drainage, plant size, preferred densities of the plant population, expected yields, harvest dates, and production during development of the planting. The nursery may be a patent holder or have a license to sell a patented variety (if applicable) and may demand a royalty on the patented variety. Government-funded varieties are not patented.

County agents are part of the cooperative extension program mentioned above. They are a reliable source of information on production and management, including varietal selection, soils, water, irrigation efficiencies, fertilization problems, cultivation practices, test plantings, and disease control.

Federal marketing orders are voluntary commodity programs initiated by producer groups. Market orders are enacted only when approved by referendums in which all producers within the proposed production area are eligible to vote. Grower organizations are self-governed under legislation established in the 1937 Agricultural

Marketing Agreement Act and, for the most part, are self-financed. Marketing orders are intended for specialty crops and are not permitted for livestock or basic field crops. A marketing order allows an agricultural crop industry to seek orderly marketing of its production. A federal marketing order sets up a mechanism for all producers of a crop in a given area to exercise control over selected aspects of the marketing of their crop and still be exempt from antitrust prosecution.

Each crop marketing order is developed by and for the particular needs of the commodity group seeking the marketing order. One or all of the following provisions may be included in a crop marketing order:

- specification of grades, size, quality, or maturity
- advertising, promotion, market development, and research
- allotting the amount each processor may handle or purchase
- establishing how much may be marketed during a set period
- establishing methods of determining surpluses as well as their control and disposition
- establishing a reserve product pool
- inspection of the product
- fixing the size, capacity, weight, dimensions, or packing of the containers used in marketing
- prohibiting unfair competition and unfair trade practices
- requiring processors to file their selling prices and to not sell below the prices filed

Only those marketing tools included in the marketing order may be used by that commodity group. Any one tool or a combination of tools described above may be written into a marketing order.

To start a marketing order, an order proposal must be submitted to the US Secretary of Agriculture along with a request for hearings on the order. If sufficient grower support is shown, the secretary holds public hearings on the proposal. An opportunity for written comments follows the hearings. Then the secretary makes a decision as to whether or not to submit a proposed marketing order to a vote of all the growers. The marketing order is established if two-thirds of the voting growers vote in favor of the marketing order or if those growers representing two-thirds of the total production vote for it. Marketing orders are ended when more than half of the growers or those representing more than half the production vote against the order. An order may be amended through a procedure similar to that followed to establish the order.

A marketing order is administered by an elected board of growers and processors and a public member who is elected by the other board members. Board members, other than the public member, are elected by those they represent on a one person-one vote basis. The US Secretary of Agriculture oversees board actions to make sure the board does not act beyond its authority as specified in the marketing order.

Under the marketing order, it is the processors who are regulated. Assessments for enacting the order are collected from processors or first handlers. However, they can pass these costs on to buyers or deduct them in making their purchases from growers. Imports are not regulated under a marketing order.

The advantages of a marketing order include industry self-control through use of selected marketing tools. A marketing order provides a means for all growers and processors to join together for various marketing activities. The disadvantage of a marketing order is that it is compulsory for all operators in the defined area. Marketing orders are attractive because they carry only modest administrative costs.

For example, the Almond Board of California administers Federal Marketing Order No. 91 over all handlers and marketers of almonds. This agency assesses growers per pound of production and provides research, quality control, promotional services, and marketing in both domestic and foreign markets.

Statistical data on bearing and nonbearing planted acreage, average production, and average price is very useful to appraisers of permanent plantings in identifying and projecting trends in supply, demand, and commodity prices. This information is compiled and distributed by various government agencies, including the USDA Economic Research Service, cooperative extension services, state agricultural statistics offices, county/parish agricultural commissioners, and some commodity organizations.

Appraisers need to verify more comprehensive data for sales of permanent crops than for sales of less complex agricultural properties. The verification sheet should include all the information needed to analyze the comparable property in all three approaches to value. The holding period is used to estimate the typical period of ownership for particular varieties and, when derived from market evidence, can assist in developing discounted cash flow (DCF) analyses and internal rates of return. Appraisers also need to confirm the marketing periods of comparable properties to project estimated marketing periods for the subject property. The date of possession is a significant issue because "high input" permanent plantings, such as almonds, table grapes, and wine grapes, can accrue significant cultivation costs prior to harvest, and sales often postpone possession until after harvest. Also, additional monies are often paid for reimbursement of the cultivation costs incurred to date or for the actual crop planted. The net operating income (*NOI*) of comparable properties is derived from actual historical yields, commodity prices, and preharvest and harvest expenses. This figure is used in the sales comparison, income capitalization, and cost approaches. An example of a typical permanent crop sales verification sheet is shown in Figure 18.5.

The sales comparison approach employs paired data analysis in which elements of comparison are isolated and differences between the comparable properties and the subject property are analyzed. Adjustments approximating the degree of comparability or quantifiable difference between properties are applied to the sales prices of the comparable properties to reflect the subject property's physical characteristics. Paired data analyses can be used to quantify elements of comparison, but in many markets the data is limited and appraisers may not be able to isolate or quantify adjustments. Appropriate elements of comparison may be varied and differ subtly. In many permanent planting markets (e.g., citrus, olives, pistachios, and some varieties of vines), plant age may not be critical, as plants tend to be long-lived. Conversely, plantings in other markets may experience short economic lives. Varietal preference may be a moving target, with today's jewel becoming tomorrow's candidate for removal. Market conditions can be a difficult adjustment, as the semipermanence of these plantings makes them especially sensitive to cyclical change.

Climate is a critical element for some types of plantings, and appraisers must draw on the market knowledge available to operators, buyers, and sellers. In most climate-sensitive industries, historical, site-specific temperature readings may be available for review and comparison. Water is a ubiquitous factor affecting the value of almost all permanent plantings. Value may depend on the quantity, quality, and cost of available water or, conversely, on the ability to drain, dispose, or control excess water or normal runoff.

In Tables 18.2 and 18.3, 10 sales of comparable almond orchards are considered, analyzed, and adjusted to develop specific indicators

Table 18.2 Unadjusted Comparable Sales of Almond Orchards

Sale No.	Sale Date	Area in Acres	Production Lbs./Acre	Water $/Acre	Tree Age	Sale Price/Acre
1	1/1/2011	320	1,800	$1,250	12	$20,765
2	1/1/2012	240	1,500	$500	15	$19,160
3	8/15/2012	480	2,000	$1,250	10	$26,305
4	3/15/2013	80	2,000	$1,250	18	$20,665
5	5/15/2013	640	2,000	$1,250	18	$20,770
6	7/15/2013	300	1,800	$500	10	$28,370
7	12/15/2013	160	1,200	$1,250	18	$9,125
8	1/15/2014	120	1,800	$1,750	10	$21,175
9	4/15/2014	320	1,800	$1,250	12	$22,820
10	8/1/2014	240	1,500	$500	15	$20,750
Subject	9/1/2014	320	1,900	$1,125	14	

Figure 18.5 Data Collection

Verification sheet—Sale No. ____________________

APN(s): ____________________

Principal crop(s): __________ Section: __________

Township: __________

Range: __________

Total assessed acreage: __________ Planted acreage: __________ Zoning: __________

Deed/escrow date: __________ County: __________ Recorder's book: __________

Recording date: __________ Recorder's page: __________

Document number: __________

Grantor: __________ Address: __________

Grantee: __________ Address: __________

Sale verified by: (buyer/seller/broker) __________ With: (appraiser) __________

Purchase price: __________ Cash to seller: __________

1st T/D amount: $ __________ Term: __________ Property rights conveyed: __________

Years: __________ Price adjustment: __________

Rate: % __________

Amortization type: __________ Cash-equivalent price: __________ Motivation: __________

Price adjustment: __________

Holding period: __________ Marketing period: __________

Cultivation costs/crop included in sale: __________ Contributory value: __________

Permanent plantings: __________ Age/variety: __________

Spacing: __________ Yield: __________

Other land uses: __________ Contributory value: __________

Topography: __________ Soil: __________

Date of possession: __________ *NOI*: __________

Water supply/district: __________ Well: __________ Blended cost: __________

Water cost/district: __________ GPMs: __________ Cost: __________

Water requirement/acre: __________ Distribution system: __________

Comments: ____________________

Sale price/gross acreage: $ ____________________

Sale price/acre (permanent plantings): $ ____________________

of value relevant to the subject property. Adjustments were derived for time, production, water cost, and tree age. Water cost and tree age were observed to be negative adjustments.

Table 18.3 Adjustment Amounts

Sale No.	Market Conditions	Production Lbs./Acre	Water Cost/Acre	Tree Age	Total Adjustments	Indicated Value/Acre
1	$2,380	$1,500	$750	-$1,500	$3,130	$23,895
2	$1,575	$6,000	-$3,750	$750	$4,575	$23,735
3	$1,645	-$1,500	$750	-$3,000	-$2,105	$24,200
4	$915	-$1,500	$750	$3,000	$3,165	$23,830
5	$815	-$1,500	$750	$3,000	$3,065	$23,835
6	$965	$1,500	-$3,750	-$3,000	-$4,285	$24,085
7	$195	$10,500	$750	$3,000	$14,445	$23,570
8	$400	$1,500	$3,750	-$3,000	$2,650	$23,825
9	$260	$1,500	$750	-$1,500	$1,010	$23,830
10	$55	$6,000	-$3,750	$750	$3,055	$23,805

Table 18.4 Paired Data Analysis for Market Conditions Adjustment

Sale No.	1	1 > 9	9	2	2 > 10	10
Date	1/1/2011	3.2877	4/15/2014	1/1/2012	2.5836	8/1/2014
Acres	320	=	320	240	=	240
Pounds per acre	1,800	=	1,800	1,500	=	1,500
$ per acre	$1,250	=	$1,250	$2,500	=	$2,500
Age	12	=	12	15	=	15
Per acre	$20,765	$2,055	$22,820	$19,160	$1,590	$20,750
Annualized rate			2.9%			3.1%

Table 18.5 Estimation of Size Adjustment

Sale No.	4	5
Date	3/15/2013	5/15/2013
$ per acre	$20,665	$20,770
Market conditions (years)	0.1671	
Adjustment	$102.35	
Adjusted	$20,767	$20,770
Acres	80	640
Pounds per acre	2,000	2,000
$ per acre	$1,250	$1,250
Age	18	18
Indicated size adjustment		$0

Adjustments for market conditions at an annualized rate of 3% were developed, as indicated in Table 18.4 by paired analyses of Sales 1 and 9 and Sales 2 and 10, which are equal in all respects except date of sale. The two pairings indicate annualized rates of 2.9% and 3.1%, so a 3% rate is applied. Once this market conditions adjustment is made, Sale 4 can be adjusted for market conditions at the indicated rate of 3%, and Sales 4 and 5 can be compared for size, as shown in Table 18.5.

Similar analyses indicated production adjustments of $15 per pound of production, with higher production meriting higher unit prices per acre. Water cost adjustments were established at $30 per dollar of cost, with higher costs equating to lower orchard values. Tree age adjustments of $750 per year of age were developed, with older orchards receiving lower unit prices per acre. The final adjusted values indicate a relatively narrow range of per-acre values for the subject orchard.

The sales comparison approach is a useful tool, but it requires attention to detail and comprehensive analysis of the general parameters of the subject market as well as the physical characteristics of the subject property and comparable properties. Appraisers must learn to recognize which factors influence value in the market for the specific property type and develop relevant adjustments for differences in these elements of comparison that lead to meaningful value conclusions.

Cost Approach

The cost approach to value is based on the principle of substitution, which affirms that no prudent investor would pay more for a property than the cost to acquire a comparable site and establish permanent crops of equal desirability and utility, without undue delay. The reproduction cost of a permanent planting, plus site value, provides a measure against which prices for similarly improved properties can be judged.

Site Value

Contributory land value is estimated by sales comparison, using sales of land similar to the land included in the subject property. These sales may be of developed row crop farmland, native land, or land previously used for other permanent plantings.

The land investment required to develop permanent plantings can be substantial, and typical investors or growers research water costs, soil adaptability, and climate conditions before making a commitment to develop a vineyard, orchard, or grove. Good-quality farmland is adaptable to a wide variety of field, vegetable, and permanent crops. The market recognizes the ease of adapting farmland use from one year to the next, depending on the economic outlook for alternative crops such as cotton, alfalfa, grain, or vegetables. However, once land is planted to trees or vines, it becomes an orchard, grove, or vineyard and no longer has the economic versatility of open land. For example, in times of drought, irrigated farmland can be left idle or lie fallow, awaiting the return of a viable water source. If the land has been developed into a permanent planting, however, it must be maintained and the owner or operator must acquire water, which is often costly at these times. Thus, the capitalization rate, which reflects the risk of ownership, changes from the relatively low market capitalization rate for farmland, or the even lower rate for native pasture, to the higher rates associated with permanent plantings.

The contribution of the land as if vacant and available for development to its highest and best use is derived from the market by direct comparison with sales of unimproved properties that are typically adjusted only for financing and time. Appraisers estimate contributory land value from verified sales data, and that indication of land value can then be incorporated into the cost approach to value.

Contributory Value of Improvements

Cost to Establish

The cost to establish a permanent planting includes four types of expenses:

- capital costs
- preharvest expenses
- harvest expenses
- the cost of capital and entrepreneurial profit

Capital costs typically include the cost of acquiring land and preparing it for planting, purchasing and planting nursery stock, and purchasing and installing the irrigation system. As noted earlier in this chapter, appraisers rely on various public and private sources for significant cost information relating to the specific planting being appraised. The available data on land value, irrigation improvements, water cost, and management must be evaluated as to its reliability and applicability to the subject property. Sample costs to establish an almond orchard are shown in Table 18.6.

The format shown in Table 18.6 can be modified to develop the estimated cost to establish any permanent crop. This format presents detailed items of expense and offsetting income through the development period. The cost of capital should reflect the risk rates associated with the permanent crop being appraised as well as the cost of borrowing capital from private or institutional lenders. Typically, permanent plantings are perceived as high-risk enterprises, and the yield and direct capitalization rates observed in the marketplace reflect this higher risk. Entrepreneurial profit is often observed in the expansion phase of planting cycles, as demand for the product

Table 18.6 Sample Costs per Acre to Establish an Almond Orchard (San Joaquin Valley North, 2016)

Year	1	2	3	4	5
Meat pounds per acre ($2.50 per pound)			400	800	1,600
Preplanting Costs					
Orchard removal	375				
Spread chips	125				
Rip–3-foot depth root removal	117				
Rip–4-foot depth	150				
Disc and roll 2x	50				
Laser level	70				
Soil fumigate–tree row strip	1,400	1			
Make berms/tree rows (GPS)	30	5			
Smooth/float between rows	30				
Total Preplanting Costs	**2,347**				
Plant	6				
Dip roots/plant/top	1,241				
Whitewash/wrap/irrigate	62				
Total Planting Costs	**1,304**				
Cultural Costs					
Frost protection–irrigate				17	17
Pollination–beehives			90	180	360
Pests–disease/fertilize (Zn)	20	35	48	50	69
Pests–disease			60	60	60
Pests–mites	8	11	19	21	21
Pests–insects (NOW)			44	44	44
Pests–insects (ants)			3	12	12
Pests–vertebrates (gophers)	16	23	23	14	14
Pests–vertebrates (squirrels)				25	25
Irrigate	92	175	267	350	350
Irrigation labor	60	63	54	81	81
Irrigation–well/water (test/analysis)	5	5	5	5	5
Fertilize–hand (15-15-15)	187				
Fertigate (CAN17)/(UAN32)		40	52	70	93
Fertilize–foliar (Boron)		6	11	11	11
Fertilize–lab analysis (leaf/hull)			2	2	2
Prune/train/sucker/tie ropes/stack brush	76	59	141	97	57
Shred brush			20	20	20
Pests–weeds (disc middles)	9				
Pests–weeds (float middles)	9				
Pests–weeds (mow middles)	17	30	26	26	26
Pests–weeds (broadcast spray/preharvest)			29	32	32
Pests–weeds (spot spray)	5	10			
Pests–weeds, dormant (strip spray)	35	30	38	38	38
Winter sanitation–shake/blow/mow			45	109	109
PCA fee	20	20	20	35	35

Table 18.6 Sample Costs per Acre to Establish an Almond Orchard (San Joaquin Valley North, 2016) *(continued)*

Year	**1**	**2**	**3**	**4**	**5**
Pickup 1/2-ton farm use	41	55	55	55	55
ATV farm use	21	25	25	25	32
Total Cultural Costs	**621**	**585**	**1,109**	**1,441**	**1,687**
Harvest Costs					
Pole trees			46		
Shake trees				94	94
Sweep nuts				75	75
Hand rake/blow nuts			15	4	10
Pick up/haul nuts			38	38	56
Hull/shell nuts			20	40	40
Total Harvest Costs			**119**	**250**	**315**
Interest on operating capital @ 4.25%	147	10	17	20	25
Total Operating Costs/Acre	**4,419**	**595**	**1,245**	**1,711**	**2,027**
Cash Overhead Costs					
Office expense	60	60	60	60	60
Liability insurance	8	8	8	8	8
Sanitation fees	8	8	8	8	8
Environmental fees	10	10	10	10	10
Property taxes	264	264	265	297	297
Property insurance	22	22	22	25	25
Investment repairs	50	50	50	56	56
Total Cash Overhead Costs	**422**	**422**	**423**	**465**	**465**
Total Cash Costs/Acre	**4,481**	**1,017**	**1,668**	**2,176**	**2,492**
Income/Acre from Production			**1,000**	**2,000**	**4,000**
Net Cash Costs/Acre for the Year	**4,481**	**1,017**	**668**		
Profit/Acre above Cash Costs				**176**	**1,508**
Accumulated Net Cash Costs/Acre	**4,481**	**5,858**	**6,526**	**6,350**	**4,842**
Noncash Overhead					
Building (4,000 square feet)	34	34	34	34	34
Fuel tanks (2 @ 1,000 gallons)	7	7	7	7	7
Booster pump (125 HP)	13	13	13	13	13
Shop/field tools	8	8	8	8	8
Irrigation system (micro)	89	89	89	89	89
Land SJV	813	813	813	813	813
Establishment costs				420	420
Equipment	20	23	37	37	37
Total Noncash Overhead Cost/Acre	**985**	**988**	**1,001**	**1,421**	**1,421**
Total Cost/Acre for the Year	5,826	2,005	2,669	3,597	3,913
Income/Acre from Production			1,000	2,000	4,000
Total Net Cost/Acre for the Year	5,826	2,005	1,669	1,597	
Net Profit/Acre above Total Cost					87
Total Accumulated Net Cost/Acre	5,826	7,831	9,500	11,097	11,010

Source: University of California Cooperative Extension, *2016 Almond Costs & Returns Study (San Joaquin Valley-North)*

outpaces supply and growers find they can sell young, immature plantings at a profit. However, few growers actually develop new plantings with the intent of selling them at a quick profit. More often, the grower plans on holding the planting over a long term and extracting profit from the ongoing operation of the development.

A permanent planting is considered "established" or "developed" at the end of the last year of negative cash flow–that is, in the year prior to the first year in which projected annual revenues exceed total annual expenses. While the planting may continue to mature for some years past this point in time, the cost to establish–or replacement cost–of the almond orchard (excluding the cost of capital) in Table 18.6 is projected as $11,097 at the end of Year 4.

Depreciation

Permanent plantings, like most other improvements, tend to suffer loss in value, or depreciation, from physical, functional, and external causes. When sufficient verified sales data is available, the most reliable indicator of depreciation is the actual loss of value observed in the market, which may be divided into physical, functional, and external depreciation. If not enough information is available to allow appraisers to extract market-derived depreciation, appraisers may use a traditional age-life method, isolating the respective components of the property and applying appropriate levels of depreciation to each. Many properties may include short-lived items such as drip irrigation or trellis systems, which may experience accelerated depreciation. When these items are present, depreciation analysis should be performed. The various components of depreciation may be either curable or incurable. Appraisers should also attempt to determine when the living improvements actually begin to accrue physical depreciation. If the permanent planting is truly the highest and best use of the land, the living planting may not experience physical depreciation until it reaches maturity, although the ancillary infrastructure (irrigation and trellis systems) may begin to lose value from the moment of installation.

Table 18.7 illustrates the analysis of physical depreciation based on the effective ages of the permanent plantings. As the table shows, the market appears to recognize economic lives of 25 to 30 years for these types of orchards. Physical depreciation rates should be extracted from sales of plantings with ages similar to that of the subject property because rates may vary with plant age. For example, higher or lower rates may be observed in sales of younger or older trees. As noted previously, many types of permanent plantings are relatively long-lived, and the markets for these types of properties may exhibit little, if any, physical depreciation. Short-lived items such as drip irrigation systems often experience accelerated depreciation.

Table 18.7 Depreciation Rates Derived from Sales Data

Sale No.	1	2	3	4
Sale date	1/1/2014	1/1/2015	8/15/2015	3/15/2016
Sale price/acre	$35,000	$36,000	$35,750	$34,000
Less land value/acre	– $20,000	– $22,000	– $23,250	– $20,000
Contributory value of trees	$15,000	$14,000	$12,500	$14,000
Replacement cost of orchard (less land)	$30,705	$32,115	$30,750	$31,600
Less contributory value	– $15,000	– $14,000	– $12,500	– $14,000
Observed depreciation	$15,705	$18,115	$18,250	$17,600
Effective age of trees (years)	14	17	15	14
Annual depreciation ($)	$1,122	$1,065	$1,217	$1,257
Annual depreciation (%)	3.65%	3.32%	3.96%	3.98%
Indicated economic life	27.4	30.1	25.3	25.1

Table 18.8 Functional Obsolescence Estimate

Contributory orchard value, per acre	
Sale price per acre	$27,000
Less land value per acre	– $15,000
Contributory value of orchard (less land)	$12,000
Total depreciation	
Replacement cost of orchard, per acre	$31,500
Less contributory value	– $12,000
Total observed depreciation	$19,500
Physical depreciation, long-lived	
Total observed depreciation	$19,500
Less curable physical (irrigation system)	– $750
Less incurable physical, short-lived	– $2,500
Physical depreciation, long-lived	$16,250
Replacement cost of orchard less curable physical and incurable short-lived depreciation ($31,500 less $750 and $2,500)	$28,250
Effective age of trees	15 years
Estimated annual physical depreciation rate (from sales)	3.3%
Estimated physical depreciation (3.3% per year × 15 years = 49.5% × $28,250)	– $13,984
Physically depreciated replacement cost	$14,266
Additional depreciation from functional obsolescence—low production	
Total observed, less physical, depreciation ($16,250 less $13,984)	$2,266
Functional obsolescence as a percentage of physically depreciated replacement cost ($2,266 ÷ $14,266)	15.9%

Table 18.9 Functional Obsolescence Due to Change in Preference

Contributory vineyard value	
Sale price per acre	$28,500
Less land value per acre	– $15,000
Contributory vineyard value	$13,500
Total depreciation	
Replacement cost of vineyard (per acre)	$39,500
Less contributory value	– $13,500
Total observed depreciation	$26,000
Effective age of vines	15 years
Estimated annual physical depreciation	2.0%
Estimated physical depreciation	
2.0% per year × 15 years = 30.0%	
$39,500 × 30.0% = physical depreciation	– $11,850
Functional obsolescence = total dep., less physical	$14,150
Physically depreciated replacement cost	$27,650
Functional obsolescence as a percentage of physically depreciated replacement cost	51.2%

Functional obsolescence may be observed when the planting is inconsistent with contemporary market preferences in terms of the varieties planted, plant densities, or trellising systems. Advances in technology, and consequent cost savings or production increases, may make certain types of plantings obsolete. Table 18.8 illustrates the functional obsolescence observed in a comparable property with a less-than-optimum production of 1,800 pounds per acre. This obsolescence is derived from the market as shown in the table.

Table 18.9 shows an analysis of the functional obsolescence suffered by a Chenin Blanc (white wine grape) vineyard during a period when consumer preference strongly favored red wines.

Appraisers are faced with a question as to whether the functional obsolescence derived in Table 18.9 is curable. The obsolescence may be physically cured by grafting the existing vines to one of the red varieties that are currently popular, thus allowing the vineyard to command higher grape prices. The cost to graft the vines and maintain the grafted vines until full production is restored must then be measured against the resulting value of the fully productive, new variety. Solving this problem requires income capitalization approach techniques.

External obsolescence can be observed in permanent crops when conditions outside of the subject property affect value. For example, supply and demand imbalances caused by natural disaster, new varieties, or fluctuations in competitive bearing acreage can affect the market, lowering the commodity prices and values of those permanent crops. The modern market is increasingly a global market.

Income Capitalization Approach

The income capitalization approach measures the present worth of future benefits. The application of direct capitalization may be appropriate in the case of long-lived permanent plantings, such as some types of grape vineyards. The Thompson Seedless grape vine, for example, can be extremely long-lived, producing at economic levels for many decades. In the valuation of those properties, the direct capitalization approach can be a valid indicator of value. Appraisers must, however, have sufficient knowledge of the plantings being appraised and the market's treatment of those plantings to determine the appropriate method of valuation.

Use of the income capitalization approach in the valuation of many permanent plantings, however, is complicated by the permanence or semipermanence of these developments. Traditional direct capitalization using an overall rate may not be appropriate in many cases. In the valuation of young or immature plantings, or of older plantings past their peak production years, application of this methodology may be misleading or inaccurate. In those cases, DCF analysis is the proper technique. Appraisers must attempt to measure the income stream generated by the subject property over an appropriate period of time.

Young, immature plantings require the use of DCF analysis. Appraisers must measure the income stream generated by the planting during the early years of development and the years of maturing production. Appraisers must also measure the value of the development as a mature, productive asset.

In the case of older plantings, the holding period is defined by the remaining economic life of the planting. The economic life of any permanent planting can be considered to end when the planting fails to generate sufficient revenue to offset production costs. Thus, an orchard has reached the end of its economic life when total gross revenues no longer exceed total costs of production. This determination is, of course, affected by the price received for the commodity produced. A decline in the unit price for the particular commodity hastens the end of the development's economic life. As stated previously, market trend analysis is critical in the determination of probable crop prices and the resultant economic lives of the permanent plantings that produce these crops.

At the end of the planting's economic life, the residual value of the underlying land, less any costs to be incurred in restoring the land to a useful state, must be estimated as the final component of the income stream. The first years of development reflect the initial costs of establishing the planting, including the cost of the raw land in the first year. Maturity is reached some years later, as the value of production overtakes cultivation costs. Net operating income continues to increase for several more years as yields continue to expand, and a period of stabilized income is achieved during the mature years of production. Net income begins to decline after the years of peak production as the planting enters the final phase of its life. The economic life of the planting ends when the net operating income fails to offset the cultivation and harvest costs.

The application of DCF analysis to this cycle measures the income over the total life of the investment and the rate of return earned by the capital invested in the initial development. In Table 18.10, the internal rate of return for the same investment is measured as 15.37%. The reversionary value in the final year of the holding period is measured as the sum of the annual net operating income (-$145), plus the market value of the underlying land ($12,000), minus the cost to remove the obsolete trees ($750), or $11,105. This analysis serves to graphically illustrate the time value of money. While this enterprise generates a total income stream of $81,285, the present worth of this income stream is, by definition, $0 at the indicated internal rate of return of 15.37%. The critical components of the income capitalization approach are revenue (income), expenses (costs), and rate.

Revenue

The typical means for measuring revenue in the income capitalization approach is to measure the rental income potential of the subject property. This approach is frustrated, in many cases, because most permanent plantings are owner-operated rather than rented or held by investors for rental income. When plantings are rented, it is often the result of an involuntary acquisition by a lender, who then offers the property for rent rather than incur significant management and production costs. Typically, these rentals are for a short term, while the owner seeks to market the property. When permanent plantings are held for investment purposes, they are commonly leased on a cropshare basis or for a combination of a fixed minimum cash rent plus owner participation in a share of revenues above a certain level. Appraisers are obligated to survey the market and determine the rental terms typical in the specific market under examination. Given the scarcity of rentals of permanent plantings, some appraisers erroneously mix cash and cropshare rents in appraisals of permanent plantings. Typically, cropshare rents exhibit a higher rate of return on investment because the investor shares in the risk of the crop and thus demands a higher rate of return. Consequently, if both types of rent must be used, appraisers should measure the increment of return attributable to the risk factor and adjust accordingly.

Proper analysis of the revenue generated by a permanent planting may require appraisers to explore a variety of factors. In most cases,

Table 18.10 Discounted Cash Flow Analysis

Crop Year	Costs (Dev. + Cult.)	Yield (Pounds per Acre)	Price ($/Pound)	Total Gross Rev.	*NOI*	Net Cash Flow (Plus Reversion)	*PV* Factor	*PV*
1	169,095				-19,095	-19,095	0.866798	-16,551.50
2	2,205				-2,205	-2,205	0.751339	-1,656.70
3	3,685	2,000	$5.35	2,140	-1,545	-1,545	0.651259	-1,006.20
4	4,400	4,000	$5.35	4,280	-120	-120	0.564510	-67.75
5	4,400	7,500	$5.35	8,025	3,625	3,625	0.489317	1,773.75
6	4,485	9,000	$5.35	9,630	5,145	5,145	0.424139	2,182.20
7	4,565	10,000	$5.35	10,700	-6,135	6,135	0.367643	2,255.50
8	5,290	11,000	$5.35	11,770	6,480	6,480	0.318672	2,065.00
9	5,930	12,000	$5.35	12,840	6,910	6,910	0.276224	1,908.70
10	5,930	12,000	$5.35	12,840	6,910	6,910	0.239431	1,654.45
11	5,930	12,000	$5.35	12,840	6,910	6,910	0.207538	1,434.10
12	5,930	12,000	$5.35	12,840	6,910	6,910	0.179894	1,243.05
13	5,930	12,000	$5.35	12,840	6,910	6,910	0.155932	1,077.50
14	5,930	12,000	$5.35	12,840	6,910	6,910	0.135161	933.95
15	5,795	11,040	$5.35	11,815	6,015	6,015	0.117157	704.95
16	5,670	10,155	$5.35	10,870	5,195	5,195	0.101552	527.65
17	5,560	9,345	$5.35	10,000	4,440	4,440	0.088025	390.85
18	5,455	8,595	$5.35	9,200	3,745	3,745	0.076300	285.75
19	5,355	7,910	$5.35	8,465	3,105	3,105	0.066137	205.40
20	5,270	7,275	$5.35	7,785	2,515	2,515	0.057327	144.30
21	5,185	6,695	$5.35	7,165	1,975	1,975	0.049691	98.15
22	5,110	6,160	$5.35	6,590	1,480	1,480	0.043072	63.65
23	5,045	5,670	$5.35	6,065	1,020	1,020	0.037335	38.05
24	4,980	5,215	$5.35	5,580	600	600	0.032362	19.35
25	4,920	4,795	$5.35	5,130	210	210	0.028051	5.90
26	4,870	4,410	$5.35	4,720	-145	11,105*	0.024315	269.95
						81,285		0.00
Internal rate of return			15.37%					

* Includes reversionary value of land, less cost to remove trees

the historical production of the planting is the key to this estimate, but historical information must be viewed in light of many aspects of production and the value of that production. Production in any specific crop year may have been influenced by climatic conditions such as a frost at bloom time. Appraisers must determine if the recurrence of these types of events is a reasonable probability to determine if that year's production should be included, or excluded, in calculating average yields. A particular property may be located in an area susceptible to frosts or other negative events. The appraisal of that property should properly measure the effect of the adverse location on the property's income. Conversely, another property may enjoy a location that it is not significantly influenced by events that affect other properties in the general area.

The value of the crop may be determined by the size, maturity, color, or market timing of the crop produced. Many permanent planting crops are highly volatile, and a particular box of fruit may command one price in June and a markedly lower or higher price two weeks later in July. Appraisers must also isolate factors attributable to the specific management or marketing skills of the grower. Helpful sources of information include commodity groups, marketing commissions, farm advisors, university extension offices, county agricultural commissioners, county crop reports, and federal and state market news services.

When rental data is available, the gross returns to the land may be estimated by multiplying the total gross revenue generated by the property by the appropriate cropshare rental factor. A farm generating total gross revenues of $10,000 per acre in a market where a 30% cropshare is typical would yield a gross return to the land of $3,000 per acre ($10,000 × 30%).

Expenses

Expenses must be deducted from gross revenues to reflect the typical terms found in the specific market. In some markets, rental rates are absolute net, with the tenant (lessee) assuming all costs associated with the land, including property taxes and water district or special assessments. In other markets, the landowner (lessor) may be responsible for property taxes and other expenses. In the leasing of permanent plantings in some markets, it may be typical for the lessor to share in costs associated with maintenance of the plantings–e.g., in a vineyard lease, the lessor may agree to share in the cost of trellis replacement, with the lessor responsible for the direct cash costs of materials and the lessee agreeing to provide the necessary labor to retrofit those materials. Net operating income may then be established by simply deducting appropriate expenses from gross revenue.

Rate

Direct, overall rates can be calculated from sales of comparable properties using traditional methods. Appraisers select an appropriate overall rate and apply this rate directly to the net operating income developed for the subject property. In the absence of adequate market data to extract overall rates, rates may be developed as they are in the appraisal of other types of properties. Again, appraisers should take care to select rates from homogenous data–that is, from cash rents or cropshare rents but not from both. Typically, cropshare rents yield higher overall rates as the lessor demands a higher return for the increased risk associated with cropshare leases. Table 18.11 shows data on a raisin vineyard and four comparable properties, illustrating a typical application of direct capitalization.

The four properties are leased on a cropshare basis. The appraiser has developed average returns for raisins over five years prior to the sales dates and the effective date of appraisal for the subject property. Revenues, expenses, and overall rates were then calculated as shown. The lease on Sale 3 calls for the tenant to pay taxes, so no owner expenses were estimated for this sale. Professional farm management is not typical for small parcels such as these in this market, thus no management expenses were included. The analysis indicates a range of overall capitalization rates from 7.08% to 7.74%, with the highest and lowest rates indicated by the least similar sales, Sales 2 and 3. The most similar sales, Sales 1 and 4, indicate rates of 7.55% and 7.47%, so a rate of 7.5% can be selected for the subject property. The analysis indicates an estimated net operating income of $135,995 for the subject property. The indicated overall rate of 7.5% is then applied to the subject property's net income, yielding an estimate of market value by the direct capitalization of income of $1,813,265, or $37,775 per acre.

DCF analyses may be performed using similar methods to develop estimates of revenues, expenses, and net operating income. In a DCF analysis, however, an appraiser must also develop an estimate of the reversionary value of the property at the end of the holding

Table 18.11 Sales and Income Analysis of Subject and Comparable Properties

Sale	**1**	**2**	**3**	**4**	**Subject**
Price	$1,638,000	$1,862,500	$1,828,800	$1,736,500	
Size (acres)	42	50	48	46	48
Price/acre	$39,000	$37,250	$38,100	$37,750	
Revenue					
Yield (tons/acre)	2.8	2.4	2.65	2.5	2.6
Price/ton	$4,450	$4,450	$4,450	$4,450	$4,450
Gross/acre	$12,460	$10,680	$11,795	$11,125	$11,570
Share	25.00%	26.00%	25.00%	27.00%	26.00%
Share rent/acre	$3,115.25	$2,776.80	$2,948.10	$3,003.75	$3,008.20
Total revenue	$130,830	$138,840	$141,510	$138,175	$144,395
Expenses					
Expenses/acre	$170	$140	$0	$185	$175
Total expenses	$7,140	$7,000	$0	$8,510	$8,400
NOI	$123,690	$131,840	$141,510	$129,665	$135,995
Overall rate	7.55%	7.08%	7.74%	7.47%	

period. This may be accomplished by direct capitalization of the next year's anticipated income or by another relevant method.

As an example, a DCF analysis is performed for a white wine grape vineyard at the end of the first year of development (see Table 18.12). The land was acquired one year ago at a cost of $17,000 per acre, and development began immediately. The vineyard is considered to be established at the end of Year 3, when revenue first exceeds expenses, even though grape yields will continue to increase through Year 5, when the vineyard reaches full mature production. The reversionary value of the vineyard at the end of Year 5 is estimated by the direct capitalization of the anticipated mature net income of $18,620, capitalized at a terminal rate of 20%. At a discount rate of 20%, the present worth of the annual cash flows is estimated as shown in Table 18.12. The value of the property, then, is estimated as the sum of the present values of each annual cash flow, or $45,465.

Table 18.12 Discounted Cash Flow Analysis

Year of Development	**2**	**3**	**4**	**5**	
Yield (tons/acre)		3	6	9	
Price ($/ton)		$3,000	$3,000	$3,000	
Revenue	$0	$9,000	$18,000	$27,000	
Expenses	$19,600	$6,235	$7,630	$8,380	
NOI	-$19,600	$2,765	$10,370	$18,620	
Reversion				$93,100	
Net cash flow	-$19,600	$2,765	$10,370	$111,720	
Present value factor @ 20.0%	0.833333	0.694444	0.578704	0.482253	
Present value	-$16,335	$1,920	$6,000	$53,875	
Sum of present values:					$45,465

Chapter 19

The Valuation of Timberland

Timberland is considered an agricultural property because it produces a crop: merchantable timber harvested on a periodic basis. Unlike farm properties that have a crop to harvest annually or more frequently, timber crops are harvested every 20, 50, or even 80 years, depending on the tree species and site-growing conditions. Timberland comprises a total of 766 million acres in the United States (33% of the total land area), of which about 521 million acres can be used for commercial timber. Of this figure, 69% is privately owned and 31% belongs to federal, state, or county governments.[1] (See Tables 19.1 and 19.2.)

The amount of forest area in the United States remained relatively stable over the course of the twentieth century, in spite of the rapid rise in population and a natural increase in the demand for wood products (see Figure 19.1). Since 1965, lumber production in the United States has generally trended upwards, except for periods of economic slowdown such as the mid-1970s, early 1980s, and 2006-2011.[2] Lumber consumption in the United States in 2011 for all uses totaled 40.8 billion board feet (BF). This reflects a decrease of 33.7 billion BF since 2005, when lumber consumption peaked at 74.5 billion BF. The record high in 2005 also exceeded levels from the early 1900s, when lumber was the most important raw material used in the United States for construction, manufactured products, and shipping.[3]

1. USDA, *US Forest Resource Facts and Historical Trends* (Washington, DC: 2014), 7-8, 14.
2. James L. Howard and Rebecca M. Westby, *US Timber Production, Trade, Consumption, and Price Statistics, 1965-2011* (Washington, DC: USDA Forest Service, 2013), 3.
3. Ibid., 5.

Export prices have been much higher than domestic prices, and US forest product exports have grown 80% in value since 2009. The majority of this growth has been due to increased off-shore demand, as prices have remained relatively flat during this period.[4]

Definitions and Legal Concepts

As a term of art, *timber* is defined as "forest stands and their products or wood in forms that are suitable for heavy constructions." By extension, the term *timberland* is defined as "agricultural property from which merchantable timber is harvested periodically, usually every 20, 50, or 80 years depending on the species and growing conditions."[5]

Table 19.1 US Forest Land by Ownership Class (2012)

	Region			
Owner Class/Land Class	**US**	**North**	**South**	**West**
	Million acres			
All Owners	**766**	**176**	**244**	**346**
Timberland	521	167	210	144
Reserved forest	74	7	4	63
Other forest	172	2	31	139
National Forest	**145**	**12**	**13**	**120**
Timberland	98	10	12	75
Reserved forest	27	1	1	24
Other forest	20	0	0	20
Other Public	**176**	**35**	**20**	**122**
Timberland	63	29	15	19
Reserved forest	47	5	3	39
Other forest	67	0	2	65
Private Corporate	**147**	**29**	**65**	**53**
Timberland	111	29	61	21
Reserved forest	0	-	0	0
Other forest	36	0	4	32
Private Noncorporate	**298**	**100**	**147**	**51**
Timberland	249	99	121	28
Reserved forest	0	0	0	0
Other forest	48	1	25	22

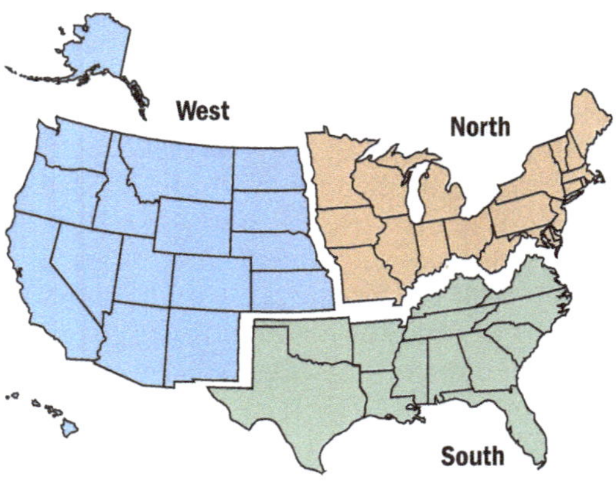

Source: USDA, *US Forest Resource Facts and Historical Trends*, 2014.

4. USDA Foreign Agricultural Service, "Money Does Grow on Trees as US Forest Product Exports Set Record," *International Agricultural Trade Report* (April 16, 2015).

5. Appraisal Institute, *The Dictionary of Real Estate Appraisal*, 6th ed. (Chicago: Appraisal Institute, 2015), 389.

Table 19.2 Land and Forest Area Trends in the United States

Category	Year	US	Region: North	South	West
		Million acres			
Land	2012	2,261	413	533	1,315
Of which:					
Forest	2012	766	176	244	346
	2007	752	172	235	346
	1997	742	170	231	341
	1987	733	165	234	333
	1977	742	164	235	343
	1963	753	166	245	342
	1953	742	161	240	341
	1940	738	159	232	346
	1920	721	149	220	353
Of which:					
Timberland	2012	521	167	210	144
	2007	514	164	204	146
	1997	504	159	201	143
	1987	485	155	195	135
	1977	491	153	198	139
	1963	515	156	209	150
	1953	509	154	205	150
	1940	502	149	200	153
	1920	482	138	194	149
Reserved forest	2012	74	7	4	63
	2007	75	6	3	65
	1997	52	8	4	40
	1987	35	7	3	25
	1977	29	6	2	21
	1963	25	4	1	19
	1953	24	4	1	19
	1940	11	4	0	6
	1920	10	3	0	6
Other forest	2012	172	2	31	139
	2007	162	2	7	134
	1997	187	3	9	157
	1987	211	4	13	173
	1977	215	5	17	182
	1963	216	5	18	173
	1953	217	3	20	172
	1940	233	6	21	187
	1920	238	7	18	197
Woodlands (nonforest)	2012	53	-	23	30
Total forest and woodlands	2012	819	176	267	376

Note: "Reserved forest" does not include some protected areas. National forest roadless areas are International Union for Conservation of Nature (IUCN) Class VI but not identified as "reserved" in Forest Inventory and Analysis program (FIA) statistics and total approximately 32 million acres. These lands are currently reported in timberland and other forest land in FIA reports. New inventories will provide more accurate data to place these lands in their proper IUCN classification.

Source: USDA, *US Forest Resource Facts and Historical Trends*, 2014.

Figure 19.1 Forest Area and Population Trends in the United States (1850-2010)

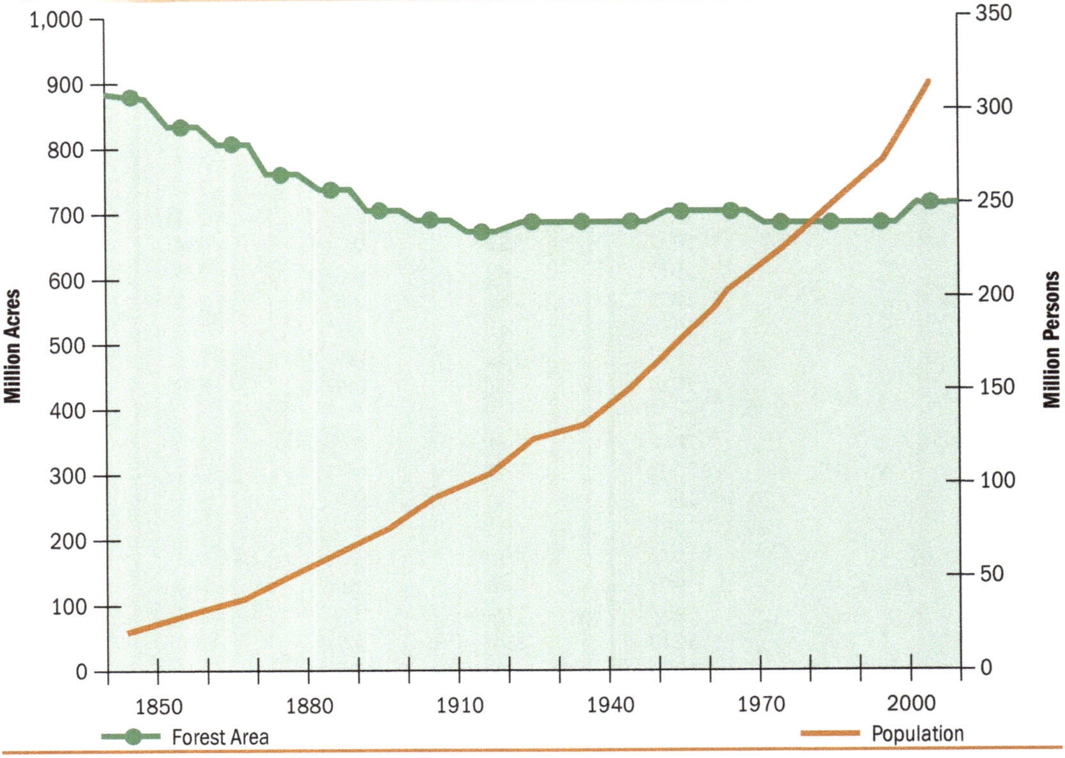

Source: USDA, *US Forest Resource Facts and Historical Trends*, 2014.

Standing timber is legally considered part of the real property rather than as a fixture or as personal property. Therefore, an appraisal of standing timber is considered a part of a real property appraisal of the fee simple interest. Standing timber can be appraised apart from the rest of the real property as well. After timber has been

Timber Terminology

merchantable timber. A stand of timber that has reached sufficient size and maturity to justify its being harvested.

premerchantable timber. Timber that is too small to economically justify its being harvested or even thinned.

old growth. Mature timber stands that remain in their natural condition; generally yield the highest volume of timber and the best quality.

residual stands. Timber stands that have been partially logged, leaving some of the original trees; tend to develop a young growth understory if not relogged.

understory. Plants growing beneath the canopy of another plant; the portion of a forest that is under, or below, the overstory or top canopy.

Figure 19.2 Forest Canopy Layers

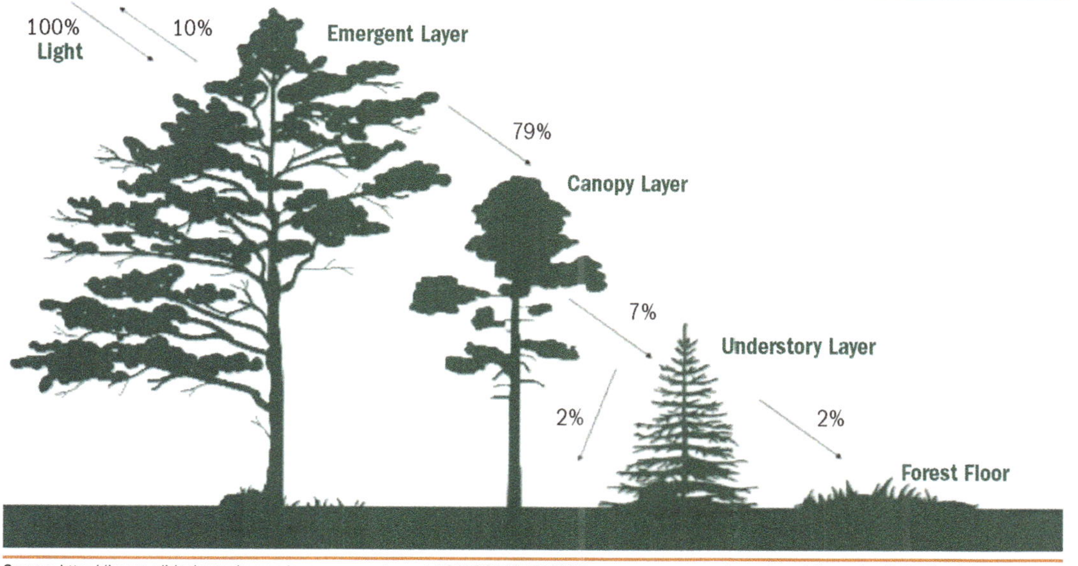

Source: http://image.slidesharecdn.com/canopymanagement-160326032522/95/canopy-management-practices-information-7-638.jpg?cb=1458963065

severed from the stump, it becomes personal property. Felled trees are referred to as *logs*.

Merchantable timber is timber that

- is of a certain size, quality, and species
- has a recognizable market at a mill
- is ready for harvest

Some states have best management practices that identify characteristics for trees and forest products that can be harvested. Sometimes merchantable timber cannot be removed from a tract of land because of its difficult location. To have value as a commodity, timber must be located in an area accessible for harvesting. This type of timber is

Table 19.3 Types of Timber Stands

Type	Characteristics
All-aged stand	Trees of all or almost all age classes, including those of exploitable age
Two-aged stand	A growing area with trees of two distinct age classes separated in age by more than 20% of rotation
Uneven-aged stand	Trees of three or more distinct age classes, either intimately mixed or in small groups
Multi-aged (multicohort) stand	A stand with two or more age classes or cohorts

Source: Society of American Foresters, *The Dictionary of Forestry* (Bethesda, MD: 1998).

sometimes referred to as *economically merchantable* timber. Federal and state regulations may restrict the removal of trees in certain areas.

In some timber-growing areas, a common practice is to *clearcut* a timber stand at harvest time. In other words, loggers remove all merchantable trees from the land. Often, state laws or local forest practice rules require that a clear-cut area be replanted after the timber is harvested.

In other areas, a common harvest procedure is *partial cutting* or *select cutting*, in which the oldest or largest trees can be marked to be cut and then removed. In operations of this type, loggers are not usually required to replant seedlings because the residual trees left after logging are assumed to be adequate to regenerate the area, or enough young trees are left to ensure that the stand continues.

When foresters use the *shelterwood* method to harvest merchantable timber, mature trees are removed in a series of cuttings. First, foresters select the trees that will remain as seed sources. Then, loggers usually remove the *overstory* (the older, larger trees or top canopy), allowing seed trees to rejuvenate the forest and then be removed later.

Some trees might be removed before a timber stand's harvest time to help other trees in the stand reach their largest and most valuable size. The common practice of *thinning* involves removing trees from a growing timber stand so that the remaining trees have adequate space and light to continue to grow.

In logging operations, some trees may be left in a timber stand to act as roadside screens or stream protection, or trees may be left due to restrictions on cutting resulting from federal or state regulations.

The Concept of Stumpage

In forestry, the term *stumpage* refers to standing timber. The term is most often encountered as a modifier of *value*. *Stumpage value* can be defined as "the value of timber as it stands uncut in terms of an amount per unit, e.g., per thousand board feet (MBF), per ton, per cord."[6] Appraisers often analyze two separate components of total property value: (1) the market value of the land and (2) the stumpage value of the economically merchantable timber as it stands in the forest attached to the ground. Stumpage value is not equivalent to the market value of logs that have been felled and delivered to a lumber mill, because stumpage value includes consideration of the cost of cutting and hauling the timber to the mill and the entrepreneurial profit realized by the timber buyer and the logger.

Stumpage can be considered real property, whereas felled logs are generally considered personal property. The assistance of a qualified, experienced forester may be required to estimate both the amount of wood product available in a stand (i.e., a timber cruise) and the stumpage value that the likely wood product would achieve in the market, depending on the appraiser's level of expertise and the scope of work of the assignment.

Note that an appraiser may be asked for the stumpage value of standing timber on a site as well as the market value of the land and any improvements on the land in an appraisal of timberland, and valuing those separate property interests are appropriate appraisal activities. However, the market value of the real property as a whole is not simply the sum of those two property interests, as will be discussed in more detail later in the chapter.

6. Appraisal Institute, *The Dictionary of Real Estate Appraisal*, 6th ed. (Chicago: Appraisal Institute, 2015), 388.

Special Considerations for Appraisal

Standing timber is used as a crop for various purposes in different parts of the country. In the West, most of the standing timber is used as *saw timber*. Logs are run through a mill and converted to dimensional lumber to be used for construction. The crop and its processing also develop a number of by-products that are valuable to the wood fiber industry. These include Christmas trees, firewood, cedar salvage, and small logs and wood chips for biomass from the *slash* (remnants left in the woods after completion of timbering operations that are sometimes scattered in the woods or piled and burned rather than sold) or from the mill and green material (moss and salal) that might be sold. In the South and Northeast, many timber stands are used for saw timber and for pulp to make paper and fiber products. The hardwoods are often used for furniture.

There is both a domestic and export market for logs. Each of these markets has its own characteristics and laws that regulate the extent of timber usage. For example, at many timber plants in the United States, thin layers are shaved from peeler logs used to make layers of plywood. However, these logs are also exported to foreign countries for processing and can then be returned to the United States as usable construction material.

Table 19.4 Uses of Timber

Type of Use	Characteristics
Saw timber	Trees or logs cut from trees with minimum diameter and length and with stem quality suitable for conversion to lumber
Peelers	High-grade logs from which veneer is peeled on a lathe or sliced for the production of plywood
Pulpwood (pulp logs)	Low-quality wood products used in the pulp or paper industry

Source: Society of American Foresters, *The Dictionary of Forestry* (Bethesda, MD: 1998) and Appraisal Institute, *The Dictionary of Real Estate Appraisal*, 6th ec. (Chicago: 2015).

Figure 19.3 Saw Timber

Source: https://s-media-cache-ak0.pinimg.com/564x/a2/b7/8b/a2b78b7bdca0b4e89cd45033a19e2f5b.jpg

Measuring Timber Stands

The size of a timber stand, whether in board feet, cords, cubic feet, or tons of wood product, is generally measured by a licensed forester or professional, based on a timber cruise. (Another forestry professional–the scaler–generally estimates the volume of logs loaded on a truck.) The Forest Service, various universities, and the timber industry have developed standards and methodologies for use in the proper measurement of timber stand size.

A *timber cruise* is an analysis of timberland performed to inventory standing timber and to estimate the quantity of the timber by species, products, size, quality, and other characteristics. Foresters undertake a timber cruise of a timber stand using a sampling process. For example, in a fixed-area plot cruise, a percentage of trees within a stand can be sampled (i.e., counted, measured, and characterized), and then statistical methods can be applied to project the total size of the stand.[7] Timber cruises can be subject to the statistical error probabilities found in all sample projections. The accuracy of a timber cruise depends on the selection of plots (whether they are truly representative), the number of plots (the higher the number, the less probability for error), and the skill of the timber cruiser in measuring the timber and size of each plot in the sample.

In a fixed-area cruise, timber cruisers map the plots on the various stands in the tracts being cruised. The plots are selected to provide representative samples of the overall tract being cruised. The trees in each of these individual but representative plots are then physically measured on the ground, and the data from these samples are then applied to the total tract to estimate the amount of timber thought to be on the entire tract.

Using an alternative technique–a variable-radius plot cruise–timber cruisers can more quickly estimate basal area (a measure of the average amount of area in a stand occupied by tree stems) and volume per acre because the timber cruiser does not need to physically measure the diameter of the trees in the plot. Rather, observation of the trees in the study area through a prism or angle gauge allows the timber cruiser to tally the number of trees that are "in" or "out" of the tree count.

The ultimate objective of a timber cruise, regardless of the technique used, is an accurate estimate of the amount of standing timber likely to be usable as saw timber expressed in board feet. A *board foot* equals the volume of a board that is 12 × 12 × 1 inches. Actual measurements vary and may be smaller because logs are round while *board foot* is a square measurement. An estimate of board footage, after processing, will change depending on the *log rule* used, and it is important to be consistent in applying the estimates of board footage for the subject property and comparable properties. When logs are processed, only about half of the wood is converted to board footage in the log rule measurement.

In the United States, the three most common log rules used to estimate board feet are

- the Doyle rule
- the Scribner rule
- the International ¼-inch log rule

7. Rarely is every tree measured in a timber cruise because of the high cost of what is known as a *100% cruise.*

Figure 19.4 One Board Foot

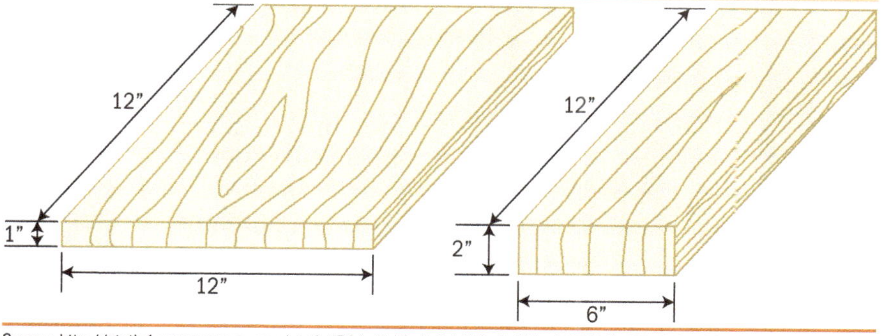

Source: http://static1.squarespace.com/static/544c4b42e4b0ddfd0e5e67ba/t/545dba25e4b0ca233d3c3e77/1415428645654

Table 19.5 Log Rules

Log Size	Doyle	International ¼-inch	Scribner
12" d x 16'	64 BF	95 BF	80 BF
20" d x 16'	256 BF	290 BF	280 BF

The estimates yielded by applying these scales are close for logs of medium or large diameter, but the Doyle rule underestimates the footage of smaller logs. As an illustration of the difference between the scales, consider 12-inch and 20-inch diameter logs, measured on the small end, that are each 16 feet in length. The board footage (BF) reported under each of the rules is illustrated in Table 19.5. Given the differences in board footage between the various log rules, appraisers must understand timber measurement and ensure that the timber estimates being employed use the same rule in all the data considered, especially if the timber cruise includes trees of smaller diameter. In the timber industry, it is rare for more than one log rule to be used in any region.

In the pulpwood market, timber is generally bought and sold on a *cord* or *ton* measure. A cord is equal to a stack of logs that is 4 × 4 × 8 feet. Pulpwood is also measured in cubic feet or in tons, typically on a truck weight scale, particularly when wood chips are the commodity.

A timber cruise is the starting point for any timberland valuation. Given the size and timing issues in discounted cash flow analysis, a forest inventory at the stand or polygon level is needed. This requires unique estimates of stand age, soil productivity, and species volumes for every productive stand on the property. As described earlier, a timber cruise is a set of plots that sample the stand and result in estimates of volume- and tree-level data for the stand. Despite the variety of methods of conducting timber cruises, the results of the

Figure 19.5 One Cord of Wood

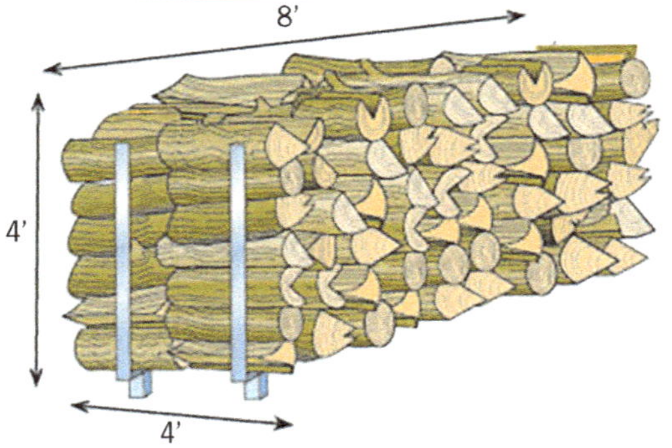

Source: http://extension.missouri.edu/explore/images/g05450art03.jpg

process should be a unique estimate for the timber stand that is statistically relevant. Most large timberland properties already have a forest inventory (i.e., cruise estimates on all their stands) that is maintained by forestry staff in a complex database system.

The mapping of the stand-level data, which is usually maintained in a geographic information system (GIS) along with streams, roads, and other legal boundaries, helps an analyst compute the gross and net productive areas of each timber stand. While the legal area of the timberland is important, the analysis of the net productive timbered area is key to estimating current and future net timber volumes. The GIS map facilitates the deduction of nonproductive road and stream areas and the estimation of the area influence of regulatory stream buffering. This analysis is generally referred to as "net-down" and is the computation of the net productive area that can be legally harvested. A typical timberland property has areas of delineated nonforest lands (e.g., swamps, wetlands, brush fields, rock pits and outcrops, lakes), and these stands are removed from the productive land base in the GIS. Other types of nonforest lands, such as roads, streams, and riparian buffers, are shown in the GIS using user-defined widths to buffer the different classes of line segments. With the nonforest and nonproductive areas defined and removed, the remaining area is considered legally available for growing trees. The area and volume in riparian buffers or other set-aside areas can also be computed, and these areas may or may not contribute to the overall value of the timberland.

Timber cruise data can be analyzed and forecasted in a variety of growth and yield models. These models can be simple yield tables that display the future volumes based on species, site productivity, and age. In other cases, foresters use complex programs in which cruised tree lists are input and forecast into the future. These tree-level growth and yield programs are available from the USDA Forest Service, state departments of natural resources, university forestry schools, and private developers. The selection of the program depends on tree species, geography, and the data available for input into the program. Forestry professionals with skills in forest measurements, or "biometricians," can assist in the development of the yield forecast of the stand data.

It is extremely important that rural appraisers recognize the limitations of their expertise. An appraiser who has forestry training and experience can complete a timber cruise alone. An appraiser who does not have that knowledge or expertise should associate with a recognized expert in the field so that the quantity and quality of the timber can be properly estimated by market-recognized procedures.

Factors Affecting the Value of Timberland

In analyzing the characteristics of a tract of timberland, appraisers frequently use geographic information systems (GIS) and ortho-rectified imagery to assist in observing and describing the property being appraised. What once was a process of reconciling a survey or other map source with aerial photographs from the USDA or other agencies is now an investigation of online information sources.

Using aerial imagery and other GIS map sources, appraisers or foresters can prepare a land use classification map, which shows the open areas in the property, the inoperable areas (such as cropland, wetlands, or rock with no timber), and the characteristics of the timber stand (such as age, species, quantity, quality, whether it is a natural stand or second growth, and probable date of the last harvest). The timber cruise usually shows the areas sampled, the statistical basis for the projections, and the projected tree count (trees per acre) and tree volume (by species) for the property.

Appraisers then analyze the various factors that affect the value of the stumpage at this location. The relevant factors include

- stand characteristics
- site characteristics
- location and access
- market demand

Stand Characteristics (Tree Species)

There are significant differences in tree species across geographic regions. Hardwood varieties may dominate northeastern markets, pine may dominate southern markets, and fir may dominate western markets. The growth rate between regions varies, as does the type of cruise inventory and valuation needed. In the Northwest, rotation may exceed 60 years, whereas pine plantations in the South may have rotations of less than 30 years. As a result, investment terms differ and the information needs of investors change accordingly.

Because of these factors, valuation training in each region has focused on information about local species. While regional variations exist to accommodate the specific scope of work of different types of assignments, certain characteristics of the valuation process for timber properties are essential regardless of geography.

Figure 19.6 Site Index Curves for Loblolly Pine in the Coastal Plain of Virginia, North Carolina, and South Carolina

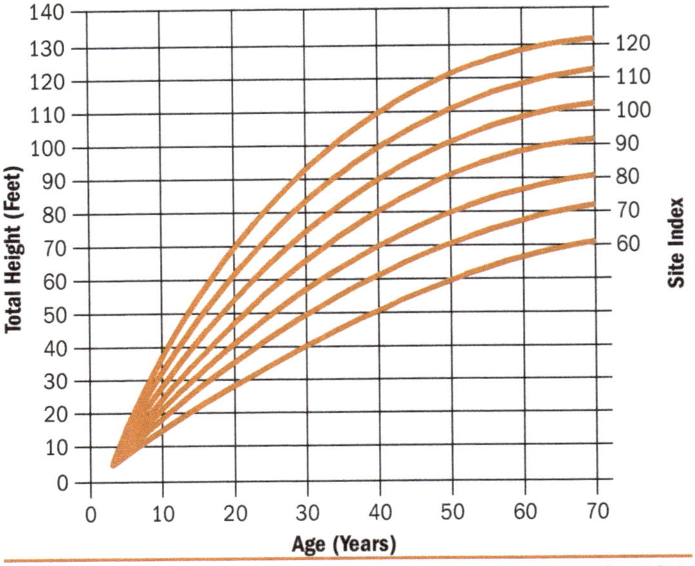

Note: The curves are based on stem analysis of 40 dominant trees in the middle and lower Coastal Plain.

Source: North Carolina Cooperative Extension Service, *Woodland Owner Notes: Forest Soils and Site Index*, http://content.ces.ncsu.edu/forest-soils-and-site-index.

Site Characteristics

Rural appraisers need to have a thorough understanding of legal descriptions of land to accurately identify a property's boundaries. Site visits to the subject property and the comparable properties should reveal the physical characteristics of the terrain, such as the direction of logging (i.e., uphill or downhill) and the steepness of the slope, that affect the cost of logging and transporting felled logs from the site. The location of better soils adjacent to a water course may create a growing site for a timber species different from the species on the adjoining hillside slopes. Elevation and climatic conditions are also noted so that the growing characteristics of trees in the various soils can be identified.

Foresters assign a site index to land to indicate the total height that the dominant trees of a given species will grow to on a given site at an index age. For example, a site index of "70 at 50 years" for loblolly pine would indicate that a tree planted today on the site in question would be expected to grow to a height of 70 feet in 50 years. Clearly, site index is closely related to timber yield. The soil factors analyzed to determine site index include topsoil depth, soil texture, the consistency of the subsoil, limiting layers, soil fertility, and internal drainage.

Location and Access

Access to the property is important. Both the distance and type of haul that the logging trucks have to make affect the property value. It may be physically possible to drive a vehicle to the property over existing roads, but it may not be legally possible to remove a large volume of timber over the same roadways. It is usually important to have a right-of-way agreement or a road easement connecting public roads to the subject property before the timber is harvested. Other arrangements based on a right-of-way charge by an intervening landowner are common. Charges are often an amount per cubic unit of fiber removed from the forest. Timber is considered inaccessible if intervening owners refuse to grant right-of-way requests.

Market Demand

The timing of the sale of stumpage or logs is important. When housing starts are low, demand for logs for lumber generally drops. It is interesting to note that, even in times of low demand for lumber and logs, there is an underlying strength to the market for good timber-growing lands with merchantable timber. This ongoing market indicates a belief in future requirements for fiber grown in the forest despite cyclical economic changes. The fact that the trees are physically growing during low demand periods also contributes to the stability of timberland prices.

The quality and quantity of timber in a specific stand also affect its marketability. Loggers generally avoid small stands of poor-quality timber. If the quality is poor, the log price received at a mill could be too low to make the logging operation economical. A similar condition results from a light timber stand where logging costs will be disproportionately high because of the small volume of material to be removed and the increase in time necessary to travel between harvestable timber. The felling and bucking time of small-diameter timber (e.g., diameter at a breast height of 20 inches or less) is slightly faster, but the volume per piece decreases faster than the working time, thus increasing costs. Both felling and bucking time increase as tree size increases. However, volume increases faster than, or at the same rate as, working time, and the cost will level or decrease. In the case of excessively large trees when special care must be taken to fell them (e.g., redwoods), the cost can be expected to rise.

Likewise, a timber stand with a steep slope will be less marketable than a stand where traveling and working is easier and less time-consuming because costs would be expected to rise on more difficult terrain. The distances of a timber stand to landing areas (where logs are sorted and graded) and then to saw mills (where logs are processed) are also considerations in the analysis of marketability.

Another consideration is topography and elevation. Lower, flood-prone properties are generally identified as sites suitable only for dry weather harvests, meaning the timber can only be

Figure 19.7 Diameter at Breast Height

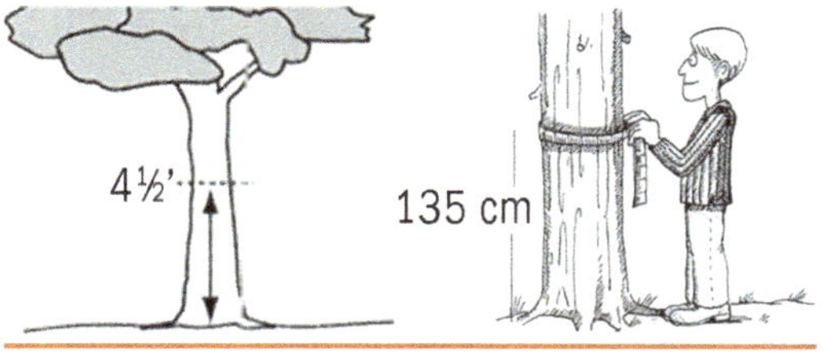

Sources: www.portlandonline.com/shared/cfm/image.cfm?id=514880 and https://archive.epa.gov/esd/land-sci/lcb/nrb/VFRDB/web/gif/image6.gif

harvested when it is dry. This influences the price paid for the timber because the timber is not as readily harvestable as it would be at an all-weather site (where the timber can be harvested at any time).

The location in relation to the mills also has a strong bearing on the prices that will be paid for standing timber. If loggers experience a long haul to a lumber mill, the price paid for the lumber is lower because of the higher hauling cost. Alternatively, if a site is near an operating mill, the price could be influenced positively.

Valuation Procedures

When valuing timberland used for timber harvesting purposes, real estate appraisers usually provide an opinion of the total property value including both land and timber. Although the valuation process is the basis for timberland valuation, the nature of the timber crop and its cyclical harvest necessitate some adaptation in applying the approaches to value.

Elements of the cost approach can be employed in the valuation of timberland, although the cost approach itself is generally not considered applicable because there are no building improvements on the land and because of the long amount of time often required to recapture expenses such as planting costs.

Whether an appraisal problem requires an opinion of the total property value or an opinion of value for the land and timber separately, appraisers employ certain basic valuation techniques unique to the timber field.

Sales Comparison Approach

The use of comparable sales is important in valuing timberland. Appraisers must consider many variables and details to accurately describe timber properties and their complex differences, and the application of the sales comparison approach relies on appraisers having sufficient information about comparable properties.

The analysis of sales of comparable properties requires the same identification of characteristics that is required for the subject property, which includes species identification, quantity and quality estimates, recognition of sources of value, and recognition of all available opportunities for the use of the crop on the land.

However, appraisers will rarely have the same level of information about all the comparable sales and the property being appraised. The availability of expensive timber cruises is a significant factor. Appraisal clients will often not pay for a timber cruise on the property being appraised, much less for timber cruises on the comparable properties. Sometimes, an appraiser will have access to a timber cruise on the subject property being appraised but will not have access to timber cruises on the relevant comparable properties, and

Highest and Best Use and the Recreational Use of Timberland

In many regions, the recreational use of forested land has had an influence on the timberland market. Historically, the traditional methods of appraising timberland have been employed and are still valid techniques to form an opinion of the value of a tract of timberland. However, recreational timberland is a new use in the marketplace since the mid- to late 1990s.

Demand for these recreational properties is from outside the timber market. Different factors influence their value, such as recreational appeal, proximity to the market (buyers), topography, access, and the presence of improvements and additional on-site improvements (e.g., interior roads, lakes, cabins, and lodges). Thorough market analysis and highest and best use analysis are particularly important in the identification of how the timberland would be marketed if it were presented to potential buyers or investors (i.e., for recreational use or for timber production). For land with a highest and best use as recreational timberland, the value of the standing timber might be allocated separately from the value of the land. However, adding the value of the land and the stumpage value of the timber for an overall value indication could very well produce an inflated opinion of value for timberland used for recreational purposes rather than for harvesting. Recreational use is ultimately a separate use that needs to be considered in the highest and best use analysis because recreational use has its own values, which are typically higher than the values for the forest only.

the client will not want to spend the money to have five or six comparable tracts cruised as a part of the appraisal assignment. In fact, appraisers may discover that comparable properties were purchased without a timber cruise at all. When experienced investors in timber tracts visit a timberland site, they are often able to assess the quality, species, and quantity of timber on a tract just by walking a portion of it, and they may make their buying decision based on that observation, without the benefit of a timber cruise. Similarly, if experienced timberland appraisers do not have access to timber cruises on the subject property or the comparable properties, they should be able after visiting the subject property and each comparable property to have a general idea of the timber quality and quantity on each tract visited and make appropriate adjustments. This expertise includes a general understanding of timber cruise techniques, the importance of quality (e.g., tree branching patterns) on a tree's value, the ability to identify species of trees on the tract, and other factors.

In addition to the physical characteristics on the respective properties, appraisers must recognize and consider the time between the dates of sale and the valuation date as well as the changes occurring in the expectations of the timber market. Appraisers must also consider the differences in logging and delivery costs for the logs, as well as the effect of harvest and management plans or other restrictions associated with the respective properties.

In comparing the subject property with comparable properties, appraisers first consider the market conditions variable–that is, the change in timber prices and timberland market conditions over the time interval between the sale date and valuation date. Variations in the quality and quantity of the timber crop and in all related costs necessary to convert and deliver the crop to the log delivery point are then considered, and market conditions adjustments may be appropriate for differences in those characteristics. Appraisers need to be careful to avoid making market conditions adjustments based on prices of timber delivered to the mill because those prices can fluctuate significantly, as shown in Figure 19.8. Productive timberland is often purchased with long-term horizons in mind, so the prices paid

Figure 19.8 Hardwood Lumber Prices in Indiana (Dollars per Thousand Board Feet, One Inch Thick)

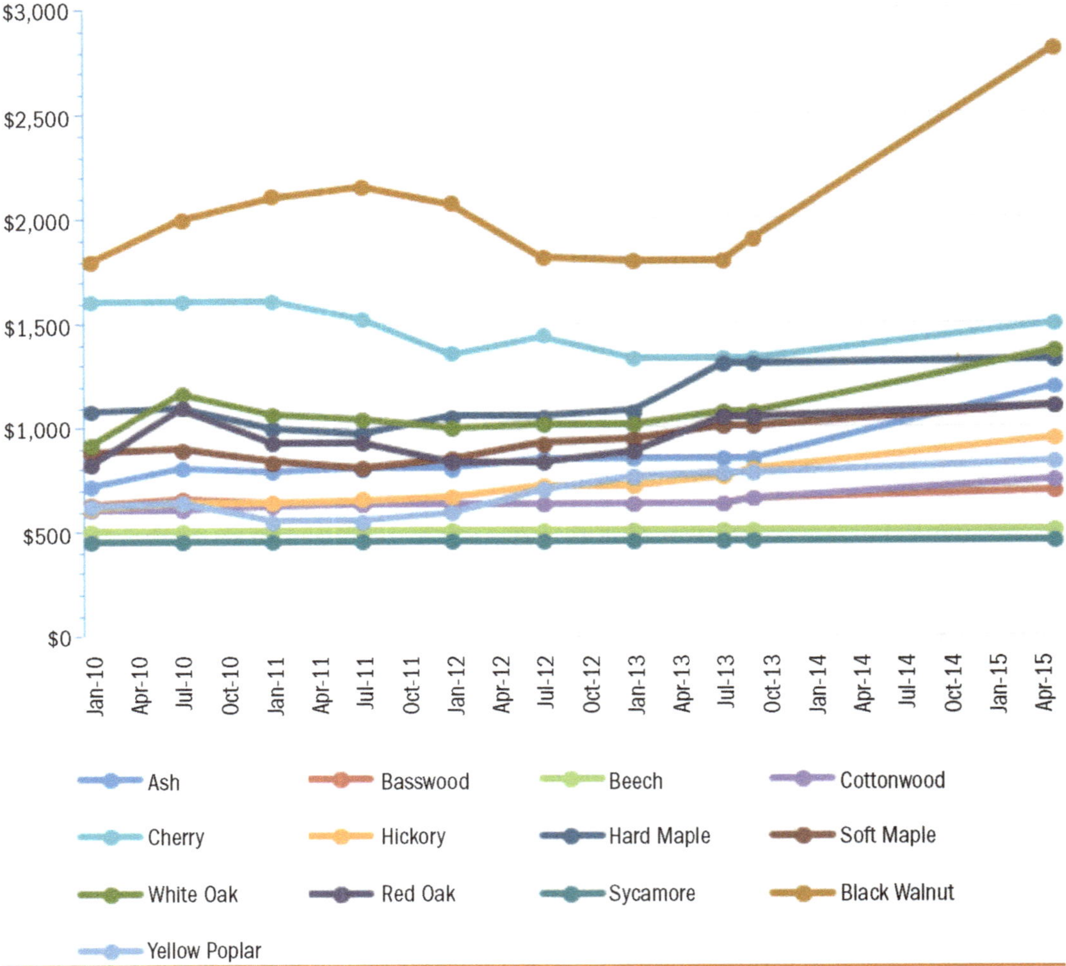

Data source: *Hardwood Market Report*

Table source: www.in.gov/dnr/forestry/files/fo-Spring_2015_Price_Report.pdf

for these property types are not usually affected and do not move in concert with short-term prices of timber delivered to the mill.

When the valuation is of the total property, comparisons must also reflect the variations in land characteristics such as aspect, slope, water courses, elevation, tree cover, and productivity. The age class structure of trees is also an important driver of timberland value.

Appraisers may analyze the land and timber separately, using the same comparison requirements, and then combine the two value contributions into a single indication of value for the total property. In this case, it is essential that an appraiser recognize (1) the relationship of the two elements, (2) that the land is not available for other uses until the present timber is removed, and (3) that this is actually how the market would accept this real estate. Adding the two represents summation, which may or may not represent market behavior. In considering the land separately, appraisers recognize that the land is still part of a unit property, made up of both land and standing timber. In developing an opinion of land value, appraisers typically rely on comparable sales of cutover land on which standing or residual timber is of nominal consequence. Appraisers must also consider if the land sales include any residual timber volumes that may have affected the sale price.

Although application of the sales comparison approach would ideally be based on the analysis and comparison of sales of properties made up of both land and timber at the time of sale, that type of comparable sales data is not often available. When seeking sales of comparable properties, appraisers are often confronted with variations in the composition of crop species, quantity, and quality, so that meaningful comparisons are either cumbersome or too diverse to be valid.

Adjustments in the sales comparison approach can be made with the use of property allocation. That is, appraisers can develop market-based rates (unit/area or unit/volume) for the different components of a timberland property. Generally, properties can be broken down into (1) merchantable acres and volume by species and (2) premerchantable acres divided into age groups. Appraisers can develop stumpage rates for the merchantable volume or area and then apply discounting techniques to develop an opinion of the present value of the premerchantable acres by age class. These values and bare land value can be used to allocate the values of the components of comparable properties, and then those rates can be applied to the subject property to develop an indication of value for that property.

Income Capitalization Approach

The income capitalization approach has become the most applicable valuation technique for large timberland properties because of the long-term nature of investment in the properties and the ability of income capitalization techniques to forecast the present value of future benefits. Despite the historical resistance to assumptions necessary to apply discounted cash flow (DCF) analysis, the technique has become more prevalent among sophisticated investors and analysts in the timberland market since the 1980s.

A DCF analysis can be used for larger timbered properties on which annual or regular harvests take place that can be predicted. It can also be used for estimating the value of timberland by discounting the expected future regular harvests back to the present, which is called a "soil expectation" analysis.

In the valuation of large timberland properties (more than, say, 5,000 acres), the wood flow forecast and timing of the timber harvest become complex. The use of wood flow models and growth-and-yields programs becomes necessary. Large timberland properties cannot be harvested in a single year because of physical and regulatory restrictions, which is why the timing of the timber harvest and the cost of capital necessitate the use of DCF modeling. The timber industry and investors recognize the need for sophisticated timber analysis by qualified forestry professionals.

A wood flow forecast model for timberland can be a complex system using expensive linear programming models or simulation programs. Small forests with a limited number of stands can be modeled in spreadsheets. DCF analysis of timberland is different from the analysis of other types of real property because of the long nature of tree growth. DCF models are built and forecasted for at least one rotation of the tree crop, which could be 50 years in the Pacific Northwest and 25 years in the South. While shorter forecast periods may be reviewed within the model based on investment lengths and exit strategies, the value of timberland is based on all the stands, regardless of age, that contribute value to the property.

The wood flow model accounts for harvest timing of all the stands during the planning horizon and replanting of new stands. The wood flow model incorporates the timber cruise and GIS data and then forecasts the data into the future based on the growth of the trees and the forest as a whole. The wood flow model can be used to develop several options for timber harvest from the property. In smaller properties with few age classes of timber, the wood flow forecast may be to harvest the stands over time as they become available (i.e., reach harvest age). Larger properties can have many options, depending on the age class structure of the property, with young forests having limited up-front timber harvests and older forests having excess volume that can be harvested more heavily in the early years. The wood flow forecast should examine legal and likely buyer harvest scenarios to help develop a market value for the timberland.

Drivers of the DCF Model

The DCF model used to compute the value of the income generated by timberland can be completed within a complex modeling system or by loading the resulting harvest forecast into a detailed spreadsheet. Both methods have advantages, but typically wood flow forecasts are loaded into spreadsheets along with all the other underlying financial assumptions.

The basic methods of DCF modeling apply in timberland valuation: (a) discounting the annual net operating income to present value and (b) computing a terminal value at the end of the planning horizon and discounting that value to the present. Despite the criticisms of DCF analysis and the many assumptions needed to develop a value indication using a DCF model, the timber industry recognizes that DCF modeling is one of the best methods of valuing timberlands given the long-term nature of growing trees and the differences in timing that the age class structure can have on timber harvests.

The wood flow model is based on a forecast of the timber cruise data into the future using many harvest scenarios. The forecast includes volume flow by species, harvested area, area classified by silvicultural treatment (such as premerchantable and commercial thinning, spraying, or forest management operations), and total forest inventory over time. All this information is input into the DCF financial model, and the associated costs and values can be applied. Because timberland is inherently an income-producing property, most wood flow models rely on the analysis of maximum net present value based on an estimated discount rate. That is, all the financial assumptions are loaded into the wood flow model, and the harvest decisions can be based on maximizing the value for the total property. To maximize net present value, wood flow constraints are typically applied to model legal forest management requirements, such as green up and maximum harvest size openings. Typically, several wood flow scenarios are developed by applying different harvest flow constraints to reflect how likely buyers of the property would manage the property.

Log prices are needed to value the proposed wood flows and compute revenues from timber harvests. These values are a combination of starting log prices and a forecast of prices. The prices can be stumpage prices or log (mill) prices. Stumpage prices already include a deduction for the cost of logging and hauling, so those costs would need to be subtracted from log prices. Logging costs can be variable due to terrain, and those costs can be modeled in the wood flow model and incorporated in the computation of net log values over time. The timber cruise and local market knowledge will help in the development of log prices that reflect the species and qualities of the logs. Prices are typically derived by species at a minimum. The forecasted prices can be derived from (a) services that provide price forecasts for the forest products industry and (b) the analysis of historical price levels and future trends in housing and other macroeconomic drivers. The price forecast can be performed in nominal or real dollars but must be consistent with cost and discount rate assumptions.

Along with log prices, management costs for the forestry operation must be accounted for. At a minimum, management costs would include road building and maintenance, planting and site preparation, other silvicultural treatments, and land management costs (i.e., overhead). Some of the costs are applied on a harvest volume basis, such as costs for logging, hauling, and roads, while others are based on harvested or treated area. Also, overhead can be based on total property size or a split between harvest volume ($/volume) and property size ($/property area).

A key driver in all DCF models is the discount rate. Discount rates can be nominal or real, but the trend in forestry investment modeling is the use of real rates. Also, as in most DCF models, the going-in rate can be different than the terminal discount rate. In forest modeling, given the long forecast lengths, the terminal discount rate is typically the same as the *NOI* discount rate. The rate used can be developed from a variety of sources with one of the best methods being to derive rates from past sales. When DCF models have been developed for a property for sale and after closing, the final discount rate can be derived. These derived rates should be for similar properties in similar geographic regions and log markets. Other good sources for discount rates are industry surveys conducted by investors or other appraisers. These surveys typically include rates by geographic regions and other investment parameters used in DCF modeling like price appreciation rates. Because timberland is recognized as an institutional-grade investment, examination of historical trends and alternative investments, such as CBD offices or corporate bond rates, can be helpful.

The terminal value of the DCF model is important. Because of the wood flow forecast, the value can vary greatly at the final terminal period. Typical DCF models would discount the terminal value at "stabilization." However, terminal values of timberland need to be reviewed because of widely varying harvest rates and remaining inventory. If the wood flows in the ending periods and the total forest inventory levels are stable, then the forest would be considered at "stabilization." But if harvest levels are lower or higher than would be assumed at stabilization because of the uneven structure of the age classes in the timberland, then a review would be necessary. An alternative method of deriving the terminal value would be to consider the land value, the value of the remaining forest volume, and the age class area value.

Because of the long length of the DCF analysis, the terminal value is typically less than 10% of the total value (25-50 years). Institutional investors may be looking at much shorter investment periods (7-10 years). While overall values of timberland are based on long-term investment, investors often incorporate shorter-term holding length and terminal value exit strategies.

DCF Conclusions

The scope of work and the client's objectives determine the depth of analysis and type of value developed in a DCF analysis. Reliance on "market value" assumptions in the analysis requires that the wood flow forecasts be similar to those found in the market by likely buyers of the property. It can be helpful to examine the existing age class structure of the property and look at how many acres have been harvested in the last decade and how many acres are currently in excess of the minimum cutting age. These past trends and current availability will be good indications of short-term harvest options. Investors may be seeking investment value or liquidation value rather than market value. In that case, the scope of work of the appraisal would involve the analysis of much different wood flow forecasts and should be described as such. But given the scope of work of a typical market value appraisal, several possible harvest scenarios will likely need to be evaluated in a DCF model.

As part of DCF modeling, sensitivity analysis of the key drivers is essential and is an important check on the credibility of the overall model. Appraisers may want to develop several wood flow forecasts to examine the physical possibilities of the property and demonstrate the unique makeup of the property and species mix. Because the discount rate is a key driver in a DCF model, testing the effect of 50-basis-point changes in the discount rate on total value can be helpful. While costs are a direct deduction to log prices, the log price forecast should also be tested for sensitivity. Small changes to the log price forecast or price appreciation rates can have large effects on values in long-term investment models like timberland DCF analysis. The development of a sensitivity matrix will be helpful in the final reconciliation of the income capitalization approach and the final value opinion.

In presenting DCF analysis, appraisers should examine and discuss the strengths and weaknesses of the models. The wood flow model is built on the timber cruise and GIS mapping of the property. The quality and analysis conducted at this primary level should be discussed in the appraisal report–e.g., how current the data is and how the field review aligns with the computer data. The type and kind of data and programs used in the growth and yield modeling are important and demonstrate the strength of the wood flow forecast. The forecast model used can also add credibility to the forecast of the overall resulting wood flow. Information on the source of the prices and costs used in the DCF model along with discussion of the forecast and historical prices will create a framework for the reader to understand the modeling assumptions. A review of discount rates recently extracted from the market and an overview of the current macroeconomic situation lay the foundation for a DCF timberland model. A final review of the sensitivity matrix, along with the discussion described above, will produce credible value opinions using DCF analysis for timberland valuation.

Cost Approach

The conversion return technique has been used in the valuation of timberland for decades. In the absence of truly comparable sales of timberland properties, some appraisers may employ this methodology, which begins with the value of the timber in logs or finished lumber (although the subject property still has standing timber). An estimate of the myriad of lumber products and prices is not required, nor is an analysis of mill operating costs, sales expense, or profit margin.

The analysis starts with the tree price of logs delivered to the mill, assuming that the buyer at the mill has already considered the price at which the lumber produced from the logs can be sold at a profit. Costs are developed through investigation and interviews with persons or organizations performing the various services necessary to sell and deliver the logs to the sawmill. These include loggers, log truck operators, logging supervisors, and others associated with the timber industry. The costs must also include profit, supervision, management, right-of-way charges (if appropriate), preparation of harvest plans, cost of restocking and site cleanup (if appropriate), and tax consequences (if any).

After all costs necessary to fell the tree and deliver the logs to a mill are deducted from the market price of the logs delivered to the sawmill, an indication of the stumpage value of the timber can be made.

Table 19.6 Prices for Conversion Return

Species	Sort	Grade	%	Delivered Log Price	Weighted Log Price
Douglas fir	Export	Japan	28%	$745	$209
		Korea	28%	$625	$175
		China	18%	$600	$108
	Domestic	2 saw	12%	$510	$61
		3 saw	8%	$600	$48
		CNS	4%	$580	$23
		Pulp	2%	$265	$5
	Total		100%		$629
Hemlock	Export	Korea	49%	$555	$272
	Domestic	3 saw	23%	$400	$92
		CNS	23%	$450	$104
		Pulp	5%	$290	$15
	Total		100%		$483
Red alder	Saw	12"-14"+	10%	$790	$79
		10"-11"	27%	$735	$198
		8"-9"	23%	$675	$155
		Pulp	40%	$240	$96
	Total		100%		$528

Land values can be considered separately and valued through comparison based on analysis of cutover land sales, considering physical and economic variations in the respective land areas. Appraisers must also consider the fact that the subject land is not yet harvested and therefore not in a cutover condition on the valuation date.

Table 19.7 Costs for Conversion Return

Cost Items	$/MBF
Road use fee	$1.50
Road construction (12 stations @ $1,300/1,009 MBF)	$15.46
Logging (on-board truck)	$95.00
Hauling	$85.00
Fire protection	$4.00
Engineering, surveying, and permits	$5.00
Timber excise tax	$17.00
Total	$223

Table 19.8 Valuation Using Conversion Return

Species	Volume (Net MBF)	Price ($/MBF)	Cost ($/MBF)	Net ($/MBF)	Value Total
Douglas fir	319	$629	$223	$406	$129,622
Hemlock	39	$483	$223	$260	$10,140
Red alder	651	$528	$223	$305	$198,555
Gross timber value					$338,317
Less entrepreneurial profit @ 10%					– $33,831
Merchantable timber value					$304,486

The conversion return method is generally applied to smaller properties with fewer stands and where complex harvest timing is not an issue. Conversion return analysis works well when the timber stand is considered merchantable or harvestable. In properties where older and younger timber exist, a combination of conversion return and present value analysis may be required.

Tables 19.6, 19.7, and 19.8 illustrate a series of simple conversion return calculations yielding the market value of the merchantable timber on several stands of different species in the Pacific Northwest.

Reconciliation

If an appraiser has used more than one procedure for developing a value indication for a property, these value indications must be reconciled into a single value opinion. An appraiser who has developed a value indication for the property by a specialized technique, valuing the timber by comparison to timber sales contracts or by a conversion return analysis, and then separately valuing the land as if cutover, must be careful in blending these indications of value.

Sometimes marketing conditions enable appraisers to simply add the value of the land as if cutover to the value of the timber. At other times, however, the quantity of timber involved or the marketing conditions for the species and the location of the timber require that additional adjustments be made. For example, suppose the subject property has a considerably larger volume of timber than comparable properties. The discount rate used may need to be adjusted to account for the holding costs and other associated costs as well as any advantages or disadvantages in taking three or four seasons to remove the timber from the comparable properties. Sometimes a discount may be required. At other times, the strong competitive spirit of the market for this type of property may offset the need for a discount. In any event, appraisers should consider the effect on the discount rate and decide according to existing market conditions.

When an appraiser has employed more than one means of analyzing the value of the total property, more than one indication of value usually results. In the reconciliation, appraisers must address the validity, reliability, significance, and relevance of the various elements employed in the respective analyses for the total property. Appraisers consider the significance of the data, the means of analysis and its validity in the market, and the extent to which each analysis is a true and proper representation of activity in the market for the type of property being appraised.

After a thorough review of the work product and the reports commissioned by other professionals such as foresters, appraisers develop a final opinion of value for the total property, giving the greatest weight to those elements and methods of analysis that are used in the timberland market.

Index